Numerical Approximation Methods

Harold Cohen

Numerical Approximation Methods

$$\pi \approx \frac{355}{113}$$

Harold Cohen
Department of Physics and Astronomy
California State University
Los Angeles, California
USA
twohco@yahoo.com

ISBN 978-1-4899-9159-1   ISBN 978-1-4419-9837-8 (eBook)
DOI 10.1007/978-1-4419-9837-8
Springer New York Dordrecht Heidelberg London

Mathematics Subject Classification (2010): 65L-xx, 65M-xx, 65N-xx, 65Q-xx, 65R-xx, 65T-xx

Cover design: The cover of this text contains a highly accurate rational fraction approximation of π.
This approximation was initially discovered in the 5th century by a Chinese mathematician named
Zu Chongzi (see, for example, Solomon, R., 2008, p. 15). It was rediscovered by the author in 2005
during a day of extreme boredom.

Printed on acid-free paper

Springer is part of Springer Science+Business Media (www.springer.com)

This book is dedicated to the most important people in my life.
Foremost among these are:
my wife Helen,
my daughters Lisa and Allyson,
my son David
and my son-in-law Jeff.

It is also dedicated to the memory of my parents.

I also dedicate this work to the memory of Dr. Charles Pine
whose kindness and academic nurturing helped me to recognize
the beauty of mathematics and its applications to scientific problems.

I am.
Therefore I think.

Preface

Approximate solutions to many problems that cannot be solved analytically, arise from numerical and other approximation techniques. In this book, the numerical methods and algorithms are developed for approximating such solutions. It also includes some methods of computation that yield exact results.

This book is appropriate as a text for a course in computational methods as well as a reference for researchers who need such methods in their work. The book contains a presentation of some well-known approximation methods that are scattered throughout existing literature as well as techniques that are obscure such as Chio's method for evaluating a determinant in ch. 5 and Namias' extension of the Stirling approximation to the gamma function for large argument in Appendix 3. Material that seems to be original is also presented; such as evaluating integrals using Parseval's theorem for periodic functions, beta and gamma functions, and Heaviside operator methods, along with solving integral equations with singular kernels and numerical evaluation of Cauchy principle value integrals.

To provide the reader with concrete applications of these methods, the book relies heavily on illustrative examples. I have provided a table of examples (like a table of contents) with descriptive titles to give the reader quick access to these illustrations.

This manuscript was prepared using Microsoft Word and MathType. MathType is an equation editor developed and marketed by Design Science Co. of Long Beach, California. Because of constraints encountered using these programs, it is sometimes necessary to position a mathematical expression in a sentence that does not fit on the same line as the text in that sentence. Such expressions have been placed on a separate line, centered on the page. They should be read as if they were text within the sentence. These expressions are distinguished from equations in that they are in the center of the page, they do not contain an "equal" ($=$), "not equal" ($\neq$), or "inequality" ($>$, $\leq$, etc.) symbol and are not designated with an equation number. Equations are displayed starting close to or at the left margin, have one or more equal, not equal or inequality symbols, and are identified by an equation number. An example of this "out of line" part of a sentence can be found at the bottom of page 133.

My thanks are expressed to those students in the Mathematical Methods of Physics courses that I taught over the years at California State University, Los Angeles. Their curiosity and questions helped in the development of this material. My motivation to write this book came from a group of physics, chemistry, and mathematics students who sat in on a course that I taught at Shandong University in Jinan, Peoples Republic of China. I thank them all for that motivation. I am particularly grateful to Mr. Li Yuan (Paul) for his help in that course.

Los Angeles, CA   Harold Cohen

Contents

Chapter 1
INTERPOLATION and CURVE FITTING

Discrete data

Often one is presented with numerical values of a function $f(x)$ at specified values of x. Experimental results are often presented in a table as a set of discrete data points. When data is presented in this way, the values of the function at points not given in the table must be found by some numerical technique.

Table 1.1 is a numerical representation of a function $f(x)$ at five different values of x.

x	$f(x)$
0.0	0.50
1.1	1.10
1.8	2.10
2.4	2.90
3.7	4.00

Table 1.1 Sample data table

Values of $f(x)$ at points $x < 0.0$ and $x > 3.7$ are called *extrapolated values*. For points $0.0 \leq x \leq 3.7$, values of $f(x)$ are called *interpolated values* of the data. Methods of finding interpolated values of $f(x)$ that we will develop in this chapter can also be used to predict extrapolated values of $f(x)$, but extrapolated values are often unreliable.

Graphical interpolation

A most straightforward approach to interpolation is to construct an approximate graph of the function and read the values of $f(x)$ from the graph.

H. Cohen, *Numerical Approximation Methods*, DOI 10.1007/978-1-4419-9837-8_1,
© Springer Science+Business Media, LLC 2011

Example 1.1: Interpolated values by graphing

The graph of Fig. 1.1 is a reasonable estimate of the data of Table 1.1.

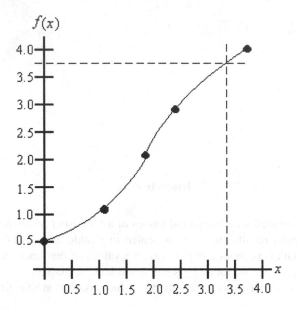

Fig. 1.1 Graph of the data in Table 1.1

We see from the graph that

$$f(3.4) \simeq 3.75 \tag{1.1}\square$$

1.1 Lagrange and Lagrange-Like Interpolation

Polynomial interpolation

One widely used analytic method of interpolating is by means of a polynomial. This technique involves approximating $f(x)$ by a polynomial over the range of x values. If one has N data points, one can interpolate by approximating $f(x)$ by a polynomial of order $N-1$.

The method consists of expressing $f(x)$ as a polynomial with N undetermined coefficients. As such, the polynomial has N terms in powers of x; $x^0, x, x^2, \ldots, x^{N-1}$.

$$f(x) = A_0 + A_1 x + \ldots + A_{N-1} x^{N-1} \tag{1.2}$$

Then, at each given value of $x = x_k$ in the data table, one substitutes the values of x_k and $f(x_k)$ into eq. 1.2. This results in a set of N equations in the N undetermined coefficients $\{A_0, \ldots, A_{N-1}\}$.

Example 1.2: Polynomial interpolation

For the five data points in Table 1.1, we approximate $f(x)$ by the fourth order polynomial

$$f(x) = A_0 + A_1 x + A_2 x^2 + A_3 x^3 + A_4 x^4 \qquad (1.3)$$

substituting the five data points into the expression yields

$$A_0 = 0.50 \qquad (1.4a)$$

$$A_0 + 1.1 A_1 + (1.1)^2 A_2 + (1.1)^3 A_3 + (1.1)^4 A_4 = 1.10 \qquad (1.4b)$$

$$A_0 + 1.8 A_1 + (1.8)^2 A_2 + (1.8)^3 A_3 + (1.8)^4 A_4 = 2.10 \qquad (1.4c)$$

$$A_0 + 2.4 A_1 + (2.4)^2 A_2 + (2.4)^3 A_3 + (2.4)^4 A_4 = 2.90 \qquad (1.4d)$$

and

$$A_0 + 3.7 A_1 + (3.7)^2 A_2 + (3.7)^3 A_3 + (3.7)^4 A_4 = 4.00 \qquad (1.4e)$$

Solving this set of equations for $\{A_0, A_1, A_2, A_3, A_4\}$, we obtain

$$f(x) \simeq 0.50 - 0.67x + 1.57x^2 - 0.47x^3 + 0.04x^4 \qquad (1.5)$$

from which

$$f(3.4) \simeq 3.81 \qquad (1.6)\square$$

It is possible to achieve the same order polynomial interpolation as that of eq. 1.2 without having to solve for coefficients of various powers of x. To accomplish this, we consider the $N-1$ order polynomial

$$\mu_k(x) \equiv \frac{(x - x_1)(x - x_2)...(x - x_{k-1})(x - x_{k+1})...(x - x_N)}{(x_k - x_1)(x_k - x_2)...(x_k - x_{k-1})(x_k - x_{k+1})...(x_k - x_N)} \qquad (1.7)$$

where $\{x_1, \ldots, x_N\}$ is the set of N different x values in the data table.

We see that

$$\mu_k(x_n) = \begin{cases} 1 & n = k \\ 0 & n \neq k \end{cases} \qquad (1.8a)$$

This can be expressed as a quantity called the *Kroenecker delta symbol* which is defined by

$$\delta_{kn} = \begin{cases} 1 & k = n \\ 0 & k \neq n \end{cases} \qquad (1.8b)$$

Thus, $\mu_k(x_n)$, as given in eq. 1.7, satisfies

$$\mu_k(x_n) = \delta_{kn} \tag{1.8c}$$

Using this property,

$$\mu_k(x_n)f(x_n) = f(x_n)\delta_{kn} = \begin{cases} f(x_k) & k = n \\ 0 & k \neq n \end{cases} \tag{1.9}$$

Writing

$$f(x) = \sum_{k=1}^{N} \mu_k(x)f(x_k) \tag{1.10}$$

we see that this expression guarantees that

$$\lim_{x \to x_n} f(x) = \sum_{k=1}^{N} \mu_k(x_n)f(x_k) = \sum_{k=1}^{N} \delta_{kn}f(x_k) = f(x_n) \tag{1.11}$$

the condition that is required to find the coefficients of the polynomial of eq. 1.2. The polynomial interpolation of eq. 1.10 is called a *Lagrange polynomial interpolation*.

Example 1.3: Lagrange interpolation

Labeling the points in Table 1.1 by x_1 through x_5, we have

$$\mu_1(x) = \frac{(x - 1.10)(x - 1.80)(x - 2.40)(x - 3.70)}{(0.00 - 1.10)(0.00 - 1.80)(0.00 - 2.40)(0.00 - 3.70)} \tag{1.12a}$$

$$\mu_2(x) = \frac{(x - 0.00)(x - 1.80)(x - 2.40)(x - 3.70)}{(1.10 - 0.00)(1.10 - 1.80)(1.10 - 2.40)(1.10 - 3.70)} \tag{1.12b}$$

$$\mu_3(x) = \frac{(x - 0.00)(x - 1.10)(x - 2.40)(x - 3.70)}{(1.80 - 0.00)(1.80 - 1.10)(1.80 - 2.40)(1.80 - 3.70)} \tag{1.12c}$$

$$\mu_4(x) = \frac{(x - 0.00)(x - 1.10)(x - 1.80)(x - 3.70)}{(2.40 - 0.00)(2.40 - 1.10)(2.40 - 1.80)(2.40 - 3.70)} \tag{1.12d}$$

and

$$\mu_5(x) = \frac{(x - 0.00)(x - 1.10)(x - 1.80)(x - 2.40)}{(3.70 - 0.00)(3.70 - 1.10)(3.70 - 1.80)(3.70 - 2.40)} \tag{1.12e}$$

From eqs. 1.12 we obtain

$$\{\mu_k(3.4)\} = \{-0.06,\ 0.63,\ -1.63,\ 1.54,\ 0.53\} \tag{1.13}$$

and, eq. 1.11 yields

$$f(3.4) \simeq 3.81 \tag{1.14}\square$$

It is occasionally useful to approximate a function by polynomial interpolation using *special polynomials*, the properties of which are well studied and appear in the literature. Examples of such polynomials are:

- Legendre polynomials denoted by $P_N(x)$, defined for $-1 \leq x \leq 1$.
- Laguerre polynomials denoted by $L_N(x)$, defined for $0 \leq x \leq \infty$.
- Hermite polynomials denoted by $H_N(x)$, defined for $-\infty \leq x \leq \infty$.

(For properties of these and other *special functions*, see, for example, Cohen, H., 1992, Table 6.2, p. 281, pp. 288–386.)

Let $Z_N(x)$ be a special polynomial. Then

$$\mu_k(x) \equiv \frac{Z_N(x)}{(x - x_k)Z_N'(x_k)} \tag{1.15}$$

is a polynomial of order $N-1$ that has the property

$$\mu_k(x_n) = \delta_{kn} \tag{\textbf{\textit{1.8c}}}$$

It is straightforward to see that the polynomial of eq. 1.15 is identical to that given in eq. 1.7.

Lagrange-like (general functional) interpolation

Let the x dependence of $f(x)$ arise via a function $q(x)$. For example,

$$f(x) = \frac{1}{x^{3/2}} \tag{1.16}$$

can be written in terms of

$$q(x) = \frac{1}{x} \tag{1.17a}$$

as

$$f(x) = q^{3/2}(x) \tag{1.17b}$$

One can generalize the Lagrange polynomial interpolation to interpolation over a function $q(x)$ as

$$v_k(x) \equiv \frac{[q(x) - q(x_1)]...[q(x) - q(x_{k-1})][q(x) - q(x_{k+1})]...[q(x) - q(x_N)]}{[q(x_k) - q(x_1)]...[q(x_k) - q(x_{k-1})][q(x_k) - q(x_{k+1})]...[q(x_k) - q(x_N)]} \quad (1.18)$$

with the condition that $q(x_k) \neq q(x_m)$ if $x_k \neq x_m$.

We see that as the functions $\mu_k(x)$ of eq. 1.7,

$$v_k(x_n) = \delta_{kn} \quad (1.19)$$

Therefore, the Lagrange-like interpolation of $f(x)$ using $q(x)$ is given by

$$f(x) = f[q(x)] = \sum_{k=1}^{N} v_k(x)f(x_k) \quad (1.20)$$

Example 1.4: Lagrange-like interpolation

We consider the interpolation of

$$f(x) = \frac{1}{x^{3/2}} \quad (1.16)$$

by a Lagrange polynomial interpolation and by a Lagrange-like interpolation using eqs. 1.17, with

$$q(x) = \frac{1}{x} \quad (1.17a)$$

The data table for $f(x)$ for various values of x is shown in Table 1.2.

x	$f(x) = x^{-3/2}$
0.3	6.08581
1.2	0.76073
2.7	0.22540
4.9	0.09220

Table 1.2   Values of $x^{-3/2}$ for various values of x

The Lagrange polynomial interpolation of this data is

$$\frac{1}{x^{3/2}} \simeq \sum_{k=1}^{4} \mu_k(x)f(x_k) \quad (1.21a)$$

with $\mu_k(x)$ given by eq. 1.7. The Lagrange-like interpolation is written as

$$\frac{1}{x^{3/2}} \simeq \sum_{k=1}^{4} v_k(x)f(x_k) \quad (1.21b)$$

with $v_k(x)$ given by eq. 1.18 with $q(x) = 1/x$.

The interpolated results at $x = 0.2$ and $x = 3.5$ are given in Table 1.3.

x	Lagrange polynomial	Lagrange-like in $1/x$	Exact values of $1/x^{3/2}$
0.2	7.03069	9.49406	11.18034
3.5	1.33971	0.15332	0.15272

Table 1.3 Interpolated values of $1/x^{3/2}$ using polynomial and $1/x$ interpolations

As can be seen, for both large and small values of x, the Lagrange-like interpolation over $1/x$ yields a better estimate of the values of $1/x^{3/2}$ than does the Lagrange polynomial interpolation. □

This example illustrates that one must have a sense of how the data behaves (something like $1/x$ in this example) to determine the function $q(x) \neq x$ over which the preferred interpolation is generated.

1.2 Spline Interpolation

If a set of points do not lie on a straight line, any straight line can contain no more than two data points. A flexible device called a *spline*, which is used to make architectural and engineering drawings, can be bent so that more than two noncolinear points can be connected by placing those noncolinear points along the edge of the spline.

To describe the mathematical equivalent of this, is called *spline interpolation*, we begin by dividing the *range* of points $[x_1, x_N]$ into *intervals* of widths $x_{k+1} - x_k$ and defining specified groups of intervals as *segments* (Fig. 1.2).

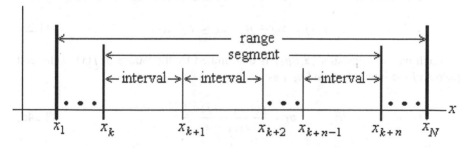

Fig. 1.2 Definitions of range, segment, and interval

Spline interpolations are obtained by fitting a function over a specified segment by a Lagrange or Lagrange-like interpolation. To illustrate, we will use a set of five points as an example.

Cardinal spline interpolation

The simplest spline interpolation involves approximating $f(x)$ by some constant over each interval between consecutive points. Thus, each segment consists of one interval.

One common choice is to take $f(x)$ to be its value at the midpoint of each interval;

$$f(x) \simeq f\left(\tfrac{1}{2}(x_k + x_{k+1})\right) \equiv f\left(x_{k+\frac{1}{2}}\right) \quad x_k \leq x \leq x_{k+1} \tag{1.22a}$$

A second common option is to take the value of the function to be the average of $f(x)$ over each interval;

$$f(x) \simeq \tfrac{1}{2}[f(x_k) + f(x_{k+1})] \quad x_k \leq x \leq x_{k+1} \tag{1.22b}$$

This approximation of $f(x)$ by a constant over each interval is the simplest interpolation one can construct. It is called a *cardinal spline*. It results in a graph like the one shown in Fig. 1.3a.

One obvious problem with the cardinal spline is that it defines $f(x)$ as a function that is discontinuous at each of the data points $\{x_k\}$. Therefore, for this interpolation, the first derivative of $f(x)$ is undefined.

Linear spline interpolation

The next higher order spline interpolation approximates $f(x)$ by a straight line between two adjacent points (Fig. 1.3b). This is described mathematically by

$$f(x) = \alpha_k x + \beta_k \quad x_k \leq x \leq x_{k+1} \tag{1.23}$$

Each pair of constants $\{\alpha_k, \beta_k\}$ is determined by the values of $f(x)$ at the end points of the interval, x_k and x_{k+1} as

$$\alpha_k = \frac{f(x_{k+1}) - f(x_k)}{x_{k+1} - x_k} \tag{1.24a}$$

and

$$\beta_k = f(x_k) - x_k \frac{f(x_{k+1}) - f(x_k)}{x_{k+1} - x_k} = \frac{x_{k+1} f(x_k) - x_k f(x_{k+1})}{x_{k+1} - x_k} \tag{1.24b}$$

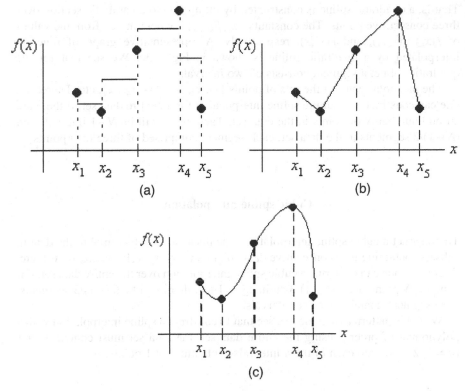

Fig. 1.3 (a) Cardinal spline interpolation. (b) Linear spline interpolation. (c) Quadratic spline interpolation

Clearly $f(x)$ is continuous at each point x_k, but its first derivative is discontinuous at these points.

Since a cardinal and a linear spline interpolations involve connecting two neighboring points x_k and x_{k+1}, each interval is defined by these two neighboring points. Therefore, these spline interpolations can be constructed no matter how many points are in the data set.

Quadratic spline interpolation

The *quadratic spline interpolation* is constructed by writing the function as a second order polynomial between sets of three successive points. This is described mathematically by

$$f(x) = \alpha_k x^2 + \beta_k x + \gamma_k \quad x_k \leq x \leq x_{k+2} \tag{1.25}$$

That is, a quadratic spline is constructed by fitting $f(x)$ to a parabolic section over three consecutive points. The constants $\{\alpha_k, \beta_k, \gamma_k\}$ are determined from the values of $f(x_k)$, $f(x_{k+1})$, and $f(x_{k+2})$, respectively. A representative graph of function interpolated by a quadratic spline is shown in Fig. 1.3c. We see that for the quadratic spline, a segment consists of two intervals.

The parabolas connect the sets of points $\{x_1, x_2, x_3\}$, $\{x_3, x_4, x_5\}$, etc. Therefore, one can construct a quadratic spline interpolation for an entire data set, if there are an odd number N of points in the data set. Then, there will be $N-1$ intervals and $(N-1)/2$ segments in the data set, each segment comprised of three data points.

Cubic spline interpolation

To construct a cubic spline interpolation, one fits a cubic polynomial to the data in subsets containing four consecutive data points, $\{x_1, x_2, x_3, x_4\}$, $\{x_4, x_5, x_6, x_7\}$, etc. Therefore, one can construct a cubic spline interpolation over the entire data set, if it contains N points with $(N-1)/3$ an integer. Then, there will be $(N-1)/3$ segments, each segment containing three intervals.

With this pattern as a guide, we see that to construct a spline interpolation with a polynomial of order p using the entire data set, the data set must contain $pn+1$ ($n=1, 2, \ldots$), with each of the n intervals containing $p+1$ points.

Higher order splines

We can see that the higher the order of the spline interpolation, and therefore the higher the order of the polynomial one uses to fit a subset of points, the more difficult it becomes to solve for the coefficients of that polynomial.

An easier way to generate the interpolating polynomial is to use the Lagrange interpolation over each segment. For example, for a quadratic spline interpolation over the segment defined by x_k, x_{k+1}, and x_{k+2}, instead of writing

$$f(x) = \alpha_k x^2 + \beta_k x + \gamma_k \quad x_k \leq x \leq x_{k+2} \qquad (1.25)$$

in this interval $x_k \leq x \leq x_{k+2}$, one could express $f(x)$ as

$$f(x) = \mu_k(x)f(x_k) + \mu_{k+1}(x)f(x_{k+1}) + \mu_{k+2}(x)f(x_{k+2}) \qquad (1.26)$$

where

$$\mu_k(x) = \frac{(x - x_{k+1})(x - x_{k+2})}{(x_k - x_{k+1})(x_k - x_{k+2})} \qquad (1.27a)$$

$$\mu_{k+1}(x) = \frac{(x - x_k)(x - x_{k+2})}{(x_{k+1} - x_k)(x_{k+1} - x_{k+2})} \qquad (1.27b)$$

and

$$\mu_{k+2}(x) = \frac{(x - x_k)(x - x_{k+1})}{(x_{k+2} - x_k)(x_{k+2} - x_{k+1})} \qquad (1.27c)$$

Or, if one knows that the x dependence of $f(x)$ is in the form $f[q(x)]$, one can construct a Lagrange-like quadratic spline interpolation as

$$f(x) = v_k(x)f(x_k) + v_{k+1}(x)f(x_{k+1}) + v_{k+2}(x)f(x_{k+2}) \qquad (1.28)$$

for $x_k \le x \le x_{k+2}$, with

$$v_k(x) = \frac{[q(x) - q(x_{k+1})][q(x) - q(x_{k+2})]}{[q(x_k) - q(x_{k+1})][q(x_k) - q(x_{k+2})]} \qquad (1.29a)$$

$$v_{k+1}(x) = \frac{[q(x) - q(x_k)][q(x) - q(x_{k+2})]}{[q(x_{k+1}) - q(x_k)][q(x_{k+1}) - q(x_{k+2})]} \qquad (1.29b)$$

and

$$v_{k+2}(x) = \frac{[q(x) - q(x_k)][q(x) - q(x_{k+1})]}{[q(x_{k+2}) - q(x_k)][q(x_{k+2}) - q(x_{k+1})]} \qquad (1.29c)$$

Example 1.5: Spline interpolation

Using the five data points in Table 1.1,

(a) a linear spline interpolation is given by

$$f(x) = 0.54546x + 0.50000 \quad 0.0 \le x \le 1.1 \qquad (1.30a)$$

$$f(x) = 1.42857x - 0.47143 \quad 1.1 \le x \le 1.8 \qquad (1.30b)$$

$$f(x) = 1.33333x - 0.30000 \quad 1.8 \le x \le 2.4 \qquad (1.30c)$$

and

$$f(x) = 0.84615x + 0.86923 \quad 2.4 \le x \le 3.7 \qquad (1.30d)$$

(b) a quadratic spline interpolation is given by

$$f(x) = 0.49062x^2 - 0.00577x + 0.50000 \quad 0.0 \le x \le 1.8 \qquad (1.31a)$$

and

$$f(x) = -0.25641x^2 + 2.41026x - 1.40770 \quad 1.8 \leq x \leq 3.7 \qquad (1.31b)$$

Using the linear spline approximation, we find

$$f(1.5) = 1.67170 \qquad (1.32a)$$

and with the quadratic spline approximation

$$f(1.5) = 1.59524 \qquad (1.32b)\square$$

1.3 Interpolation by Pade Approximants

If a function $f(x)$ is analytic over some domain, it can be approximated by a Pade (pronounced "Pah-day") approximant developed in Appendix 1;

$$f^{[N,M]}(x) = \frac{p_0 + p_1(x - x_0) + p_2(x - x_0)^2 + ... + p_N(x - x_0)^N}{1 + q_1(x - x_0) + q_2(x - x_0)^2 + ... + q_M(x - x_0)^M} \qquad (A1.9)$$

A particular Pade approximation is defined by the determination of the $M + N + 1$ coefficients $\{p_k, q_k\}$. As noted in Appendix 1, the most accurate Pade approximations to $f(x)$ are those for which $M = N$ or $M = N - 1$. These contain $2N + 1$ (for $M = N$) or $2N$ (for $M = N-1$) coefficients to be determined. Therefore, if a data set contains $2N + 1$ (an odd number of) data points, the function $f(x)$ that the data describes is best approximated by $f^{[N,N]}(x)$. For a data set with $2N$ (an even number of) points, the best approximation to $f(x)$ will be achieved using $f^{[N,N-1]}(x)$.

Since there are N roots of the denominator polynomial, there are at most N singularities (infinities) of the Pade Approximant which may not be singularities of the original function. In that case, in order for the Pade Approximant to be a good approximation to $f(x)$, it is essential that the data set not contain any values of x that are near the singularities of the Pade Approximant.

Example 1.6: Interpolation by Pade Approximant

Table 1.4 contains discrete data points for the exponential function.

x	$f(x) = e^x$
−1.2	0.301
0.5	1.649
1.4	4.055

Table 1.4 Data for e^x

To fit the data in Table 1.4, we note that there are only three points in the data set. Thus, taking $x_0 = 0$, we fit $f(x)$ to the diagonal Pade Approximant

$$f^{[1,1]}(x) = \frac{p_0 + p_1 x}{1 + q_1 x} \tag{A1.5}$$

To determine p_0, p_1, and q_1, we require that the Pade Approximant and the function have the same value at the three values of x in the table. Therefore, we have

$$f(-1.2) = 0.301 = \frac{p_0 - 1.2p_1}{(1 - 1.2q_1)} \tag{1.33a}$$

$$f(0.5) = 1.649 = \frac{p_0 + 0.5p_1}{(1 + 0.5q_1)} \tag{1.33b}$$

and

$$f(1.4) = 4.055 = \frac{p_0 + 1.4p_1}{(1 + 1.4q_1)} \tag{1.33c}$$

Solving for p_0, p_1, and q_1, we obtain

$$[e^x]^{[1,1]} = \frac{1.062 + 0.514x}{1 - 0.401x} \tag{1.34a}$$

For example, this predicts

$$[e^1]^{[1,1]} = 2.628 \tag{1.34b}$$

which differs from the exact value by about 3%. Considering that this Pade Approximant is of fairly low order, this is a very reasonable result. □

1.4   Operator Interpolation for Equally Spaced Data

If the values of x are spaced equally, then we define

$$h \equiv x_{k+1} - x_k \tag{1.35}$$

where h is the same constant for all k. For equal spacing, we can develop an *operator interpolation* by defining two operators.

The *raising operator E* is defined such that

$$Ef(x_k) = f(x_{k+1}) \tag{1.36a}$$

and the *difference operator* Δ is defined by

$$\Delta f(x_k) = f(x_{k+1}) - f(x_k) \tag{1.36b}$$

From eq. 1.36a, we have

$$\Delta f(x_k) = Ef(x_k) - f(x_k) = (E - 1)f(x_k) \tag{1.37}$$

Thus, the difference operator is related to the raising operator by

$$\Delta = E - 1 \tag{1.38}$$

Consider successive applications of E operating on $f(x_1)$. We have

$$Ef(x_1) = f(x_2) = f(x_1 + h) \tag{1.39a}$$

$$E^2 f(x_1) = Ef(x_2) = f(x_3) = f(x_1 + 2h) \tag{1.39b}$$

$$E^3 f(x_1) = f(x_1 + 3h) \tag{1.39c}$$

and so on. Thus, for any integer $n > 1$,

$$E^n f(x_1) = f(x_1 + nh) \tag{1.39d}$$

Since n is an integer, $f(x_1 + nh)$ is one of the values of the function given in the data set. To interpolate to a point that is not in the data set, we replace n by a non-integer value α and define the value of x at which we wish to find $f(x)$ by

$$x = x_1 + \alpha h \tag{1.40}$$

Then, generalizing from eq. 1.39d and using the relation between E and Δ, we have

$$f(x) = f(x_1 + \alpha h) = E^\alpha f(x_1) = (1 + \Delta)^\alpha f(x_1) \tag{1.41a}$$

We then expand the operator $(1 + \Delta)^\alpha$ in a binomial series to obtain

$$f(x) = \left(1 + \alpha\Delta + \frac{\alpha(\alpha - 1)}{2!}\Delta^2 + \frac{\alpha(\alpha - 1)(\alpha - 2)}{3!}\Delta^3 + \ldots\right)f(x_1) \tag{1.41b}$$

This expansion cannot be carried out indefinitely. To determine the number of terms in the expansion can be used, let us assume that the data set contains N values $f(x_1), \ldots, f(x_N)$. We then consider

$$\Delta f(x_1) = f(x_2) - f(x_1) \tag{1.42a}$$

$$\Delta^2 f(x_1) = \Delta f(x_2) - \Delta f(x_1) = f(x_3) - 2f(x_2) + f(x_1) \qquad (1.42\text{b})$$

$$\Delta^3 f(x_1) = f(x_4) - 3f(x_3) + 3f(x_2) - f(x_1) \qquad (1.42\text{c})$$

and so on.

We see that for each k, $\Delta^k f(x_1)$ contains $f(x_{k+1})$. Thus, $\Delta^N f(x_1)$ contains $f(x_1)$, ..., $f(x_{N+1})$. Since the values of $f(x)$ at $x > x_N$ are not specified, we cannot keep powers of Δ that yield $f(x_p)$ with $p > N$. Thus, k cannot be larger than $N-1$. Therefore, we approximate $f(x)$ by

$$f(x) \simeq \left(1 + \alpha\Delta + \frac{\alpha(\alpha-1)}{2!}\Delta^2 + \dots \frac{\alpha(\alpha-1)\dots(\alpha-(N-2))}{(N-1)!}\Delta^{N-1}\right)f(x_1) \quad (1.43)$$

The patterns developed in eqs. 1.42 also give us a *recurrence relation* for determining $\Delta^k f(x_1)$. We note that the coefficients of the various $f(x_k) \equiv f_k$ in eqs. 1.42 are the coefficients of the binomial expansion. As such, we consider the binomial expansion of $(f-1)^k$. Then, for each power p, we make the replacement

$$f^p \rightarrow f_{p+1} \qquad (1.44)$$

This yields the expression for $\Delta^k f_1$.
   For example, with $f^0 = 1$

$$(f-1)^2 = f^2 - 2f + f^0 \rightarrow f_3 - 2f_2 + f_1 = \Delta^2 f_1 \qquad (1.45\text{a})$$

and

$$(f-1)^3 = f^3 - 3f^2 + 3f + f^0 \rightarrow f_4 - 3f_3 + 3f_2 - f_1 = \Delta^3 f_1 \qquad (1.45\text{b})$$

Comparing eqs. 1.45 with eqs. 1.42b and 1.42c, we see that this recurrence relation yields the correct expressions for $\Delta^k f(x_1)$.

Example 1.7: Operator interpolation

We consider the values of the exponential function at equally spaced values of x, given in Table 1.5.

x	$f(x) = e^x$
0.0	1.00000
0.1	1.10517
0.2	1.22140
0.3	1.34986
0.4	1.49182

Table 1.5 Tabulated values of e^x at equally spaced values of x

To find $e^{0.33}$ we note that $x_1 = 0$ and $h = 0.1$. With

$$x = 0.33 = x_1 + \alpha h \tag{1.46}$$

we obtain

$$\alpha = 3.30 \tag{1.47}$$

Therefore, we must expand $(1 + \Delta)^{3.3}$ in a binomial expansion. Since there are five points in the data set, we keep only terms up to Δ^4 in the expansion. Thus, the operator interpolation of $e^{0.33}$ is given by

$$e^{0.33} \simeq \left[1 + 3.3\Delta + \frac{(3.3)(2.3)}{2!}\Delta^2 + \frac{(3.3)(2.3)(1.3)}{3!}\Delta^3 \right.$$
$$\left. + \frac{(3.3)(2.3)(1.3)(0.3)}{4!}\Delta^4 \right] f_1 \tag{1.48}$$

With

$$f_1 = 1.00000 \tag{1.49a}$$

$$\Delta f_1 = f_2 - f_1 = 0.10517 \tag{1.49b}$$

$$\Delta^2 f_1 = f_3 - 2f_2 + f_1 = 0.01106 \tag{1.49c}$$

$$\Delta^3 f_1 = f_4 - 3f_3 + 3f_2 - f_1 = 0.00116 \tag{1.49d}$$

and

$$\Delta^4 f_1 = f_5 - 4f_4 + 6f_3 - 4f_2 + f_1 = 0.00012 \tag{1.49e}$$

we obtain from eq. 1.48 that

$$e^{0.33} \simeq 1.39097 \tag{1.50}$$

which agrees with the exact value to five decimal places. □

Operator interpolation using a Taylor series

Defining

$$D^k f_1 \equiv \left. \frac{d^k f(x)}{dx^k} \right|_{x=x_1} \tag{1.51}$$

The Taylor series expansion of $f(x_1 + \alpha h)$ can be written as

$$f(x) \equiv f(x_1 + \alpha h) = \sum_{k=0}^{\infty} \frac{\alpha^k h^k}{k!} D^k f_1 \tag{1.52}$$

To express the differential operator D in terms of the raising and lowering operators E and Δ, we expand

$$Ef_k = f_{k+1} = f(x_k + h) = \sum_{n=0}^{\infty} \frac{h^n}{n!} D^n f_k \tag{1.53}$$

or

$$Ef_k = \sum_{n=0}^{\infty} \frac{h^n}{n!} D^n f_k \tag{1.54}$$

Therefore, the raising operator can be written as

$$E = \sum_{n=0}^{\infty} \frac{h^n}{n!} D^n = e^{hD} \tag{1.55a}$$

from which

$$D = \frac{1}{h} \ell n(E) = \frac{1}{h} \ell n(1 \mid \Delta) = \frac{1}{h} \left(\Delta - \frac{\Delta^2}{2} + \frac{\Delta^3}{3} \cdots \right)$$

$$= \frac{1}{h} \sum_{n=1}^{\infty} (-1)^{n+1} \frac{\Delta^n}{n} \tag{1.55b}$$

Referring to eq. 1.52, the Taylor series involves various powers of D operating on f_1. As argued earlier, since the powers of D are expressed as a sums of terms involving powers of Δ, we can only keep terms in Δ up to a power determined by the number of entries in the data table. For a data set with N entries, we only keep powers of D, and therefore powers of Δ, in the expansion of eq. 1.55b, so that each term in the Taylor expansion only has terms up to and including Δ^{N-1}.

Example 1.8: Interpolation using a Taylor series with operators to approximate derivatives

We again consider the data given in Table 1.5 for $f(x) = e^x$. Since this data set contains five entries, we will keep terms in eq. 1.55b up to Δ^4. These terms will come from the terms in a Taylor expansion up to D^4. Therefore, we first approximate the expression in eq. 1.52 as

$$f(x) \simeq \sum_{k=0}^{4} \frac{\alpha^k h^k}{k!} D^k f_1 = \left(1 + \alpha h D + \frac{\alpha^2 h^2}{2!} D^2 + \frac{\alpha^3 h^3}{3!} D^3 + \frac{\alpha^4 h^4}{4!} D^4 \right) f_1 \tag{1.56}$$

Referring to eq. 1.55b, we keep powers of Δ up to Δ^4. Then, the terms in eq. 1.56 can be written as

$$hDf_1 \simeq \left(\Delta - \frac{\Delta^2}{2} + \frac{\Delta^3}{3} - \frac{\Delta^4}{4}\right)f_1 \qquad (1.57a)$$

$$h^2D^2f_1 \simeq \left(\Delta - \frac{\Delta^2}{2} + \frac{\Delta^3}{3} - \frac{\Delta^4}{4}\right)^2 f_1 \simeq \left(\Delta^2 - \Delta^3 + \frac{11}{12}\Delta^4\right)f_1 \qquad (1.57b)$$

$$h^3D^3f_1 \simeq \left(\Delta - \frac{\Delta^2}{2} + \frac{\Delta^3}{3} - \frac{\Delta^4}{4}\right)^3 f_1 \simeq \left(\Delta^3 - \frac{1}{2}\Delta^4\right)f_1 \qquad (1.57c)$$

and

$$h^4D^4f_1 \simeq \left(\Delta - \frac{\Delta^2}{2} + \frac{\Delta^3}{3} - \frac{\Delta^4}{4}\right)^4 f_1 \simeq \Delta^4 f_1 \qquad (1.57d)$$

From these, we obtain

$$f(0.33) = e^{0.33} \simeq$$
$$f_1 + \alpha\left(\Delta f_1 - \frac{1}{2}\Delta^2 f_1 + \frac{1}{3}\Delta^3 f_1 - \frac{1}{4}\Delta^4 f_1\right) + \frac{\alpha^2}{2}\left(\Delta^2 f_1 - \Delta^3 f_1 + \frac{2}{3}\Delta^4 f_1\right)$$
$$+ \frac{\alpha^3}{6}\left(\Delta^3 f_1 - \frac{3}{2}\Delta^4 f_1\right) + \frac{\alpha^4}{24}\Delta^4 f_1 \simeq 1.39080 \qquad (1.58)$$

which differs from the exact value by 0.01%. $\square$

1.5   Curve Fitting by the Method of Least Squares

Curve fitting is another method of determining a functional form for a given set of data. When the data is expected to fit a specific function based on theory, one can determine the best fit of the data to that expected function using the *method of least squares*.

For example, let the theory of a particular phenomenon predict that $f_{th}(x)$ describes a straight line. Because there are always uncertainties in measuring $f(x)$, the measured data may not lie on a straight line. Figure 1.4 is a representation of such data. The method of least squares is a technique for determining the "best" expected curve that represents the data. The line in Fig. 1.4 represents the "best" linear fit to that data.

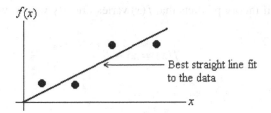

Fig. 1.4 Data for $f(x)$ vs. x that is predicted by theory to be a *straight line*

Least squares method

Let $f_{th}(x)$ describe the curve predicted by theory. Then, theory predicts that at $x = x_k$, $f(x) = f_{th}(x_k)$. Let f_k be the measured value of $f(x)$ at the point x_k. The deviation (or error) of f_k from $f(x_k)$ is

$$\varepsilon_k \equiv f(x_k) - f_k \tag{1.59}$$

which, in general, will not be zero (Fig. 1.5).

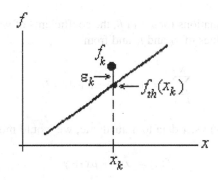

Fig. 1.5 Error between $f(x_k)$ predicted by theory and the measured value f_k

The method of finding the "best" curve is to construct the curve $f(x)$ that is expected from theory using undetermined constants. Those coefficients are then adjusted so that the *square of the rms error*

$$E \equiv \sum_{k=1}^{N} \varepsilon_k^2 = \sum_{k=1}^{N} [f(x_k) - f_k]^2 \tag{1.60}$$

is a minimum. This is accomplished by setting the partial derivatives of E with respect to each of the undetermined coefficients to zero. This is the method of least squares.

For example, if theory predicts that $f(x)$ varies linearly with x, we would model this system as

$$f(x) = \alpha x + \beta \qquad (1.61)$$

then minimize

$$E = \sum_{k=1}^{N} [\alpha x_k + \beta - f_k]^2 \qquad (1.62)$$

with respect to α and β. We do this by setting

$$\frac{\partial E}{\partial \alpha} = 2 \sum_{k=1}^{N} x_k [\alpha x_k + \beta - f_k] = 2 \left[\alpha \sum_{k=1}^{N} x_k^2 + \beta \sum_{k=1}^{N} x_k - \sum_{k=1}^{N} x_k f_k \right] = 0 \quad (1.63a)$$

and

$$\frac{\partial E}{\partial \beta} = 2 \sum_{k=1}^{N} [\alpha x_k + \beta - f_k] = 2 \left[\alpha \sum_{k=1}^{N} x_k + \beta \sum_{k=1}^{N} 1 - \sum_{k=1}^{N} f_k \right] = 0 \qquad (1.63b)$$

These are the linear equations for α and β, the coefficients of which are determined from the measured values of x_k and f_k and from

$$\sum_{k=1}^{N} 1 = 1 + 1 + \ldots + 1 = N \qquad (1.64)$$

If we were to fit $f(x)$ vs. x data to a quadratic, we would model $f(x)$ as

$$f(x) = \alpha x^2 + \beta x + \gamma \qquad (1.65)$$

and minimize

$$E = \sum_{k=1}^{N} [\alpha x_k^2 + \beta x_k + \gamma - f_k]^2 \qquad (1.66)$$

with respect to α, β, and γ. This would result in a set of three equations which are linear in these three parameters.

Example 1.9: Least squares method for a mass–spring system

From Hooke's law for a mass M on a spring with spring constant k,

$$Mg = kx \qquad (1.67a)$$

theory predicts that the displacement from equilibrium of an object of mass M suspended on a spring will increase linearly with increasing mass (Fig. 1.6);

$$x = \frac{g}{k}M \qquad (1.67b)$$

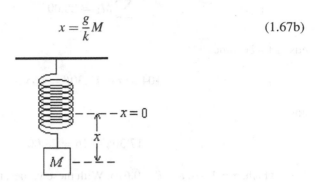

Fig. 1.6 Displacement from equilibrium of a mass suspended on a spring

The data of Table 1.6 represent the measured values of masses needed to cause corresponding displacements of an object suspended on a spring

M	x
2.1	1.2
4.2	2.3
9.4	5.7
12.3	8.1

Table 1.6 Data for displacement of a mass on a spring

To fit this data to a straight line, we minimize

$$E = \sum_{k=1}^{4} [\alpha x_k + \beta - M_k]^2 \qquad (1.68)$$

Then, with

$$\sum_{k=1}^{4} x_k^2 = 104.83 \qquad (1.69a)$$

$$\sum_{k=1}^{4} x_k = 17.30 \qquad (1.69b)$$

$$\sum_{k=1}^{4} x_k M_k = 165.39 \qquad (1.69c)$$

and

$$\sum_{k=1}^{4} M_k = 28.00 \qquad (1.69\text{d})$$

eqs. 1.64 become

$$104.83\alpha + 17.30\beta = 165.39 \qquad (1.70\text{a})$$

and

$$17.30\alpha + 4\beta = 28.00 \qquad (1.70\text{b})$$

from which, $\alpha = 1.476$ and $\beta = 0.616$. With these values, the square of the rms error for the linear fit is

$$E_{linear} = \sum_{k=1}^{4} [1.48x_k + 0.62 - M_k]^2 = 0.330 \qquad (1.71)\square$$

We note two differences between a least squares fit of data to a specific curve and the various interpolation approximations that were introduced earlier.

- With a Lagrange, Lagrange-like, spline interpolation or Pade Approximant, $f(x)$ is forced to have the measured values $f(x_k)$ at each x_k in the data set.
  For a least squares fit to a function $f(x)$, none of the measured values $f(x_k)$ are required to lie on the curve defined by $f(x)$.
- For a spline interpolation, one must have a specific number of points in the data set in order to achieve a given spline interpolation. For example, for a quadratic spline, one must have an odd number of points in the data set. The number of points in the data set determines the spline interpolation used to model $f(x)$. For a Lagrange polynomial or Lagrange-like interpolation, the number of factors in $\mu_k(x)$ (eq. 1.7) or $v_k(x)$ (eq. 1.18) is determined by the number of points in the data set.
  Using the least squares criterion, one chooses the type of curve that the data is to fit and that curve is independent of the number of points in the data set.

Justification for the least squares method

As stated above, the correct method for fitting data to a particular function $f(x)$ is to minimize

$$E \equiv \sum_{k=1}^{N} \varepsilon_k^2 = \sum_{k=1}^{N} [f(x_k) - f_k]^2 \qquad (\textit{1.61})$$

with respect to coefficients used to define $f(x)$.

Usually, a "best" curve is one for which there will be as much positive error as there is negative error. That is, a general rule for a "best" curve is

$$E_0 = \sum_{k=1}^{N} \varepsilon_k = \sum_{k=1}^{N} [f(x_k) - x_k] = 0 \tag{1.72}$$

Thus, adding errors does not tell us anything about how large or small an overall error is for a given curve.

The method of least squares removes the negative signs on individual errors by squaring each error. But there are other ways to eliminate these negative signs. For example, one could define an error as

$$E_1 \equiv \sum_{k=1}^{N} |\varepsilon_k| = \sum_{k=1}^{N} |f(x_k) - f_k| = \sum_{k=1}^{N} \sqrt{[f(x_k) - f_k]^2} \tag{1.73a}$$

or

$$E_4 = \sum_{k=1}^{N} \varepsilon_k^4 = \sum_{k=1}^{N} [f(x_k) - f_k]^4 \tag{1.73b}$$

again, minimizing each expression for E with respect to the parameters used to define $f(x)$.

To get a sense of the correct approach, we consider the simple problem of determining the average value of a set of N numbers $\{f(x_k)\}$. The "curve" defined by the average value of $f(x)$ is the horizontal straight line given by

$$f(x) = constant = f_{avg} \tag{1.74}$$

Then, the constant f_{avg} is the only parameter defining the horizontal line. If we use eq. 1.73a to find the best horizontal line fit to $f(x) = f_{avg}$, we minimize

$$E_1 = \sum_{k=1}^{N} |f(x_k) - f_{avg}| = \sum_{k=1}^{N} \sqrt{[f(x_k) - f_{avg}]^2} \tag{1.75a}$$

with respect to f_{avg}. Then

$$\frac{\partial E_1}{\partial f_{avg}} = -\sum_{k=1}^{N} \frac{f(x_k) - f_{avg}}{\sqrt{[f(x_k) - f_{avg}]^2}} = -\sum_{k=1}^{N} sign[f(x_k) - f_{avg}] = 0 \tag{1.75b}$$

Clearly, this is not how one defines the average value of a set of data points.

If we minimize

$$E_4 = \sum_{k=1}^{N} \left[f(x_k) - f_{avg} \right]^4 \tag{1.76}$$

we obtain the expression

$$\frac{\partial E_4}{\partial f_{avg}} = -4 \sum_{k=1}^{N} \left[f(x_k) - f_{avg} \right]^3 = 0 \tag{1.77a}$$

With eq. 1.65, this yields f_{avg} as the solution to the cubic equation

$$f_{avg}^3 N - 3f_{avg}^2 \sum_{k=1}^{N} f(x_k) + 3f_{avg} \sum_{k=1}^{N} f^2(x_k) - \sum_{k=1}^{N} f^3(x_k) = 0 \tag{1.77b}$$

This too is an incorrect definition of f_{avg}.
   However, minimizing

$$E = \sum_{k=1}^{N} \left[f(x_k) - f_{avg} \right]^2 \tag{1.78}$$

we obtain

$$\frac{\partial E}{\partial f_{avg}} = -2 \sum_{k=1}^{N} \left[f(x_k) - f_{avg} \right] = -2 \left(\sum_{k=1}^{N} f(x_k) - f_{avg} \sum_{k=1}^{N} 1 \right)$$

$$= -2 \left(\sum_{k=1}^{N} f(x_k) - f_{avg} N \right) = 0 \tag{1.79}$$

This results in

$$f_{avg} = \frac{1}{N} \sum_{k=1}^{N} f(x_k) \tag{1.80}$$

which is the correct definition of f_{avg}.

Fitting data to non-polynomial curves

It is not necessary to model a function in terms of a polynomial. Just as with interpolations, one can use any appropriate function $q(x)$ to define a least squares fit.

If one knows that the x dependence of $f(x)$ is introduced in the form of a function $q(x)$, one could model $f(x)$ in the form

$$f(x) = \alpha q(x) + \beta q^2(x) + \ldots \tag{1.81}$$

and minimize

$$E = \sum_{k=1}^{N} \left[\alpha q(x_k) + \beta q^2(x_k) + \ldots - f_k \right]^2 \tag{1.82}$$

with respect to the parameters $\alpha, \beta, \ldots$

Example 1.10: Least squares fit with a general function $q(x)$

Theory predicts that the data in Table 1.7 decreases with increasing x as a quadratic in $1/x$.

x	$f(x)$
1.3	5.42
2.2	4.28
3.7	3.81
4.9	3.62

Table 1.7 Data with inverse power of x decrease

To find the best fit of this data, we define

$$f(x) = \alpha + \beta \frac{1}{x} + \gamma \frac{1}{x^2} \tag{1.83}$$

Minimizing

$$E = \sum_{k=1}^{4} \left[\alpha + \beta \frac{1}{x_k} + \gamma \frac{1}{x_k^2} - f_k \right]^2 \tag{1.84}$$

by

$$\frac{\partial E}{\partial \alpha} = 2 \sum_{k=1}^{4} \left[\alpha + \beta \frac{1}{x_k} + \gamma \frac{1}{x_k^2} - f_k \right] = 0 \tag{1.85a}$$

$$\frac{\partial E}{\partial \beta} = 2 \sum_{k=1}^{4} \frac{1}{x_k} \left[\alpha + \beta \frac{1}{x_k} + \gamma \frac{1}{x_k^2} - f_k \right] = 0 \tag{1.85b}$$

and

$$\frac{\partial E}{\partial \gamma} = 2 \sum_{k=1}^{4} \frac{1}{x_k^2} \left[\alpha + \beta \frac{1}{x_k} + \gamma \frac{1}{x_k^2} - f_k \right] = 0 \tag{1.85c}$$

we obtain

$$\alpha + \beta \sum_{k=1}^{4} \frac{1}{x_k} + \gamma \sum_{k=1}^{4} \frac{1}{x_k^2} = \sum_{k=1}^{4} f_k \tag{1.86a}$$

$$\alpha \sum_{k=1}^{4} \frac{1}{x_k} + \beta \sum_{k=1}^{4} \frac{1}{x_k^2} + \gamma \sum_{k=1}^{4} \frac{1}{x_k^3} = \sum_{k=1}^{4} \frac{f_k}{x_k} \tag{1.86b}$$

and

$$\alpha \sum_{k=1}^{4} \frac{1}{x_k^2} + \beta \sum_{k=1}^{4} \frac{1}{x_k^3} + \gamma \sum_{k=1}^{4} \frac{1}{x_k^4} = \sum_{k=1}^{4} \frac{f_k}{x_k^2} \tag{1.86c}$$

This results in

$$4.000\alpha + 1.698\beta + 0.913\gamma = 17.130 \tag{1.87a}$$

$$1.698\alpha + 0.913\beta + 0.577\gamma = 7.883 \tag{1.87b}$$

and

$$0.913\alpha + 0.577\beta + 0.400\,\gamma = 4.520 \tag{1.87c}$$

from which $\alpha = 3.261$, $\beta = 1.480$, and $\gamma = 1,722$. With these values

$$E = \sum_{k=1}^{4} \left[3.261 + \frac{1.480}{x_k} + \frac{1.722}{x_k^2} - f_k \right]^2 = 0.001 \tag{1.88} \square$$

Problems

1. Table P1.1 lists three values of $f(x) = x^2 - 2e^{-x}$. Estimate the value of $f(0.65)$ using a Lagrange polynomial interpolation.

x	$f(x) = x^2 - 2e^{-x}$
0.50	-0.963
0.75	-0.382
1.00	0.264

Table P1.1 Selected values of $x^2 - 2e^{-x}$

2. Tables P1.2 contain selected data for an exponential function, a sine function, and a fractional power function.

(a)		(b)		(c)	
x	$e^{-x^2/2}$	x	$\sin(x/2)$	x	$x^{3/2}$
0.50	0.8825	$\pi/5$	0.3090	1.1	1.1537
1.20	0.4868	$\pi/3$	0.5000	2.3	3.4881
1.81	0.1944	$\pi/2$	0.7071	4.5	9.5459
2.25	0.0796	π	1.0000	6.2	15.4349

Table P1.2 Values of common functions at selected values of x

(a) For each data set, using just the numerical values, approximate the function by a Lagrange (polynomial) interpolation. Use that approximation to estimate the value of the function at the midpoints of each of the intervals (x_1, x_2), (x_2, x_3), and (x_3, x_4).

(b) For each data set, using just the numerical values, approximate the function by a Lagrange-like interpolation over the function specified below.

For the data in Table (a): construct two Lagrange-like interpolations; one with $q(x) = e^{-x}$, the second with $q(x) = e^{-x^2}$.
For Table (b): construct a Lagrange-like interpolation with $q(x) = \sin x$.
For Table (c): construct a Lagrange-like interpolation using $q(x) = x^2$.
Use each approximation to estimate the value of the function at the midpoints of each of the intervals (x_1, x_2), (x_2, x_3), and (x_3, x_4).

3. Tables P1.3 contain selected data for an exponential function, a sine function, and a fractional power function.

(a) For each table, approximate the function by a linear spline interpolation. Use that approximation to estimate the value of the function at the midpoints of each of the intervals (x_1, x_2), (x_2, x_3), (x_3, x_4), and (x_4, x_5).

(b) For each table, approximate the function by a quadratic spline interpolation. Use that approximation to estimate the value of the function at the midpoints of each of the intervals (x_1, x_2), (x_2, x_3), (x_3, x_4), and (x_4, x_5).

(c) Give a brief explanation of the reason why the data in these tables CANNOT be approximated using a cubic spline interpolation.

(a)		(b)		(c)	
x	$e^{-x^2/2}$	x	$\sin(x/2)$	x	$x^{3/2}$
0.50	0.885	$\pi/5$	0.3090	1.1	1.1537
1.20	0.4868	$\pi/3$	0.5000	2.3	3.4881
1.81	0.1944	$\pi/2$	0.7071	4.5	9.5459
2.25	0.0796	$2\pi/3$	0.8660	5.6	13.2520
2.70	0.0261	π	1.0000	6.2	15.4349

Table P1.3 Values of common functions at selected values of x

4. Tables P1.4 contain selected data for an exponential function, a sine function, and a fractional power function.

   (a) For each table, approximate the function by an operator interpolation. Use that approximation to estimate the value of the function at the midpoints of each of the intervals (x_1, x_2), (x_2, x_3), and (x_3, x_4).
   (b) For each table, approximate the function by a Taylor series interpolation using operators to estimate derivatives. Use that approximation to estimate the value of the function at the midpoints of each of the intervals (x_1, x_2), (x_2, x_3), and (x_3, x_4).

(a)		(b)		(c)	
x	$e^{-x^2/2}$	x	$\sin(x/2)$	x	$x^{3/2}$
0.50	0.8825	$\pi/6$	0.2588	1.5	1.8371
1.00	0.6065	$5\pi/6$	0.9659	2.5	3.9529
1.50	0.3247	$9\pi/6$	0.7071	3.5	6.5479
2.00	0.1353	$13\pi/6$	-0.2588	4.5	9.5459

Table P1.4 Values of common functions at selected values of x

5. There are four data points in Table P1.4c. What is the most accurate Pade Approximant for representing this data? For this Pade Approximant, write out the equations satisfied by the undetermined coefficients.
6. What is the best Pade Approximant for representing the three data points in Table P1.1? Find the values of the coefficients of this Pade approximants and use this result to determine $f(0.6)$ and $f(0.9)$.
7. Construct an operator interpolation approximation for each of the functions for which numerical data is given in Tables P1.4. Using this interpolation, estimate the value of each function at the midpoint of the interval between x_2 and x_3.
8. Construct a Taylor series interpolation with operator approximation of the derivatives for each of the functions for which numerical data is given in Tables P1.4. Using this interpolation, estimate the value of each function at the midpoint of the interval between x_2 and x_3.

9. Fit the data in Table P1.5 to a parabola in t. Determine the square of the rms error.

t	$S(t) = 2t + 3t^2$
1.5	9.95
2.5	24.25
4.1	56.90
6.0	118.35

Table P1.5 Distance and time data for the motion in one dimension of a particle with constant acceleration

10. The data in Table P1.6 are supposed to represent the straight line defined by $f(x) = 4.5x - 1.5$. Determine the best linear fit to this data. Compare the slope and intercept to the correct values.

x	$f(x)$
−1.2	−7.01
−0.3	−2.95
0.9	2.62
2.1	7.83

Table P1.6 Data for points on a straight line $f(x) = 4.5x - 1.5$

11. Table P1.7 contains data for $f(x) = 2\sin(x\pi/2) + 1$.

x	$f(x)$
0.6	−0.895
1.1	0.011
1.4	2.990
1.9	0.691

Table P1.7 Data for $f(x) = 2\sin(x\pi/2) + 1$

Determine the best fit to this data in the form $f(x) = \alpha\sin(x\pi/2) + \beta$ and find the squared rms error for this fit.

Chapter 2
ZEROS of a FUNCTION

In this chapter, we present methods for finding the zeros of $f(x)$

- when $f(x)$ is a polynomial. By the time one has finished high school, the methods for finding the roots of first- and second order polynomials have been learned. It is well known that it is not possible to solve for the roots of a polynomial in $f(x)$ in terms of the coefficients of x for a polynomial of order $N \geq 5$. (This was first proven by Niels Henrik Abel in 1826. For a version of Abel's proof in English, see Pesic, P., 2003.) As such, we restrict this discussion to polynomials of third and fourth order.
- when $f(x)$ is presented in tabular form giving specific values of f at specified values of x.
- when $f(x)$ is given in closed form (which is applicable to both non-polynomial functions and polynomials of any order).

2.1 Roots of a Cubic Polynomial

There are N *roots* of a polynomial of order N. If two or more of the roots have the same value, these are *repeated* roots.

Let

$$x^3 + px^2 + qx + r = 0 \qquad (2.1)$$

Factoring

Grouping the first and second terms of eq. 2.1, we write this equation as

$$x^2(x+p) + q\left(x + \frac{r}{q}\right) = 0 \qquad (2.2)$$

H. Cohen, *Numerical Approximation Methods*, DOI 10.1007/978-1-4419-9837-8_2,
© Springer Science+Business Media, LLC 2011

If

$$r = pq \tag{2.3}$$

then eq. 2.2 can be written in factored form as

$$\left(x^2 + q\right)(x + p) = 0 \tag{2.4}$$

If we group the first and third terms, eq. 2.2 can be expressed in the form

$$x\left(x^2 + q\right) + p\left(x^2 + \frac{r}{p}\right) = 0 \tag{2.5}$$

Again, if

$$r = pq \tag{2.3}$$

eq. 2.5 can again be expressed as

$$\left(x^2 + q\right)(x + p) = 0 \tag{2.4}$$

Thus, to find the roots of a cubic polynomial by this method of factoring, one must only ascertain that, with the coefficient of $x^3 = 1$, eq. 2.3 is satisfied. Then, the roots of eq. 2.4 are found straightforwardly.

This factoring approach can be applied to higher order polynomials. However, the constraints [as that of eq. 2.3] become more unwieldy and more unlikely. We leave such analysis to the reader.

Cardan's method

Cardan's method for finding the solutions to eq. 2.1 begins by making the substitution

$$x = y - \frac{p}{3} \tag{2.6}$$

This results in *the reduced cubic equation*

$$y^3 + Py + Q = 0 \tag{2.7}$$

where

$$P = q - \frac{p^2}{3} \tag{2.8a}$$

and

$$Q = \frac{2p^3}{27} - \frac{pq}{3} + r \tag{2.8b}$$

Defining

$$y = u + v \tag{2.9}$$

eq. 2.7 becomes

$$u^3 + v^3 + (u+v)(3uv + P) + Q = 0 \tag{2.10}$$

If we substitute

$$v = -\frac{P}{3u} \tag{2.11}$$

into eq. 2.10, then we obtain

$$u^6 + Qu^3 - \frac{P^3}{27} = 0 \tag{2.12a}$$

Likewise, if we substitute for u in eq. 2.10, we find

$$v^6 + Qv^3 - \frac{P^3}{27} = 0 \tag{2.12b}$$

Eqs. 2.12 are quadratic equations for u^3 and v^3, respectively, the solutions to which are

$$u^3 = \frac{1}{2}\left(-Q \pm \sqrt{Q^2 + \frac{4P^3}{27}}\right) \tag{2.13a}$$

and

$$v^3 = \frac{1}{2}\left(-Q \pm \sqrt{Q^2 + \frac{4P^3}{27}}\right) \tag{2.13b}$$

Since u and v are not the same, we take

$$u^3 = \frac{1}{2}\left(-Q + \sqrt{Q^2 + \frac{4P^3}{27}}\right) \tag{2.14a}$$

and

$$v^3 = \frac{1}{2}\left(-Q - \sqrt{Q^2 + \frac{4P^3}{27}}\right) \tag{2.14b}$$

Then

$$u_n = \omega_n \left[\frac{1}{2} \left(-Q + \sqrt{Q^2 + \frac{4P^3}{27}} \right) \right]^{\frac{1}{3}} \tag{2.15a}$$

and

$$v_n = \omega_n \left[\frac{1}{2} \left(-Q - \sqrt{Q^2 + \frac{4P^3}{27}} \right) \right]^{\frac{1}{3}} \tag{2.15b}$$

where

$$\omega_n = e^{2\pi i n/3} \tag{2.16}$$

with $n = 0, 1, 2$ are the three cube roots of 1. They are

$$\omega_0 = 1 \tag{2.17a}$$

$$\omega_1 = \frac{-1 + i\sqrt{3}}{2} = \omega_2^* \tag{2.17b}$$

and

$$\omega_2 = \frac{-1 - i\sqrt{3}}{2} = \omega_1^* \tag{2.17c}$$

Therefore, the three solutions to eqs. 2.14 are

$$u_k = \omega_k \left[\frac{1}{2} \left(-Q + \sqrt{Q^2 + \frac{4P^3}{27}} \right) \right]^{\frac{1}{3}} \tag{2.18a}$$

and

$$v_k = \omega_k \left[\frac{1}{2} \left(-Q - \sqrt{Q^2 + \frac{4P^3}{27}} \right) \right]^{\frac{1}{3}} \tag{2.18b}$$

The roots of eq. 2.1 are given by

$$x = u + v - \frac{p}{3} \tag{2.19}$$

Since there are three solutions for u and three solutions for v, there are nine different combinations of $u + v$. To determine which expressions for u and v

form the solutions of eq. 2.1, we see from eqs. 2.11 and 2.18 that the product $u_m v_k$ must satisfy

$$u_m v_k = -\omega_m \omega_k \frac{P}{3} = -\frac{P}{3} \qquad (2.20)$$

Thus, we must choose those cube roots of 1 that satisfy

$$\omega_k \omega_m = 1 \qquad (2.21)$$

Since only

$$\omega_0 \omega_0 = 1 \qquad (2.22a)$$

and

$$\omega_1 \omega_2 = \omega_2 \omega_1 = 1 \qquad (2.22b)$$

the three roots of the cubic equation of eq. 2.1 are

$$x_{00} = u_0 + v_0 - \frac{p}{3} \qquad (2.23a)$$

$$x_{12} = u_1 + v_2 - \frac{p}{3} \qquad (2.23b)$$

and

$$x_{21} = u_2 + v_1 - \frac{p}{3} \qquad (2.23c)$$

We see from eqs. 2.18 that if the discriminant satisfies

$$Q^2 + \frac{4P^3}{27} = 0 \qquad (2.24)$$

then $u_1 = v_1$ and $u_2 = v_2$, from which $x_{21} = x_{12}$. That is, if the discriminant is zero, there are only two distinct roots. If all three roots are repeated, then in addition to requiring the discriminant to be zero, we also require

$$u_0 = v_0 = u_1 = v_1 = u_2 = v_2 \qquad (2.25)$$

Since $\omega_0 \neq \omega_1 \neq \omega_2$, this also requires

$$Q = 0 \qquad (2.26a)$$

With the discriminant being zero, this leads to

$$P = 0 \tag{2.26b}$$

and

$$u_k = v_k = 0 \tag{2.27}$$

Then

$$x_{00} = x_{12} = x_{21} = -\frac{p}{3} \tag{2.28}$$

This result can also be obtained by noting that with eqs. 2.8, $P = Q = 0$ allows us to write the cubic equation in the form

$$\left(x + \frac{p}{3}\right)^3 = 0 \tag{2.29}$$

If p, q, and r (and therefore P and Q) are real, and the discriminant is negative, eqs. 2.15 can be written as

$$u_n = \omega_n \left[\frac{1}{2}\left(-Q + i\sqrt{\left|Q^2 + \frac{4P^3}{27}\right|}\right)\right]^{\frac{1}{3}} \tag{2.30a}$$

and

$$v_n = \omega_n \left[\frac{1}{2}\left(-Q - i\sqrt{\left|Q^2 + \frac{4P^3}{27}\right|}\right)\right]^{\frac{1}{3}} = u_n^* \tag{2.30b}$$

Then

$$x_{00} = u_0 + v_0 - \frac{p}{3} = u_0 + u_0^* - \frac{p}{3} = 2\operatorname{Re}(u_0) - \frac{p}{3} \tag{2.31a}$$

$$x_{12} = \omega_1 u_0 + \omega_2 v_0 - \frac{p}{3} = \omega_1 u_0 + \omega_1^* u_0^* - \frac{p}{3} = 2\operatorname{Re}(\omega_1 u_0) - \frac{p}{3} \tag{2.31b}$$

and

$$x_{21} = \omega_2 u_0 + \omega_1 v_0 - \frac{p}{3} = \omega_2 u_0 + \omega_2^* u_0^* - \frac{p}{3} = 2\operatorname{Re}(\omega_2 u_0) - \frac{p}{3} \tag{2.31c}$$

Thus, when the discriminant is negative, all three roots are real.

If the P and Q are real and the discriminant is positive, u_0 and v_0 are real. Therefore,

$$x_{00} = u_0 + v_0 - \frac{p}{3} \tag{2.32a}$$

$$x_{12} = \omega_1 u_0 + \omega_2 v_0 - \frac{p}{3} = \omega_1 u_0 + \omega_1^* v_0 - \frac{p}{3} \tag{2.32b}$$

and

$$x_{21} = \omega_2 u_0 + \omega_1 v_0 - \frac{p}{3} = \omega_1^* u_0 + \omega_1 v_0 - \frac{p}{3} = x_{12}^* \tag{2.32c}$$

That is, when the discriminant is positive, the polynomial has one real root and two complex roots, which are conjugates of one another.

Example 2.1: Roots of a cubic polynomial using Cardan's method

The solutions to

$$x^3 - 3x^2 - 10x + 24 = 0 \tag{2.33a}$$

are $x = \{2, -3, 4)$. Referring to eq. 2.6, we obtain the reduced equation

$$y^3 - 13y + 12 = 0 \tag{2.33b}$$

Referring to eqs. 2.13, the discriminant for this cubic equation is

$$Q^2 + \frac{4P^3}{27} = 12^2 - \frac{(4)(13)^3}{27} = -181.48148 < 0 \tag{2.34}$$

Therefore, the polynomial of eq. 2.33a has three real roots. From eqs 2.31, they are

$$x_{00} = 2\text{Re}(u_0) - \frac{p}{3} = 2\text{Re}\left[(-6 + i6.73575)^{\frac{1}{3}} \right] + 1 \tag{2.35a}$$

$$x_{12} = 2\text{Re}(\omega_1 u_0) - \frac{p}{3} = 2\text{Re}\left[\left(-\frac{1}{2} + i\frac{\sqrt{3}}{2} \right)(-6 + i6.73575)^{\frac{1}{3}} \right] + 1 \tag{2.35b}$$

and

$$x_{21} = 2\text{Re}(\omega_3 u_0) - \frac{p}{3} = 2\text{Re}\left[\left(-\frac{1}{2} - i\frac{\sqrt{3}}{2} \right)(-6 + i6.73575)^{\frac{1}{3}} \right] + 1 \tag{2.35c}$$

Since $\text{Re}(-6 + i6.73575) < 0$ and $\text{Im}(-6 + i6.73575) > 0$, the argument of this complex number is in the second quadrant. Therefore, we write

$$(-6 + i6.73575)^{\frac{1}{3}} = 2.08167 e^{\frac{i}{3}\left[\pi - \tan^{-1}\left(\frac{6.73575}{6}\right)\right]} = 1.50000 + i1.44338 \qquad (2.36)$$

From this we obtain the correct results $\{x_{00}, x_{12}, x_{21}\} = \{4, -3, 2\}$. $\square$

Hyperbolic function method

Following Namias, V., (1985), we begin with the identity

$$4\sinh^3(\theta) + 3\sinh(\theta) - \sinh(3\theta) = 0 \qquad (2.37)$$

which has the same form as the reduced cubic equation

$$y^3 + Py + Q = 0 \qquad (2.8)$$

Therefore, we take

$$y^3 + Py + Q = \lambda\left(4\sinh^3(\theta) + 3\sinh(\theta) - \sinh(3\theta)\right) \qquad (2.38)$$

and identify

$$y^3 = 4\lambda\sinh^3(\theta) \qquad (2.39a)$$

$$Py = 3\lambda\sinh(\theta) \qquad (2.39b)$$

and

$$Q = -\lambda\sinh(3\theta) \qquad (2.39c)$$

From the ratio of eqs. 2.39a and 2.39b, we have

$$y^2 = \frac{4P}{3}\sinh^2\theta \qquad (2.40a)$$

from which

$$y = \pm 2\sqrt{\frac{P}{3}}\sinh(\theta) \qquad (2.40b)$$

Taking the positive square root, we substitute this into eq. 2.39b to obtain

$$\lambda = 2\sqrt{\frac{P^3}{27}} \tag{2.41a}$$

from which eq. 2.39c yields

$$-\sinh(3\theta) = \sinh(3\theta \pm i\pi) = \frac{Q}{2}\sqrt{\frac{27}{P^3}} \tag{2.41b}$$

If we take the negative square root, we have

$$\lambda = -2\sqrt{\frac{P^3}{27}} \tag{2.42a}$$

and

$$\sinh(3\theta) = \frac{Q}{2}\sqrt{\frac{27}{P^3}} \tag{2.42b}$$

With θ found from either eq. 2.41b or 2.42b, y is given by eq. 2.40b.

Since $\sinh(3\theta)$ and $\sinh(3\theta \pm i\pi)$ are unchanged by adding $2k\pi i$, for some integer k, to the arguments of these hyperbolic sine functions, the values of y obtained from eq. 2.41b are given by

$$y_k = 2\sqrt{\frac{P}{3}}\sinh\left(\theta + \frac{(2k+1)i\pi}{3}\right) \tag{2.43a}$$

The values of y found using eq. 2.42b are

$$y_k = 2\sqrt{\frac{P}{3}}\sinh\left(\theta + \frac{2ki\pi}{3}\right) \tag{2.43b}$$

Thus, we have six possible solutions to the cubic equation. By substituting each solution into the cubic equation, the three solutions that are not roots are eliminated trivially.

Example 2.2: Roots of a cubic polynomial using a hyperbolic function

We consider the reduced cubic equation

$$x^3 + x + 10 = 0 \tag{2.44}$$

which has solutions $x = -2, 1 \pm 2i$.

With $P = 1$, $Q = 10$, eqs. 2.41b and 2.42b yield,

$$\theta = \frac{\sinh^{-1}\left(5\sqrt{27}\right)}{3} + \frac{i\pi}{3} \qquad (2.45a)$$

and

$$\theta = \frac{\sinh^{-1}\left(5\sqrt{27}\right)}{3} \qquad (2.45b)$$

respectively. Then, from eq. 2.43a, we obtain

$$x_0 = -\frac{2}{\sqrt{3}}\sinh(\theta) = -2 \qquad (2.46a)$$

and

$$x_1, \ x_1^* = \frac{\sinh(\theta)}{\sqrt{3}} \pm i\cosh(\theta) = 1 \pm 2i \qquad (2.46b)$$

Using eq. 2.43b, we find

$$x_0 = \frac{2}{\sqrt{3}}\sinh(\theta) = 2 \qquad (2.47a)$$

and

$$x_1, \ x_1^* = -\frac{\sinh(\theta)}{\sqrt{3}} \pm i\cosh(\theta) = -1 \pm 2i \qquad (2.47b)$$

It is trivial to demonstrate that the values of eqs. 2.47 are not the roots of the polynomial of eq. 2.44 and that those of eqs. 2.46 are those roots. $\square$

Trigonometric function method

If all three roots of a cubic polynomial are known to be real, it is possible to determine them using the trigonometric identity

$$4\cos^3(\theta) - 3\cos(\theta) - \cos(3\theta) = 0 \qquad (2.48)$$

This too has the same form as the reduced cubic equation

$$y^3 + Py + Q = 0 \qquad (2.8)$$

Therefore, if we set

$$y^3 + Py + Q = \lambda\left(4\cos^3(\theta) - 3\cos(\theta) - \cos(3\theta)\right) \tag{2.49}$$

we can make the associations

$$y^3 = \lambda 4\cos^3(\theta) \tag{2.50a}$$

$$Py = -3\lambda\cos(\theta) \tag{2.50b}$$

and

$$Q = -\lambda\cos(3\theta) \tag{2.50c}$$

Taking the square root of ratio of eqs. 2.50a and 2.50b we obtain

$$y = \pm 2\sqrt{\frac{-P}{3}}\cos(\theta) \tag{2.51}$$

With the positive square root, we substitute this into eq. 2.50b to obtain

$$\lambda = -2\sqrt{\frac{-P^3}{27}} \tag{2.52a}$$

from which eq. 2.50c yields

$$\cos(3\theta) - \frac{Q}{2}\sqrt{\frac{-27}{P^3}} \tag{2.52b}$$

If we take the negative square root, we have

$$\lambda = 2\sqrt{\frac{-P^3}{27}} \tag{2.53}$$

and

$$-\cos(3\theta) = \cos(\pi - 3\theta) = \frac{Q}{2}\sqrt{\frac{-27}{P^3}} \tag{2.54}$$

We find the values of θ from eqs. 2.52b and 2.54. Then, for each θ, we determine y from eq. 2.51.

Adding $2k\pi$ ($k = 0, 1, 2$) to the arguments of $\cos(3\theta)$ and $\cos(\pi - 3\theta)$ does not change their values. Thus, we can add integer multiples of $2\pi/3$ to the values of θ found above to obtain the y values

$$y_k = 2\sqrt{\frac{-P}{3}}\cos\left(\theta + \frac{2\pi k}{3}\right) \tag{2.55a}$$

for the positive square root in eq. 2.51. For the negative square root, we have

$$y_k = 2\sqrt{\frac{-P}{3}}\cos\left(\frac{\pi}{3} - \theta + \frac{2\pi k}{3}\right) = 2\sqrt{\frac{-P}{3}}\cos\left(\frac{(2k+1)\pi}{3} - \theta\right) \tag{2.55b}$$

As with the analysis using hyperbolic functions, we have six possible solutions to the cubic equation, three of which are eliminated trivially by substitution.

If θ is real, then $-1 \le \cos(3\theta), \cos(\pi - 3\theta) \le 1$ which requires

$$-1 \le \frac{Q}{2}\sqrt{\frac{-27}{P^3}} \le 1 \tag{2.56a}$$

from which

$$Q^2 + \frac{4P^3}{27} \le 0 \tag{2.56b}$$

The left side of this inequality is the discriminant in the Cardan solution and this condition results in all three roots being real.

If this discriminant is zero, then

$$\frac{Q}{2}\sqrt{\frac{-27}{P^3}} = \pm 1 \tag{2.57}$$

from which we can take

$$\theta + \frac{2\pi k}{3} = \left\{0, \frac{2\pi}{3}, \frac{4\pi}{3}\right\} \tag{2.58a}$$

or

$$\frac{(2k+1)\pi}{3} - \theta = \left\{0, \frac{2\pi}{3}, \frac{4\pi}{3}\right\} \tag{2.58b}$$

Referring to eq. 2.51, these results yield

$$y_k = \{2P, -P, -P\} \tag{2.59}$$

Example 2.3: Roots of a cubic polynomial using a trigonometric function

We consider

$$x^3 - 3x^2 - 10x + 24 = 0 \tag{2.33a}$$

which has solutions 2, −3, and 4. The reduced cubic equation for this is

$$y^3 - 13y + 12 = 0 \qquad (2.33b)$$

Therefore,

$$\cos(3\theta + 2\pi k) = \frac{12}{2}\sqrt{\frac{-27}{-13^3}} \qquad (2.60a)$$

and

$$\cos((2k+1)\pi - 3\theta) = \frac{12}{2}\sqrt{\frac{-27}{-13^3}} \qquad (2.60b)$$

From the three values of θ obtained from eq. 2.60a, we find from eq. 2.55a that $x = \{5, -2, 0\}$ which do not satisfy eq. 2.33a. Using the values of θ given by eq. 2.60b, we obtain $x = \{4, -3, 2\}$ which are the correct roots. □

2.2 Roots of a Quartic/Biquadratic Polynomial

The general form of the *quartic* (also called the *biquadratic*) equation is

$$x^4 + px^3 + qx^2 + rx + s = 0 \qquad (2.61)$$

Such an equation can have four real roots, two real and two complex roots, or four complex roots.

Ferrari's method

Ferrari's method for finding the roots of a quartic polynomial is like the method of completing the square. One determines the values of α, β, and γ such that

$$x^4 + px^3 + qx^2 + rx + s + (\alpha x + \beta)^2 = \left(x^2 + \frac{p}{2}x + \gamma\right)^2 \qquad (2.62)$$

It is straightforward to show that α, β, and γ satisfy

$$\alpha^2 - 2\gamma = \frac{p^2}{4} - q \qquad (2.63a)$$

$$2\alpha\beta - p\gamma = -r \tag{2.63b}$$

and

$$\beta^2 = \gamma^2 - s \tag{2.63c}$$

Equating the product $\alpha^2\beta^2$ from eqs. 2.63a and 2.63c to $\alpha^2\beta^2$ obtained from eq. 2.63b, we find

$$\gamma^3 - \frac{q}{2}\gamma^2 + \left(\frac{pr}{4} - s\right)\gamma + \left(\frac{qs}{2} - \frac{(sp^2 + r^2)}{8}\right) = 0 \tag{5.64}$$

Three values of γ are then obtained using one of the methods described in Sect. 2.1. For each of these values, we obtain corresponding values of α and β from eqs. 2.63a and 2.63c.

Adding $(\alpha x + \beta)^2$ to both sides of eq. 2.61, we obtain

$$\left(x^2 + \frac{p}{2}x + \gamma\right)^2 = (\alpha x + \beta)^2 \tag{2.65a}$$

the solutions to which are given by

$$x^2 + \frac{p}{2}x + \gamma = \pm(\alpha x + \beta) \tag{2.65b}$$

Not all solutions to eq. 2.65b will be solutions to the original quartic equation. Each value of x must be tested in eq. 2.61.

Example 2.4: Roots of a quartic polynomial by Ferrari's method

For the quartic equation

$$x^4 - 2x^3 - 12x^2 + 10x + 3 = 0 \tag{2.66}$$

(see Conkwright, N., 1941, p. 79) we write

$$x^4 - 2x^3 - 12x^2 + 10x + 3 + (\alpha x + \beta)^2 = \left(x^2 - x + \gamma\right)^2 \tag{2.67}$$

where γ satisfies

$$\gamma^3 + 6\gamma^2 - 8\gamma - 32 = 0 \tag{2.68}$$

The values γ that satisfy eq. 2.68 are

$$\gamma = -2, -2\left(1 \pm \sqrt{5}\right) \tag{2.69a}$$

and for each of these three values

$$\alpha = \pm\sqrt{12 + 2\gamma} \tag{2.69a}$$

and

$$\beta = \pm\sqrt{\gamma^2 - 3} \tag{2.69b}$$

We obtain

$$\gamma = -2, \ \alpha = \pm 3, \ \beta = \pm 1 \tag{2.70a}$$

$$\gamma = -2\left(1 + \sqrt{5}\right), \ \alpha = \pm\sqrt{9 - 4\sqrt{5}}, \ \beta = \pm\sqrt{21 + 8\sqrt{5}} \tag{2.70b}$$

and

$$\gamma = -2\left(1 - \sqrt{5}\right), \ \alpha = \pm\sqrt{9 + 4\sqrt{5}}, \ \beta = \pm\sqrt{21 - 8\sqrt{5}} \tag{2.70c}$$

Since there are many values of α, β, and γ, there are many values of x. Those that satisfy eq. 2.66 are

$$x = -3, \ 1, \ \left(2 \pm \sqrt{5}\right) \tag{2.71}$$

These are the roots of the polynomial of eq. 2.66. $\square$

Descarte's method

To use the *Descarte method* for determining the roots of a fourth order polynomial, we begin by obtaining the *reduced quartic* equation, the quartic equation without a y^3 term. This is achieved by substituting

$$x = y - \frac{P}{4} \tag{2.72}$$

in eq. 2.61 to obtain

$$y^4 + Py^2 + Qy + R = 0 \tag{2.73}$$

We then write this reduced quartic equation in factored form

$$y^4 + Py^2 + Qy + R = \left(y^2 + y\sqrt{\gamma} + \alpha\right)\left(y^2 - y\sqrt{\gamma} + \beta\right) = 0 \tag{2.74a}$$

Multiplying this factored form, we have

$$y^4 + (\alpha + \beta - \gamma)y^2 + \sqrt{\gamma}(\beta - \alpha)y + \alpha\beta = 0 \qquad (2.74b)$$

Comparing this to eq. 2.73, we obtain

$$\alpha + \beta - \gamma = P \qquad (2.75a)$$

$$\sqrt{\gamma}(\beta - \alpha) = Q \qquad (2.75b)$$

and

$$\alpha\beta = R \qquad (2.75c)$$

from which

$$\alpha = \frac{1}{2}\left(P + \gamma - \frac{Q}{\sqrt{\gamma}}\right) \qquad (2.76a)$$

and

$$\beta = \frac{1}{2}\left(P + \gamma + \frac{Q}{\sqrt{\gamma}}\right) \qquad (2.76b)$$

Substituting these into eq. 2.75c, we obtain

$$\gamma^3 + 2P\gamma^2 + \left(P^2 - 4R\right)\gamma - Q^2 = 0 \qquad (2.77)$$

which we solve by one of the methods described earlier. We determine α and β for each value of γ, then, from eq. 2.74a, we find the values of y that satisfy

$$y^2 + \sqrt{\gamma}y + \alpha = 0 \qquad (2.78a)$$

and

$$y^2 - \sqrt{\gamma}y + \beta = 0 \qquad (2.78b)$$

Each value of y is then tested in eq. 2.73 to determine if it is a root of the reduced quartic equation.

Example 2.5: Roots of a quartic polynomial by Descarte's method

The quartic equation

$$x^4 - 2x^2 + 8x - 3 = 0 \qquad (2.79)$$

(see Conkwright, N., 1941, pp. 83, 84) is in reduced form. Eq. 2.74 for this equation is

$$\gamma^3 - 4\gamma^2 + 16\gamma - 64 = 0 \qquad (2.80)$$

which has solutions

$$\gamma = 4 \qquad (2.81a)$$

and

$$\gamma = \pm 4i = 4e^{\pm i\pi/2} \qquad (2.81b)$$

from which we obtain

$$\gamma = 4, \ \alpha = -1, \ \beta = 3 \qquad (2.82a)$$

$$\gamma = 4i, \ \alpha = \left(-1 + 2i - 2e^{-i\pi/4}\right), \ \beta = \left(-1 + 2i + 2e^{-i\pi/4}\right) \qquad (2.82b)$$

and

$$\gamma = -4i, \ \alpha = \left(-1 - 2i - 2e^{i\pi/4}\right), \ \beta = \left(-1 - 2i + 2e^{i\pi/4}\right) \qquad (2.82c)$$

The solutions to the quadratic equations of eqs. 2.78 are then found for each set of values of α, β, and γ. Each solution is then tested to see if it satisfies the quartic equation. We find the solutions to eq. 2.79 to be

$$x = -1 \pm \sqrt{2}, \ 1 \pm i\sqrt{2} \qquad (2.83)\square$$

2.3 Regula Falsi and Newton–Raphson Methods

The *regula falsi* (false position) and the *Newton–Raphson methods* are iterative techniques that approximate a zero of any function $f(x)$ given in closed form.

Regula falsi method

For the regula falsi method, one begins by testing $f(x)$ to identify two points $x = a$ and $x = b$ such that $f(a)$ and $f(b)$ have opposite signs indicating that there is at least one zero between a and b. $f(x)$ is then approximated by the linear expression

$$f(x) = \alpha^{(0)}x + \beta^{(0)} \qquad (2.84)$$

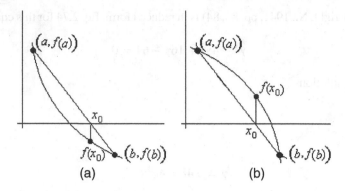

Fig. 2.1 Approximation to the zero of $f(x)$

where

$$\alpha^{(0)} = \frac{f(b) - f(a)}{b - a} \qquad (2.85a)$$

and

$$\beta^{(0)} = \frac{af(b) - bf(a)}{b - a} \qquad (2.85b)$$

The first estimate of a zero, found by setting $f(x) = 0$ in eq. 2.84, is given by

$$x_0 = -\frac{\beta^{(0)}}{\alpha^{(0)}} = -\frac{af(b) - bf(a)}{f(b) - f(a)} \qquad (2.86)$$

If $f(x_0)$ and $f(a)$ have opposite signs as in Fig. 2.1a, we then take

$$a' = a \qquad (2.87a)$$

and

$$b' = x_0 \qquad (2.87b)$$

If $f(x_0)$ and $f(b)$ are of opposite sign as in Fig. 2.1b, we set

$$a' = x_0 \qquad (2.87c)$$

and

$$b' = b \qquad (2.87d)$$

We then apply the regula falsi method with a' and b' to obtain the next iteration of the zero as

$$x_1 = -\frac{a'f(b') - b'f(a')}{f(b') - f(a')} \tag{2.88}$$

After p iterations of this process we have

$$x_p = -\frac{a^{(p)}f\left(b^{(p)}\right) - b^{(p)}f\left(a^{(p)}\right)}{f\left(b^{(p)}\right) - f\left(a^{(p)}\right)} \tag{2.89}$$

The process is continued until two successive approximations, x_p and x_{p+1}, have the same value to a required level of accuracy.

Example 2.6: The regula falsi method

For

$$f(x) = e^x - 3 \tag{2.90}$$

the exact solution (to five decimal places) is

$$x = \ell n(3) = 1.09861 \tag{2.91}$$

We note that

$$f(0.5) = e^{0.5} - 3 = -1.35128 \tag{2.92a}$$

and

$$f(1.5) = e^{1.5} - 3 = 1.48169 \tag{2.92b}$$

are of opposite sign indicating that a zero of $f(x)$ lies in the interval $0.5 \le x \le 1.5$. Starting with $a = 0.5$ and $b = 1.5$, the first approximation to the zero of $f(x)$ is obtained from

$$x_0 = -\frac{0.5f(1.5) - 1.5f(0.5)}{f(1.5) - f(0.5)} = 0.97698 \tag{2.93}$$

Since $f(0.97698) = -0.34357 < 0$ and $f(1.5) = 1.48169 > 0$, we take

$$a' = x_0 = 0.97698 \tag{2.94a}$$

and

$$b' = b = 1.5 \tag{2.94b}$$

From this, we generate the second approximation

$$x_1 = -\frac{0.97698f(1.5) - 1.5f(0.97698)}{f(1.5) - f(0.97698)} = 1.07543 \qquad (2.95)$$

The correct value, 1.09861, accurate to five decimals, is obtained after seven iterations of this method. □

Newton–Raphson method

Like the regula falsi technique, the Newton–Raphson method approximates $f(x)$ by a linear expression. But for the Newton–Raphson method, that linear expression is given by the first two terms in its Taylor series expansion about some starting point a chosen by the user;

$$f(x) \simeq f(a) + f'(a)(x - a) \qquad (2.96)$$

The first approximation to the zero is found by setting $f(x) = 0$ which yields

$$x_0 = a - \frac{f(a)}{f'(a)} \qquad (2.97a)$$

Replacing a by x_0, the next iteration is given by

$$x_1 = x_0 - \frac{f(x_0)}{f'(x_0)} \qquad (2.97b)$$

or, after p iterations,

$$x_{p+1} = x_p - \frac{f(x_p)}{f'(x_p)} \qquad (2.98)$$

Again, the process is continued until x_p and x_{p+1} have the same value to some predefined level of accuracy.

A potential problem can occur using the Newton–Raphson method. If the one of the iterations produces a value x_k that is at or near an extremum of $f(x)$, then at that point

$$f'(x_k) \simeq 0 \qquad (2.99)$$

If $f(x_k) \neq 0$ in eq. 2.98, the estimate of x_{k+1} becomes a very large number, and convergence of the method to the zero of $f(x)$ is destroyed.

A modification of the Newton–Raphson method that avoids this problem is to select a starting point a that is far from an extremum of $f(x)$ so that $f'(a)$ is distinctly different from zero. Then eq. 2.98 is replaced by

$$x_{p+1} = x_p - \frac{f(x_p)}{f'(a)} \tag{2.100}$$

This tends to converge to the zero of $f(x)$ more slowly than that given in eq. 2.98, but it avoids the problem.

Example 2.7: The Newton–Raphson method

As in example 2.6, we again consider finding the zero of

$$f(x) = e^x - 3 \tag{2.90}$$

Using eq. 2.98, this involves iterations of

$$x_{p+1} = x_p - \frac{e^{x_p} - 3}{e^{x_p}} \tag{2.101}$$

With a starting value $a = 1$, we obtain

$$x_0 = a - \frac{e^a - 3}{e^a} = 1 - \frac{e - 3}{e} = 1.10364 \tag{2.102a}$$

from which

$$x_1 = x_0 - \frac{e^{x_0} - 3}{e^{x_0}} = 1.10364 - \frac{e^{1.10364} - 3}{e^{1.10364}} = 1.09862 \tag{2.102b}$$

After four iterations, we obtain the correct result to five decimal places, 1.09861. Using the less precise eq. 2.100 with $a = 1$, we generate approximations to the zero of $f(x)$ from

$$x_{p+1} = x_p - \frac{e^{x_p} - 3}{e^a} = x_p - \frac{e^{x_p} - 3}{e} \tag{2.103}$$

we find that 63 iterations of eq. 2.103 are needed to obtain agreement with the exact value to five decimal places. $\square$

2.4   Zeros by Interpolation for Discrete Data

In ch. 1, it was noted that when $f(x)$ is represented by discrete data points (presented as a table of data), an approximation to the function can be created from such data by interpolation.

If $f(x)$ changes sign at least once over the entire range of x, this indicates that there is at least one zero of $f(x)$. To form an interpolated function, we pick two points in the range of x, x_1, and x_N, and form a polynomial representation of $f(x)$ given by

$$f(x) = \sum_{k=1}^{N} \mu_k(x) f(x_k) \qquad (1.11)$$

where $\mu_k(x)$ is the $(N-1)$th order polynomial

$$\mu_k(x) \equiv \frac{(x-x_1)...(x-x_{k-1})(x-x_{k+1})...(x-x_N)}{(x_k-x_1)...(x_k-x_{k-1})(x_k-x_{k+1})...(x_k-x_N)} \qquad (1.7)$$

If it is known (for example, from theory) that the x dependence of $f(x)$ arises through an expression $q(x)$, then $f(x)$ can be interpolated by

$$f(x) = f[q(x)] = \sum_{k=1}^{N} \mu_k[q(x)] f(x_k) \qquad (1.20)$$

where $\mu_k[q(x)]$ is the $(N-1)$th order polynomial in q

$$\mu_k[q(x)] \equiv \frac{[q(x)-q(x_1)]...[q(x)-q(x_{k-1})][q(x)-q(x_{k+1})]...[q(x)-q(x_N)]}{[q(x_k)-q(x_1)]...[q(x_k)-q(x_{k-1})][q(x_k)-q(x_{k+1})]...[q(x_k)-q(x_N)]}$$
$$(1.18)$$

If there is an indication of one zero, x_1 and x_N should be the first and last x values in the table. If there are multiple (two or more) zeros, each zero is found separately and x_1 and x_N should be chosen as values of x on opposite sides of each zero.

To determine each zero of the polynomial form of $f(x)$, we define

$$y = f(x) \qquad (2.104a)$$

When the x dependence of $f(x)$ arises through the expression $q(x)$, we write

$$y = f[q(x)] \qquad (2.104b)$$

The values of x that are the zeros of $f(x)$ are found by setting $y = 0$ in eqs. 2.104.

The interpolations of the inverses of these definitions are given by

$$x = f^{-1}(y) = \sum_{k=1}^{N} \mu_k(y) f^{-1}(y_k) = \sum_{k=1}^{N} \mu_k(y) x_k \qquad (2.105a)$$

for the polynomial $f(x)$, and, if it is known that the x dependence of $f(x)$ arises through $q(x)$,

$$q(x) = f^{-1}(y) = \sum_{k=1}^{N} \mu_k(y) f^{-1}(y_k) = \sum_{k=1}^{N} \mu_k(y) q(x_k) \qquad (2.105b)$$

where

$$\mu_k(y) = \frac{(y-f_1)(y-f_2)...(y-f_{k-1})(y-f_{k+1})...(y-f_{N-1})(y-f_N)}{(f_k-f_1)(f_k-f_2)...(f_k-f_{k-1})(f_k-f_{k+1})...(f_k-f_{N-1})(f_k-f_N)} \qquad (2.106)$$

The zero of $f(x)$ is found by setting $y = 0$ in an interpolation representation given in one of the eqs. 2.105. For the polynomial interpolation

$$x_0 = f^{-1}(0) = \sum_{k=1}^{N} \mu_k(0) x_k =$$

$$(-1)^{N-1} \sum_{k=1}^{N} \frac{f_1...f_{k-1} f_{k+1}...f_N}{(f_k-f_1)...(f_k-f_{k-1})(f_k-f_{k+1})...(f_k-f_N)} x_k \qquad (2.107a)$$

and, when the x dependence of $f(x)$ arises through the expression $q(x)$,

$$q(x_0) = f^{-1}(0) = \sum_{k=1}^{N} \mu_k(0) q(x_k) =$$

$$(-1)^{N-1} \sum_{k=1}^{N} \frac{f_1...f_{k-1} f_{k+1}...f_N}{(f_k-f_1)...(f_k-f_{k-1})(f_k-f_{k+1})...(f_k-f_N)} q(x_k) \qquad (2.107b)$$

Example 2.8: Single zero by interpolation

The data in Table 2.1 are values of $f(x)$, where x is an angle expressed in degrees.

The fact that x represents an angle and values of the magnitude of $f(x)$ are less than one, it is reasonable to consider that $f(x)$ somehow depends on x through $\sin(x)$ and/or $\cos(x)$. As an educated guess, we attempt to interpolate over

$$q(x) = \sin(x) \qquad (2.108)$$

From eq. 2.105b, the zero of $f(x)$ arises from

$$\sin(x_0) = \sum_{k=1}^{N} \mu_k(0) \sin(x_k) \qquad (2.109)$$

x (°)	$f(x)$
10	−0.42764
20	−0.29998
30	−0.14645
40	0.00153
50	0.17047
60	0.30593
70	0.41092
80	0.47730
90	0.50000

Table 2.1 Table of data points

$$\sin(x_0) = \sum_{k=1}^{N} \mu_k(0)\sin(x_k) =$$

$$(-1)^{N-1} \sum_{k=1}^{N} \frac{f_1 \cdots f_{k-1} f_{k+1} \cdots f_N}{(f_k - f_1) \cdots (f_k - f_{k-1})(f_k - f_{k+1}) \cdots (f_k - f_N)} \sin(x_k) \qquad (2.110)$$

From this we obtain

$$x_0 = 39.04699° \qquad (2.111)$$

Using a polynomial interpolation, eq. 2.107a yields

$$x_0 = \sum_{k=1}^{N} \mu_k(0)x_k \qquad (2.112)$$

from which we find

$$x_0 = 38.55133° \qquad (2.113)$$

Although it is assumed that we do not know it, the data in Table 2.1 was obtained from

$$f(x) = \sin^{3/2}(x) - \tfrac{1}{2} \qquad (2.114)$$

from which, the zero of $f(x)$ is given by

$$x_0 = \sin^{-1}\left(.5^{2/3}\right) = 39.04721° \qquad (2.115)$$

Comparing this exact value to the results given in eqs. 2.113 and 2.115, we see that the polynomial interpolation yields a reasonable estimate of the zero of $f(x)$, but interpolation over $\sin(x)$ results in an estimate that is more accurate. □

Example 2.9: Multiple zeros by interpolation

The data given in Table 2.2 are numerical values of

$$f(x) = e^{2x} - 7e^x + 10 = (e^x - 2)(e^x - 5) \qquad (2.116)$$

x	$f(x)$
0.25	2.66054
0.50	1.17723
0.75	−0.33731
1.00	−1.63892
1.25	−2.24991
1.50	−1.28629
1.75	2.83323
2.00	12.87476
2.25	33.60298

Table 2.2 Tabulated values of the function of $f(x) = e^{2x} - 7e^x + 10$

which has two zeros at

$$\{x_0, x_1\} = \{\ell n(2), \ell n(5)\} = \{0.69315, 1.60944\} \qquad (2.117)$$

To apply the regula falsi method, we see from Table 2.2 that the two zeros which we call x_0 and x_1. We see from Table 2.2 that these zeros occur within the intervals $x_0 \, \varepsilon \, (0.50, 0.75)$ and $x_1 \, \varepsilon \, (1.50, 1.75)$. We apply the method over the first interval to find x_0 and independently, x_1 is found by applying the method over the second interval.

Using the Newton–Raphson approach, we determine x_0 by choosing a starting point a within the first interval. A convenient choice is the data point in the table that is closest to that zero. For this example, that point would be

$$a = 0.75 \qquad (2.118)$$

Independently, x_1 is found using a starting point closest to the zero in the second interval, which for this example is

$$a' = 1.50 \qquad (2.119)$$

To obtain x_0 and x_1 using interpolation, we note that for $x_0 \, \varepsilon \, [0.25, 1.25]$ the values of $f(x)$ monotonically decrease, and for $x_1 \, \varepsilon \, [1.25, 2.25]$ $f(x)$ is increasing monotonically. Thus, $[0.25, 1.25]$ is the recommended interval over which one should interpolate to find x_0 and $[1.25, 2.25]$ should be used to determine x_1.

Since we have already discussed the application of these methods to find each zero, we leave this problem for the reader to solve (see problem 7). □

2.5   Roots of a Polynomial by Root Squaring Methods

A *real polynomial* is one for which the coefficient of each power of x is real. Such a polynomial can have both real and complex roots. For a real polynomial, the complex roots must occur in conjugate pairs. That is, only a product of factors containing complex roots of the form

$$\left(x - re^{i\theta}\right)\left(x - re^{-i\theta}\right) = x^2 - 2r\cos\theta + r^2 \equiv x^2 + px + q \qquad (2.120)$$

yields an expression with real coefficients p and q. Thus, a real polynomial that has real and complex roots can be written as

$$f(x) = R(x)\left(x^2 + p_1 x + q_1\right)^{M_1}\left(x^2 + p_2 x + q_2\right)^{M_2}...\left(x^2 + p_n x + q_n\right)^{M_n} \qquad (2.121)$$

where $R(x)$ is a real polynomial, the roots of which are all the real roots of $f(x)$. That is, the roots of each factor $(x^2 + p_k x + q_k)$ $1 \le k \le n$ are complex.

If the number of real roots of $f(x)$ is the same as the order of $f(x)$, then $f(x)$ does not have complex roots and $R(x) = f(x)$. If the number of real roots is less than the order of $f(x)$, then $f(x)$ has an even number of complex roots and $R(x) \ne f(x)$.

All roots are real

Let $f(x)$ be a polynomial of order N with real roots $\{x_1, x_2, ..., x_N\}$. Then $f(x)$ can be written in the form

$$f(x) = (x - x_1)(x - x_2)...(x - x_{N-1})(x - x_N) \qquad (2.122a)$$

which is equivalent to

$$f(x) = x^N + a_1 x^{N-1} + ... + a_{N-1}x + a_N \qquad (2.122b)$$

where a_0, the coefficient of x^N, has been taken to be 1.

It is well known that the coefficients of the various powers of x are related to the roots of $f(x)$ by

$$-a_1 = \sum_{k=1}^{N} x_k = (x_1 + x_2 + ... + x_N) \qquad (2.123a)$$

$$a_2 = \sum_{\substack{k,m=1 \\ m>k}}^{N} x_k x_m$$

$$= (x_1 x_2 + x_1 x_3 + ... + x_1 x_N + x_2 x_3 + x_2 x_4 + ... + x_3 x_4 + ... + x_{N-1}x_N) \qquad (2.123b)$$

$$-a_3 = \sum_{\substack{k,m,n=1 \\ n>m>k}}^{N} x_k x_m x_n = (x_1 x_2 x_3 + x_1 x_2 x_4 + \ldots + x_1 x_2 x_N + x_1 x_3 x_4 + x_1 x_3 x_5$$

$$+ \ldots + x_1 x_3 x_N + \ldots + x_2 x_3 x_4 + \ldots + x_{N-2} x_{N-1} x_N) \tag{2.123c}$$

$$(-1)^N a_N = \prod_{k=1}^{N} x_k \tag{2.123d}$$

(see, for example, Conkwright, N., 1941, p. 29).

The polynomial $g(x)$ with roots $\{-x_1, -x_2, \ldots, -x_N\}$ can be written as

$$g(x) = (x + x_1)(x + x_2)\ldots(x + x_{N-1})(x + x_N) = (-1)^N f(-x) \tag{2.124a}$$

or equivalently as

$$g(x) = x^N - a_1 x^{N-1} + a_2 x^{N-2} + \ldots + (-1)^N a_N \tag{2.124b}$$

The product of these polynomials can, therefore, be expressed as

$$f(x)g(x) = (x - x_1)(x + x_1)(x - x_2)(x + x_2)\ldots(x - x_N)(x + x_N)$$
$$= \left(x^2 - x_1^2\right)\left(x^2 - x_2^2\right)\ldots\left(x^2 - x_N^2\right) \tag{2.125a}$$

That is, $f(x)g(x)$ is an even polynomial of order $2N$, the roots of which are the squares of the roots of $f(x)$. This product can also be expressed as

$$f(x)g(x) =$$
$$\left(x^N + a_1 x^{N-1} + a_2 x^{N-2} + \ldots + a_N\right)\left(x^N - a_1 x^{N-1} + a_2 x^{N-2} + \ldots + (-1)^N a_N\right)$$
$$\equiv \left(x^2\right)^N + b_1 \left(x^2\right)^{N-1} + \ldots + b_{N-1}\left(x^2\right) + b_N \tag{2.125b}$$

where

$$b_1 = -\left(a_1^2 - 2a_2\right) \tag{2.126a}$$

$$b_2 = a_2^2 - 2a_1 a_3 + 2a_4 \tag{2.126b}$$

$$b_3 = -\left(a_3^2 - 2a_2 a_4 + 2a_1 a_5 - 2a_6\right) \tag{2.126c}$$

and so on. The general coefficient for the new polynomial is given by

$$b_k = (-1)^k \left(a_k^2 - 2a_{k-1}a_{k+1} + 2a_{k-2}a_{k+2} - \ldots + (-1)^k 2a_{2k} \right) \qquad (2.127)$$

It is obvious that for $2k > N$, for $k-m < 0$ and for $k + m > N$, the coefficients a_{2k}, a_{k-m}, or a_{k+m} are meaningless and should be ignored (set to zero). A simple way to write code for such a process is to define the coefficients a_n such that

$$a_0 \equiv 1 \qquad (2.128a)$$

and

$$a_n \equiv 0 \quad n<0 \ \text{ and } \ n>N \qquad (2.128b)$$

Since the roots of $f(x)g(x)$ are the squares of the roots of $f(x)$, then from eqs. 2.125 and 2.126,

$$- b_1 = \sum_{k=1}^{N} x_k^2 \qquad (2.129a)$$

$$b_2 = \sum_{\substack{k,m=1 \\ m>k}}^{N} x_k^2 x_m^2 \qquad (2.129b)$$

$$- b_3 = \sum_{\substack{k,m,n=1 \\ n>m>k}}^{N} x_k^2 x_m^2 x_n^2 \qquad (2.129c)$$

$$(-1)^N b_N = \prod_{k=1}^{N} x_k^2 \qquad (2.129d)$$

Root squaring methods involve iterations of this procedure. For example, coefficients c_k are generated from the coefficients b_k, and the fourth powers of the roots of $f(x)$ are related to the c_k by

$$- c_1 = \sum_{k=1}^{N} x_k^4 \qquad (2.130a)$$

$$c_2 = \sum_{\substack{k,m=1 \\ m>k}}^{N} x_k^4 x_m^4 \qquad (2.130b)$$

and so on.

After p iterations of this process, the sums involving the 2^p power of the roots are given by

$$-\beta_1^{(p)} = \sum_{k=1}^{N} x_k^{2^p} \tag{2.131a}$$

$$\beta_2^{(p)} = \sum_{\substack{k,m=1 \\ m>k}}^{N} x_k^{2^p} x_m^{2^p} \tag{2.131b}$$

and so on.

To allow for *multiple* or *repeated* roots, we write $f(x)$ in the form

$$f(x) = (x - x_1)^{M_1} (x - x_2)^{M_2} ... (x - x_n)^{M_n} \tag{2.132a}$$

where the *multiplicities* of the roots $M_1, M_2, ..., M_n$ satisfy

$$\sum_{k=1}^{n} M_k = N \tag{2.132b}$$

If $M_k = 1$, then x_k is called a *simple* root.

Let $\{x_1, ..., x_n\}$ be the different real roots of $f(x)$, ordered by size so that

$$|x_1| > |x_2| > ... > |x_n| \tag{2.133}$$

After p iterations of the root squaring process described in eqs. 2.125, we obtain a polynomial of the form

$$\phi(x) = \left(x^{2^p}\right)^N + \beta_1^{(p)} \left(x^{2^p}\right)^{N-1} + ... + \beta_N^{(p)} \tag{2.134}$$

which has N roots

$$\left\{ M_1 x_1^{2^p}, \ M_2 x_2^{2^p}, \ ..., \ M_n x_n^{2^p} \right\}$$

For p large enough, the sum of these roots will be dominated by the term involving the root with the largest magnitude. Thus, we can approximate this sum by

$$-\beta_1^{(p)} = \sum_{k=1}^{N} x_k^{2^p} \simeq M_1 x_1^{2^p} \tag{2.135a}$$

Therefore, the largest root is given approximately by

$$x_1 \simeq \pm \left(-\frac{\beta_1^{(p)}}{M_1} \right)^{1/2^p} \qquad (2.135b)$$

The two values, $\pm$, come from taking the real values of $(1)^{1/2^p}$.

The number of iterations needed for an acceptable level of accuracy (the rate of convergence of iteration process) will depend on the multiplicity M_k. For any $M_k \geq 1$, there will be some power p such that

$$M_k^{1/2^p} \simeq 1 \qquad (2.136)$$

to a required level of accuracy. For example, if five decimal place accuracy is required, then for each M_k in the range $2 \leq M_k \leq 10$, the number of iterations needed to satisfy eq. 2.136 is $20 \leq p \leq 22$.

We see that eq. 2.135b will yield the value of x_1, but not the value of M_1. Thus, for this approach to be useful, one must know the values of the multiplicities beforehand.

Graeffe root squaring method

If all multiplicities are 1, and thus, all roots are simple roots, the method is called the *Graeffe root squaring method* (Graeffe, H.K., 1837). When all multiplicities are 1, we see from eqs. 2.129a and 2.129b that the second largest root can be found from

$$x_2 \simeq \pm \left(-\frac{\beta_2^{(p)}}{\beta_1^{(p)}} \right)^{1/2^p} \qquad (2.137a)$$

and from eqs. 2.129b and 2.129c, we obtain

$$x_3 \simeq \pm \left(-\frac{\beta_3^{(p)}}{\beta_2^{(p)}} \right)^{1/2^p} \qquad (2.137b)$$

and so on.

Example 2.10: Simple roots by the Graeffe root squaring method

The cubic polynomial of example 2.1

$$x^3 - 3x^2 - 10x + 24 = 0 \qquad\qquad (2.33a)$$

has roots 2, –3, and 4.
    With

$$\{a_0, a_1, a_2, a_3\} = \{1, -2, -10, 24\} \qquad\qquad (2.138)$$

and taking a coefficient to be zero if its index is less than 0 or greater than 4, we refer to eq. 2.127 to obtain

$$b_1 = -\left(a_1^2 - 2a_2\right) = -29 \qquad\qquad (2.139a)$$

$$b_2 = a_2^2 - 2a_1a_3 + 2a_4 = a_2^2 - 2a_1a_3 = 244 \qquad\qquad (2.139b)$$

and

$$b_3 = -\left(a_3^2 - 2a_2a_4 + 2a_1a_5 - 2a_6\right) = -a_3^2 = -576 \qquad\qquad (2.139c)$$

If we know that all the roots are simple, the first iteration of the Graeffe process predicts the roots to be

$$x_1 \simeq \pm\sqrt{-b_1} = \pm 5.38517 \qquad\qquad (2.140a)$$

$$x_2 = \pm\sqrt{\frac{b_2}{b_1}} = \pm 2.90065 \qquad\qquad (2.140b)$$

and

$$x_3 = \pm\sqrt{-\frac{b_3}{b_2}} = \pm 1.53644 \qquad\qquad (2.140c)$$

The constants for the second iteration are

$$c_1 = -\left(b_1^2 - 2b_2\right) = -353 \qquad\qquad (2.141a)$$

$$c_2 = b_2^2 - 2b_1b_3 = 26128 \qquad\qquad (2.141b)$$

and

$$c_3 = -b_3^2 = -331776 \qquad\qquad (2.141c)$$

from which

$$x_1 \simeq \pm \sqrt[4]{-c_1} = \pm 4.33455 \qquad (2.142a)$$

$$x_2 = \pm \sqrt[4]{-\frac{c_2}{c_1}} = \pm 2.93314 \qquad (2.142b)$$

and

$$x_3 = \pm \sqrt[4]{-\frac{c_3}{c_2}} = \pm 1.88771 \qquad (2.142c)$$

As can be seen, the magnitudes of the Graeffe approximated roots are approaching the correct values.

After five iterations, we obtain

$$x_1 \simeq \pm 4.00001 \qquad (2.143a)$$

$$x_2 = \pm 2.99999 \qquad (2.143b)$$

and

$$x_3 = \pm 2.00000 \qquad (2.143c)$$

By testing all six values of x in eq. 2.33a, we find that the correct roots are $\{4.00001, -2.99999, 2.00000\}$, and we must discard the other three values $\{-4.00001, 2.99999, -2.00000\}$. □

For problems where the multiplicities of the roots are not known, we present a modified Graeffe method which will yield the root and multiplicity quickly.

Let x_1, the root with the largest magnitude, be a multiple root with a multiplicity M_1. We note that after p iterations, we have

$$\sum_{k=1}^{n} M_k x_k^{2^p} = -\beta_1^{(p)} \simeq M_1 x_1^{2^p} \qquad (2.144a)$$

and after $p + 1$ iterations,

$$\sum_{k=1}^{n} M_k x_k^{2^{p+1}} = -\beta_1^{(p+1)} \simeq M_1 x_1^{2^{p+1}} = M_1 x_1^{2 \times 2^p} \qquad (2.144b)$$

Then,

$$\frac{-\beta_1^{(p+1)}}{-\beta_1^{(p)}} \simeq \frac{M_1 x_1^{2 \times 2^p}}{M_1 x_1^{2^p}} = x_1^{2^p} \qquad (2.145a)$$

which is independent of M_1. Therefore,

$$x_1 \simeq \pm \left(\frac{-\beta_1^{(p+1)}}{-\beta_1^{(p)}} \right)^{1/2^p} \tag{2.145b}$$

Once x_1 is determined to a required level of accuracy, M_1 is found from eq. 2.144b to be

$$M_1 \simeq -\frac{\beta_1^{(p+1)}}{x_1^{2^{p+1}}} \tag{2.146}$$

which must converge to a positive integer.

Once the integer value of M_1 is established, the second largest root can be determined from the approximations

$$-\beta_1^{(p)} - M_1 x_1^{2^p} = \sum_{m=2}^{N} x_k^{2^p} \simeq M_2 x_2^{2^p} \tag{2.147a}$$

and

$$-\beta_1^{(p+1)} - M_1 x_1^{2^{p+1}} \simeq M_2 x_2^{2^{p+1}} \tag{2.147b}$$

Taking the ratio of these equations, we have

$$x_2^{2^p} \simeq \frac{-\beta_1^{(p+1)} - M_1 x_1^{2^{p+1}}}{-\beta_1^{(p)} - M_1 x_1^{2^p}} \tag{2.148a}$$

which yields

$$x_2 \simeq \pm \left(\frac{-\beta_1^{(p+1)} - M_1 x_1^{2^{p+1}}}{-\beta_1^{(p)} - M_1 x_1^{2^p}} \right)^{1/2^p} \tag{2.148b}$$

With this value, the multiplicity of x_2 is given by eq. 2.147b to be

$$M_2 \simeq \frac{-\beta_1^{(p+1)} - M_1 x_1^{2^{p+1}}}{x_2^{2^{p+1}}} \tag{2.149}$$

Example 2.11: Multiple roots by root squaring

The polynomial

$$f(x) = x^3 - 8x^2 + 21x - 18 \tag{2.150}$$

has roots $\{3, 3, 2\}$.

Referring to eqs. 2.126 or eq. 2.127, the coefficients for the first iteration are

$$\left\{\beta_1^{(1)}, \beta_2^{(1)}, \beta_3^{(1)}\right\} = \{-22, 153, -324\} \qquad (2.151a)$$

The second iteration yields coefficients

$$\left\{\beta_1^{(2)}, \beta_2^{(2)}, \beta_3^{(2)}\right\} = \{-178, 9153, -104976\} \qquad (2.151b)$$

and for the third iteration, we obtain

$$\left\{\beta_1^{(3)}, \beta_2^{(3)}, \beta_3^{(3)}\right\} = \{-13378, 4640953, -11019960576\} \qquad (2.151c)$$

Applying eqs. 2.145b and 2.146, the first iteration yields

$$\frac{-\beta_1^{(1)}}{-a_1} = \frac{x_1^2}{x_1} = x_1 \simeq \frac{22}{8} = 2.75000 \qquad (2.152a)$$

with the multiplicity of this root given by

$$M_1 = \frac{-\beta_1^{(1)}}{x_1^2} = \frac{22}{(2.75)^2} \simeq 2.90909 \qquad (2.152b)$$

The second iteration yields

$$\sqrt{\frac{-\beta_1^{(2)}}{-\beta_1^{(1)}}} = \sqrt{\frac{x_1^4}{x_1^2}} = x_1 \simeq \pm\sqrt{\frac{178}{22}} = \pm 2.84445 \qquad (2.153a)$$

with an estimated multiplicity of

$$M_1 = \frac{-\beta_1^{(2)}}{x_1^4} = \frac{178}{(2.84445)^4} \simeq 2.71910 \qquad (2.153b)$$

From the third iteration, we obtain

$$\sqrt[4]{\frac{-\beta_1^{(3)}}{c_1}} = \sqrt[4]{\frac{x_1^8}{x_1^4}} = x_1 \simeq \pm\sqrt[4]{\frac{13378}{178}} = \pm 2.94437 \qquad (2.154a)$$

with multiplicity

$$M_1 = \frac{-\beta_1^{(3)}}{x_1^8} = \frac{178}{(2.94437)^8} \simeq 2.36837 \qquad (2.154b)$$

We see that x_1 is converging to ± 3, and M_1 is converging to 2. We find that six iterations of this process yield the correct root and multiplicity accurate to five decimal places. The values of the coefficients for this iteration are

$$\left\{ \beta_1^{(6)}, \beta_2^{(6)}, \beta_3^{(6)} \right\} = \left\{ -6.86737 \times 10^{30}, 1.17902 \times 10^{61}, -2.17491 \times 10^{80} \right\}$$

(2.155)

Since these numbers are so large, one must be alert to the possibility of computer memory overflow problems. Testing the values of x_1 in eq. 2.150, we discard those that are not roots. The correct results are

$$x_1 = 3.00000$$

(2.156a)

and

$$M_1 = 2.00000$$

(2.156b)

To determine the value of the third root and its multiplicity, we set $x_1 = 3$ and $M_1 = 2$ in eqs. 2.148b and 2.149 to obtain

$$\frac{x_3^2}{x_3} = x_3 = \left(\frac{-\beta_1^{(1)} - M_1 x_1^2}{-a_1 - M_1 x_1} \right)^{1/2^0} = \left(\frac{22 - 18}{8 - 6} \right) = 2$$

(2.157a)

and

$$M_3 = \left(\frac{-\beta_1^{(1)} - M_1 x_1^2}{x_3^2} \right) = \left(\frac{22 - 18}{4} \right) = 1$$

(2.157b)

All higher iterations yield the same values of $x_3 = 2$ and $M_3 - 1$. $\square$

Complex roots

As noted above, if some of the roots of a real polynomial $f(x)$ are complex, $f(x)$ can be written in the form

$$f(x) = R(x)\left(x^2 + p_1 x + q_1\right)^{M_1} \left(x^2 + p_2 x + q_2\right)^{M_2} ... \left(x^2 + p_n x + q_n\right)^{M_n}$$

(**2.117**)

The real roots of $f(x)$ are the roots of $R(x)$. They can be found by methods we have introduced in this chapter (the Newton–Raphson method, for example). Thus, we can specify the polynomial

$$R(x) = (x - x_{n+1})...(x - x_N) = x^{N-n} + r_1 x^{N-n-1} + ... + r_{N-n}$$

(2.158)

The coefficients of $R(x)$ are determined from these real roots of $f(x)$ by

$$r_1 = -\sum_{k=n+1}^{N} x_k \qquad (2.159a)$$

$$r_2 = \sum_{\substack{k=n+1 \\ m=n+2 \\ m>k}}^{N} x_k x_m \qquad (2.159b)$$

and so on.

The complex roots of $f(x)$ are the roots of the polynomial

$$\frac{f(x)}{R(x)} = \left(x^2 + p_1 x + q_1\right)^{M_1} \left(x^2 + p_2 x + q_2\right)^{M_2} ... \left(x^2 + p_n x + q_n\right)^{M_n} \qquad (2.160)$$

If the order of $f(x)$ differs from that of $R(x)$ by two, then $f(x)$ has one conjugate pair of complex roots of multiplicity one. Then $R(x)$ can be written as

$$R(x) = (x - x_3)...(x - x_N) \equiv x^{N-2} + r_1 x^{N-3} + ... + r_{N-2} \qquad (2.161)$$

with

$$r_1 = -\sum_{k=3}^{N} x_k \qquad (2.162a)$$

$$r_2 = \sum_{\substack{k=3 \\ m=4 \\ m>k}}^{N} x_k x_m \qquad (2.162b)$$

etc. Then

$$\frac{f(x)}{R(x)} = x^2 + px + q \qquad (2.163)$$

To find p and q we note that

$$q = \frac{f(0)}{R(0)} = \frac{a_N}{r_{N-2}} \qquad (2.164a)$$

and

$$p = \frac{d}{dx}\left[\frac{f(x)}{R(x)}\right]\Bigg|_{x=0} = \frac{f'(0)}{R(0)} - \frac{f(0)}{R^2(0)}R'(0) = \frac{a_{N-1}}{r_{N-2}} - \frac{a_N}{r_{N-2}^2}r_{N-3} \tag{2.164b}$$

The conjugate pair of complex roots of $f(x)$ are then found from

$$x^2 + px + q = 0 \tag{2.165}$$

Since these roots are complex, we know that

$$p^2 - 4q < 0 \tag{2.166a}$$

Therefore, the complex roots are given by

$$x_\pm = \frac{1}{2}\left(-p \pm i\sqrt{4q - p^2}\right) \tag{2.166b}$$

Example 2.12: One conjugate pair of complex roots of multiplicity one

The polynomial

$$f(x) = x^4 - 7x^3 + 21x^2 - 37x + 30 \tag{2.167}$$

has two real roots, $\{3, 2\}$ found by one of the methods described earlier. Therefore, $f(x)$ has one conjugate pair of complex roots of multiplicity one. To find them, we form

$$R(x) = (x - 3)(x - 2) = x^2 - 5x + 6 \tag{2.168}$$

With $r_1 = -5$ and $r_2 = 6$, we see from eqs. 2.164 that

$$q = \frac{a_4}{r_2} = \frac{30}{6} = 5 \tag{2.169a}$$

and

$$p = \frac{a_3}{r_2} - \frac{a_4}{r_2^2}r_1 = \frac{-37}{6} - \frac{30}{36}(-5) = -2 \tag{2.169b}$$

Therefore, referring to eq. 2.166b, we have

$$x_\pm = \frac{1}{2}\left(2 \pm i\sqrt{20 - 4}\right) = 1 \pm 2i \tag{2.170}\square$$

If the orders of $f(x)$ and $R(x)$ differ by four, then $f(x)$ has either two different conjugate pairs of complex roots, each of multiplicity one, or it has one conjugate pair of roots of multiplicity two.

If $f(x)$ has two different conjugate pairs of complex roots of multiplicity one, it is of the form

$$f(x) = R(x)\left(x^2 + p_1 x + q_1\right)\left(x^2 + p_2 x + q_2\right) \tag{2.171a}$$

where

$$R(x) = \left(x^{N-4} + r_1 x^{N-5} + \ldots + r_{N-7}x^3 + r_{N-6}x^2 + r_{N-5}x + r_{N-4}\right) \tag{2.171b}$$

Then

$$\frac{f(x)}{R(x)} = \left(x^2 + p_1 x + q_1\right)\left(x^2 + p_2 x + q_2\right) =$$
$$x^4 + (p_1 + p_2)x^3 + (q_1 + q_2 + p_1 p_2)x^2 + (q_1 p_2 + q_2 p_1)x + q_1 q_2 \tag{2.172}$$

where

$$q_1 q_2 = \frac{f(0)}{R(0)} = \frac{a_N}{r_{N-4}} \tag{2.173a}$$

$$q_1 p_2 + q_2 p_1 = \frac{d}{dx}\left[\frac{f(x)}{R(x)}\right]\Bigg|_{x=0} = \frac{a_{N-1}}{r_{N-4}} - \frac{a_N r_{N-5}}{r_{N-4}^2} \tag{2.173b}$$

$$q_1 + q_2 + p_2 p_1 = \frac{1}{2}\frac{d^2}{dx^2}\left[\frac{f(x)}{R(x)}\right]\Bigg|_{x=0}$$
$$= \frac{a_{N-2}}{r_{N-4}} - \frac{a_{N-1} r_{N-5}}{r_{N-4}^2} - \frac{a_N r_{N-6}}{r_{N-4}^2} + \frac{a_N r_{N-5}^2}{r_{N-4}^3} \tag{2.173c}$$

and

$$p_2 + p_1 = \frac{1}{6}\frac{d^3}{dx^3}\left[\frac{f(x)}{R(x)}\right]\Bigg|_{x=0} =$$
$$\frac{a_{N-3}}{r_{N-4}} - \frac{a_{N-2} r_{N-5}}{r_{N-4}^2} - \frac{a_{N-1} r_{N-6}}{r_{N-4}^2} - \frac{a_N r_{N-7}}{r_{N-4}^2} + \frac{a_{N-1} r_{N-5}^2}{r_{N-4}^3} + \frac{a_N r_{N-6} r_{N-5}}{r_{N-4}^3} - \frac{a_N r_{N-5}^3}{r_{N-4}^4} \tag{2.173d}$$

It is very cumbersome to solve such a set of equations. (For example, by eliminating p_1, p_2, and q_2, we obtain a sixth order equation for q_1.)

If there is one conjugate pair of complex roots of multiplicity two, then $f(x)$ can be expressed as

$$f(x) = R(x)\left(x^2 + px + q\right)^2 =$$
$$\left(x^{N-4} + r_1 x^{N-5} + \dots + r_{N-5} x + r_{N-4}\right)\left(x^2 + px + q\right)^2 \qquad (2.174)$$

Then p and q are given by

$$q^2 = \frac{f(0)}{R(0)} = \frac{a_N}{r_{N-4}} \qquad (2.175a)$$

from which

$$q = \pm\sqrt{\frac{a_N}{r_{N-4}}} \qquad (2.175b)$$

and

$$2pq = \frac{d}{dx}\left[\frac{f(x)}{R(x)}\right]_{x=0} = \frac{a_{N-1}}{r_{N-4}} - \frac{a_N}{r_{N-4}^2} r_{N-5} \qquad (2.175c)$$

which yields

$$p = \pm\frac{1}{2}\sqrt{\frac{r_{N-4}}{a_N}}\left(\frac{a_{N-1}}{r_{N-4}} - \frac{a_N}{r_{N-4}^2} r_{N-5}\right) \qquad (2.175d)$$

Therefore, when $f(x)$ has one pair of conjugate complex roots, it is fairly straightforward to find that conjugate pair.

When the orders of $f(x)$ and $R(x)$ differ by four, one can use eqs. 2.173 to determine whether the complex roots are two distinct conjugate pairs of multiplicity one, or one conjugate pair of multiplicity two.

If there is one conjugate pair of multiplicity two, one of the values of p and one of the values of q found from eqs. 2.175b and 2.175d must also satisfy eqs. 2.173c and 2.173d with $p_1 = p_2 = p$ and $q_1 = q_2 = q$, respectively. That is, one of these solutions for p and one for q must also satisfy

$$2q + p^2 = \frac{a_{N-2}}{r_{N-4}} - \frac{a_{N-1} r_{N-5}}{r_{N-4}^2} - \frac{a_N r_{N-6}}{r_{N-4}^2} + \frac{a_N r_{N-5}^2}{r_{N-4}^3} \qquad (2.176a)$$

and

$$2p = \frac{a_{N-3}}{r_{N-4}} - \frac{a_{N-2} r_{N-5}}{r_{N-4}^2} - \frac{a_{N-1} r_{N-6}}{r_{N-4}^2} - \frac{a_N r_{N-7}}{r_{N-4}^2} + \frac{a_{N-1} r_{N-5}^2}{r_{N-4}^3} + \frac{a_N r_{N-6} r_{N-5}}{r_{N-4}^3} - \frac{a_N r_{N-5}^3}{r_{N-4}^4} \qquad (2.176b)$$

If one value of p and one value of q found from eqs. 2.175b and 2.175d do not satisfy eqs. 2.176, there are two conjugate pairs of complex roots of multiplicity one.

Example 2.13: One conjugate pair of complex roots of multiplicity two

We construct the sixth order polynomial

$$f(x) = x^6 - 9x^5 + 40x^4 - 114x^3 + 209x^2 - 245x + 150 \qquad (2.177)$$

by taking

$$f(x) = \left(x^2 - 2x + 5\right)^2 (x - 3)(x - 2) \qquad (2.178)$$

That is, as in example 2.12, $f(x)$ has two real roots $\{3, 2\}$ and

$$R(x) = (x - 3)(x - 2) = x^2 - 5x + 6 \qquad \mathbf{(2.127)}$$

Thus, we have

$$\{a_1, a_2, a_3, a_4, a_5, a_6\} = \{-9, 40, -114, 209, -245, 150\} \qquad (2.179a)$$

and

$$\{r_1, r_2\} = \{-5, 6\} \qquad (2.179b)$$

Then, from eqs. 2.175 with $a_0 = r_0 = 1$, with $a_k = r_k = 0$ for $k < 0$ and with $r_k = 0$ for $k > 2$,

$$q = \pm\sqrt{\frac{a_6}{r_2}} = \pm\sqrt{\frac{150}{6}} = \pm 5 \qquad (2.180a)$$

and

$$p = \pm\frac{1}{2}\sqrt{\frac{r_2}{a_6}}\left(\frac{a_5}{r_2} - \frac{a_6}{r_2^2}r_1\right) = \pm\frac{1}{2}\sqrt{\frac{6}{150}}\left(\frac{-245}{6} - \frac{150}{36}(-5)\right) = \pm 2 \quad (2.180b)$$

Referring to eq. 2.173c we have

$$\frac{a_4}{r_4} - \frac{a_5 r_1}{r_4^2} - \frac{a_6 r_0}{r_4^2} + \frac{a_6 r_1^2}{r_4^3}$$

$$= \frac{209}{6} - \frac{(-245)(-5)}{6^2} - \frac{(150)(1)}{6^2} + \frac{(150)(-5)^2}{6^3} = 14 \qquad (2.181a)$$

and

$$2q + p^2 = \pm 10 + 4 = \begin{cases} 14 \\ -6 \end{cases} \qquad (2.181b)$$

From eq. 2.173d,

$$
\begin{aligned}
&\frac{a_3}{r_4} - \frac{a_4 r_1}{r_4^2} - \frac{a_5 r_0}{r_4^2} - \frac{a_6 r_{-1}}{r_4^2} + \frac{a_5 r_1^2}{r_4^3} + 2\frac{a_6 r_0 r_1}{r_4^3} - \frac{a_6 r_1^3}{r_4^4} \\
&= \frac{(-114)}{6} - \frac{(209)(-5)}{6^2} - \frac{(-245)(1)}{6^2} - \frac{(150)(0)}{6^2} \\
&+ \frac{(-245)(-5)^2}{6^3} + 2\frac{(150)(1)(-5)}{6^3} - \frac{(150)(-5)^3}{6^4} = -4
\end{aligned}
\qquad (2.182a)
$$

and

$$
2p = \pm 4 \qquad (2.182b)
$$

The results of eqs. 2.180 and eqs. 2.181 and 2.182 tell us that there is one conjugate pair of complex roots of multiplicity two. Therefore, there are four possible quadratic factors, from which we obtain the complex roots. They are:

- $x^2 + 2x + 5$
- $x^2 - 2x + 5$
- $x^2 + 2x - 5$
- $x^2 - 2x - 5$.

Since the discriminants in the solutions of

$$
x^2 \pm 2x - 5 = 0 \qquad (2.183a)
$$

are positive, these factors do not yield complex roots. Therefore, the complex roots of multiplicity two are found from

$$
x^2 \pm 2x + 5 = 0 \qquad (2.183b)
$$

The solutions to eq. 2.183b are

$$
x = \pm 1 \pm 2i \qquad (2.184)
$$

By testing each of these solutions in the original polynomial, we find that $x = -1 \pm 2i$ are not roots of $f(x)$. Therefore, $x = 1 \pm 2i$ are the two complex roots of multiplicity two. $\square$

Problems

1. Find the solution to

   (a)
   $$
   x^3 - 4x^2 + 2x - 8 = 0
   $$

(b) $$6x^3 + 3x^2 + 2x + 1 = 0$$

by factoring.
2. Find the roots of $x^3 + 9x^2 + 23x + 15 = 0$ using Cardan's method.
3. Find the three real roots of $x^3 + 9x^2 + 23x + 15 = 0$ using the method based on the trigonometric identity of eq. 2.37.
4. (a) Derive an identity that relates $\sin(3\theta)$, $\sin(\theta)$, and $\sin^3(\theta)$ analogous to the identity given in eq. 2.37.
   (b) From the result of part (a), find the trigonometric solution to a cubic equation for which all roots are real.
5. (a) Derive an expression that relates $\cosh(3\phi)$, $\cosh(\phi)$, and $\cosh^3(\phi)$.
   (b) From the result of part (a), find the hyperbolic solution to a cubic equation for which all roots are real.
6. Find the solution to $x^4 - 6x^3 + 12x^2 - 14x + 3 = 0$ by Ferrari's method.
   Hint: One of the roots of $y^3 - 6y^2 + 18y - 20 = 0$ is $y = 2 + \sqrt{29}$
7. Find the solution to $x^4 - 3x^2 - 6x - 2 = 0$ by Descarte's method.
   Hint: One of the roots of $y^3 - 6y^2 + 17y - 36 = 0$ is $y = 2 + 2i\sqrt{2}$
8. Find the two zeros of the function represented numerically in Table 2.2 by

   (a) the regula falsi method to three decimal place accuracy
   (b) the Newton–Raphson method to three decimal place accuracy
   (c) interpolating over intervals surrounding each of the zeros.

9. All the roots of $f(x) = x^4 - 13x^3 + 62x^2 - 128x + 96$ are real. Use the root squaring method to find the roots of $f(x)$ and their multiplicities.
10. The polynomial $f(x) = x^4 - 2x^3 + 3x^2 - 2x + 2$ has no real roots. Use root squaring methods to determine the roots of $f(x)$ and their multiplicities.
11. The polynomial $f(x) = x^5 + 3x^4 + 4x^3 - 4x - 4$ has one real root of multiplicity 1. Use root squaring methods to find the roots of $f(x)$ and their multiplicities.

Chapter 3
SERIES

3.1 Definitions and Convergence

We consider a sum of terms written as

$$S(z) \equiv \sum_{n=n_0}^{N} \sigma_n(z) = \sigma_{n_0}(z) + \sigma_{n_0+1}(z) + \dots + \sigma_N(z) \tag{3.1}$$

When n_0 and N are both finite integers, $S(z)$ is a *finite sum* or simply a *sum*. If n_0 and N are not both finite integers (that is $n_0 = -\infty$ and/or $N = \infty$), then $S(z)$ is called an *infinite series* or simply a *series*.

When a series $S(z)$ is finite for some value of z, $S(z)$ is said to *converge to* (or is *convergent at*) z. If $S(z)$ is infinite at z, the series *diverges* (or is *divergent at*) z.

One requirement for the convergence of $S(z)$ at z is

$$\lim_{n \to \infty} \sigma_n(z) = 0 \tag{3.2}$$

This is not a sufficient condition, but it is a necessary one.

When the terms in $S(z)$ are of the form

$$\sigma_n(z) = c_n(z - z_0)^n \tag{3.3}$$

the series is called a power series. The coefficient c_n is independent of z, and z_0 is a specified value of z in the domain of $S(z)$.

H. Cohen, *Numerical Approximation Methods*, DOI 10.1007/978-1-4419-9837-8_3,
© Springer Science+Business Media, LLC 2011

Taking $n_0 = 0$, such a series

$$S(z) = \sum_{n=0}^{\infty} c_n (z - z_0)^n =$$

$$\sum_{n=0}^{\infty} \frac{1}{n!} \left(\frac{d^n S(z)}{dz^n} \right)_{z=z_0} (z - z_0)^n \equiv \sum_{n=0}^{\infty} \frac{S^{(n)}(z_0)}{n!} (z - z_0)^n \qquad (3.4)$$

is called a *Taylor series expanded about* z_0. A Taylor series with $z_0 = 0$ is called a *MacLaurin series*.

Two types of series that one encounters frequently are *absolute* and *alternating* series. An absolute series is one in which all terms have the same sign.[1]

An absolute series is of the form

$$S(z) = \sum_{n=0}^{\infty} |\sigma_n(z)| \qquad (3.5a)$$

An alternating series is one for which the signs of the terms alternate. It is of the form

$$S(z) = \sum_{n=0}^{\infty} (-1)^n |\sigma_n(z)| \qquad (3.5b)$$

Example 3.1: Absolute and alternating power series

The MacLaurin series for e^z is

$$e^z = \sum_{n=0}^{\infty} \frac{z^n}{n!} \qquad (3.6)$$

The series on the right side of eq. 3.6 converges (to e^z) for all finite values of z. If z is positive ($z = |z|$), then

$$e^z = \sum_{n=0}^{\infty} \frac{|z|^n}{n!} = 1 + |z| + \frac{|z|^2}{2!} + \frac{|z|^3}{3!} + \ldots \qquad (3.7a)$$

is an absolute series. If z is negative ($z = -|z|$), then

$$e^z = \sum_{n=0}^{\infty} (-1)^n \frac{|z|^n}{n!} = 1 - |z| + \frac{|z|^2}{2!} - \frac{|z|^3}{3!} + \ldots \qquad (3.7b)$$

is an alternating series.

---

[1] There is no loss of generality in taking all terms to be positive. If all terms in $S(z)$ are negative, then the series in which all terms are positive converges to $-S(z)$.

The geometric series

$$\frac{1}{1-z} = \sum_{n=0}^{\infty} z^n = 1 + z + z^2 + z^3 + \dots \qquad (3.8a)$$

is convergent for $|z| < 1$. For positive values of z, this is an absolute series. When z is negative, the series is the alternating series

$$\frac{1}{1+|z|} = 1 - |z| + |z|^2 - |z|^3 + \dots = \sum_{n=0}^{\infty} (-1)^n |z|^n \qquad (3.8b)$$

The logarithm series

$$\ell n(1+z) = \sum_{n=1}^{\infty} (-1)^{n+1} \frac{z^n}{n} = z - \frac{z^2}{2} + \frac{z^3}{3} - \qquad (3.9)$$

also converges for $|z| < 1$. It is an alternating series if z is positive, and for negative z, it is an absolute series for $-\ell n(1-|z|)$.

The series

$$\sin z = \sum_{n=0}^{\infty} (-1)^n \frac{z^{2n+1}}{(2n+1)!} = z - \frac{z^3}{3!} + \frac{z^5}{5!} - \dots \qquad (3.10)$$

and

$$\cos z = \sum_{n=0}^{\infty} (-1)^n \frac{z^{2n}}{(2n)!} = 1 - \frac{z^2}{2!} + \frac{z^4}{4!} - \dots \qquad (3.11)$$

converge for all finite values of z and are alternating series for both positive and negative values of z. $\square$

There are many tests for convergence of an infinite series discussed in the literature (see, for example, Cohen, H., 1992, pp. 125–134). Three of the most commonly used tests, which are discussed in Appendix 2, are

- the *Cauchy integral test* for an absolute series
- the *limit test* of eq. 3.2 for the alternating series
- the *Cauchy ratio test* for any series.

Referring to eq. 3.2, if a series is to converge, there is some value of the summation index n such that when $n \geq N$,

$$|\sigma_{n+1}(z)| < |\sigma_n(z)| \qquad (A2.2)$$

We write

$$S(z) = \sum_{n=n_0}^{N-1} \sigma_n(z) + \sum_{n=N}^{\infty} \sigma_n(z) \equiv S_1(z) + S_2(z) \tag{3.12}$$

Since $S_1(z)$ is a finite sum, it has a well-defined value and thus does not affect whether $S(z)$ converges or not. That is, the convergence or divergence of $S(z)$ is determined entirely by whether $S_2(z)$ is finite or not.

Cauchy integral test for an absolute series

By treating the summation index as a continuous variable, the *Cauchy envelope integral* is

$$\int_N^{\infty} |\sigma(n,z)| dn$$

As shown in Appendix 2, for an absolute series,

$$\int_N^{\infty} |\sigma(n,z)| dn < S_2(z) < |\sigma_N(z)| + \int_N^{\infty} |\sigma(n,z)| dn \tag{A2.9}$$

Therefore, if the envelope integral is finite, $S_2(z)$, and thus $S(z)$, is less than a finite value (its upper bound) and therefore converges. If this integral is infinite, $S_2(z)$ greater than ∞ (its lower bound) and thus diverges.

Limit test for an alternating series

It is also shown in Appendix 2 that for some large index N,

$$|\sigma_N(z)| - |\sigma_{N+1}(z)| < S_2(z) < |\sigma_N(z)| \tag{A2.14}$$

For $S_2(z)$, and therefore for $S(z)$ to converge, we require only that these bounds of $S_2(z)$ be finite. But a necessary condition for the convergence of a series is

$$\lim_{n\to\infty} \sigma_n(z) = 0 \tag{3.2}$$

Thus, this limit test is the requirement for convergence of an alternating series.

Cauchy ratio test for an any series

It is also shown in Appendix 2 that if the *Cauchy ratio*, defined by

$$\rho_N(z) \equiv \left| \frac{\sigma_{N+1}(z)}{\sigma_N(z)} \right| \tag{A2.15a}$$

satisfies

$$0 \leqslant \lim_{N \to \infty} \rho_N(z) < 1 \tag{3.13a}$$

$S_2(z)$, and therefore $S(z)$, converges. If

$$\lim_{N \to \infty} \rho_N(z) = 1 \tag{3.13b}$$

the convergence or divergence of $S_2(z)$, and thus $S(z)$, cannot be determined by the ratio test. If

$$\lim_{N \to \infty} \rho_N(z) > 1 \tag{3.13c}$$

$S_2(z)$, and thus $S(z)$, is divergent.

Example 3.2: Convergence tests

The *Riemann zeta function* is defined by the series

$$\varsigma(z) = \sum_{n=1}^{\infty} \frac{1}{n^z} \tag{3.14}$$

where z can be a complex number. For this discussion, we take z to be real. With

$$\sigma_n = \frac{1}{n^z} \tag{3.15}$$

the Cauchy ratio test yields

$$\lim_{N \to \infty} \frac{N^z}{(N+1)^z} = 1 \tag{3.16}$$

for all z. Thus, we cannot determine the convergence or divergence of the Riemann zeta series using the Cauchy ratio test.

Since the series is an absolute series, the Cauchy integral is

$$\int_N^{\infty} \frac{1}{n^z} dn = \begin{cases} -\left. \frac{1}{(z-1)n^{z-1}} \right|_N^{\infty} & z \neq 1 \\ \ln(n) \big|_N^{\infty} & z = 1 \end{cases} \tag{3.17}$$

From this, we deduce that if $z \leq 1$, the envelope integral is infinite and $\varsigma(z)$ diverges. For $z > 1$, the envelope integral is finite, so $\varsigma(z)$ converges.

For the alternating series

$$S(z) = \sum_{n=1}^{\infty} \frac{(-1)^n}{n^z} \tag{3.18}$$

we have

$$\lim_{N \to \infty} \frac{1}{N^z} = \begin{cases} 0 & z>0 \\ 1 & z = 0 \\ \infty & z<0. \end{cases} \tag{3.19}$$

Thus, the series converges for $z > 0$, and diverges when $z \leq 0$. $\square$

3.2 Summing a Series

To *sum* a series means to find the closed form of the function $S(z)$ that the series represents. If the terms in an infinite series do not contain a variable z, the series sums to a number. Some examples of the summing of series are shown in example 3.1.

Example 3.3: Simple summing of series

One technique for summing a series is to recognize that it has terms that are like the terms in a familiar series. Then, summing that series involves simply writing down the result

Referring to eq. 3.6,

$$\sum_{n=0}^{\infty} \frac{1}{2^n n!} = \sum_{n=0}^{\infty} \left(\frac{z^n}{n!} \right)_{z=\frac{1}{2}} = e^{\frac{1}{2}} \tag{3.20a}$$

and from eq. 3.9, we have

$$\sum_{n=1}^{\infty} (-1)^{n+1} \frac{1}{3^{\frac{n}{2}} n} = \sum_{n=1}^{\infty} (-1)^{n+1} \left(\frac{z^n}{n} \right)_{z=\frac{1}{\sqrt{3}}} = \ell n \left(1 + \frac{1}{\sqrt{3}} \right) \tag{3.20b} \square$$

If the series one is trying to sum is not in a recognizable form, it may be possible to manipulate that series into a familiar form. There is no prescription for doing this, but through examples, we will indicate some guidelines about what to look for in deciding how to manipulate the series.

Algebraic manipulation

Example 3.4: Algebraic manipulation to sum a series

We note that the series

$$S = \sum_{n=0}^{\infty} \frac{1}{(2n+3)(2n+4)} = \sum_{n=0}^{\infty} \left[\frac{1}{(2n+3)} - \frac{1}{(2n+4)} \right] \tag{3.21a}$$

$$= \frac{1}{3} - \frac{1}{4} + \frac{1}{5} - \frac{1}{6} + \frac{1}{7} - \frac{1}{8} + \ldots$$

contains terms like those in the logarithm series of eq. 3.9. Since the first terms in the logarithm series are 1 and $-1/2$, we write

$$S = 1 - \frac{1}{2} + \frac{1}{3} - \frac{1}{4} + \frac{1}{5} - \frac{1}{6} + \frac{1}{7} - \frac{1}{8} + \ldots - \left(1 - \frac{1}{2}\right)$$

$$= \left(\sum_{n=1}^{\infty} (-1)^{n+1} \frac{1}{n} \right) - \frac{1}{2} = (\ell n(1+z))_{z=1} - \frac{1}{2} = \ell n(2) - \frac{1}{2} \tag{3.21b}\square$$

Example 3.5: Algebraic manipulation to sum a series

Terms in the series

$$S(z) = \sum_{n=1}^{\infty} \frac{z^{2n+1}}{3^n n!} \tag{3.22a}$$

contain $1/n!$ which, referring to eq. 3.6, suggests that we try to manipulate $S(z)$ to look like a series for an exponential function.

Writing $S(z)$ as

$$S(z) = z \sum_{n=1}^{\infty} \frac{1}{n!} \left(\frac{z^2}{3} \right)^n = z \left[-1 + \sum_{n=0}^{\infty} \frac{1}{n!} \left(\frac{z^2}{3} \right)^n \right] \tag{3.22b}$$

we obtain

$$S(z) = z \left[e^{\frac{1}{3}z^2} - 1 \right] \tag{3.23}\square$$

Manipulations using integration

To manipulate terms in a series using differentiation and/or integration, the series must depend on a variable. In some examples in which the series does not depend on a variable, we begin with a series that does contain a variable, perform the calculus operations, and then assign a value to the variable.

When a power series contain factors involving the summation index in the denominator of the terms, one looks to integrating a known series.

Example 3.6: Integration to sum a series

The terms in the series

$$S = \sum_{n=1}^{\infty} \frac{1}{n(n+1)2^n} \tag{3.24}$$

contain terms in $1/n$ instead of $1/n!$ or $1/(2n)!$ or $1/(2n+1)!$ or terms in which the only dependence on n occurs in $1/2^n$. This suggests that we try to manipulate a logarithm series. However, referring to eq. 3.9, we see that S has a factor of $(n+1)$ in the denominator that does not exist in the terms in the logarithm series.

Replacing z by $-z$ in eq. 3.9, we see that

$$-\ell n(1-z) = z + \frac{z^2}{2} + \frac{z^3}{3} + \ldots = \sum_{n=1}^{\infty} \frac{z^n}{n} \tag{3.25}$$

Thus, to obtain a factor of $(n+1)$ in the denominator of the terms of the logarithm series, we consider

$$\int [-\ell n(1-z)]dz = \sum_{n=1}^{\infty} \frac{z^{n+1}}{n(n+1)} + C = (1-z)[\ell n(1-z) - 1] \tag{3.26}$$

We evaluate the constant of integration by noting that each term in the integrated series contains at least two factors of z. Therefore, setting $z = 0$, the series is zero and we obtain $C = -1$. Thus, eq. 3.26 can be written as

$$\sum_{n=1}^{\infty} \frac{z^n}{n(n+1)} = \frac{(1-z)}{z} \ell n(1-z) + \frac{1}{z} \tag{3.27a}$$

Setting $z = 1/2$, we find

$$S = \sum_{n=1}^{\infty} \frac{1}{n(n+1)2^n} = 2 - \ell n(2) \tag{3.27b}\square$$

Example 3.7: Integration to sum a series

The factor $1/n!$ suggests that to sum the series

$$S = \sum_{n=0}^{\infty} \frac{2^n}{(n+2)(n+4)n!} \tag{3.28}$$

we manipulate the series for e^z given in eq. 3.6.

To obtain the factor $(n+2)$ in the denominator, we must integrate z^{n+1}. Thus, we consider

$$\int z e^z \, dz = (z-1)e^z = \sum_{n=0}^{\infty} \frac{z^{n+2}}{(n+2)n!} + C_1 \tag{3.29a}$$

Again, by setting $z = 0$, we obtain $C_1 = -1$, so that

$$(z-1)e^z + 1 = \sum_{n=1}^{\infty} \frac{z^{n+2}}{(n+2)n!} \tag{3.29b}$$

We obtain the factor $(n+4)$ by integrating z^{n+3}. Therefore, we multiply eq. 3.29b by z and integrate to obtain

$$\int z[(z-1)e^z + 1] \, dz - (z^2 - 3z + 3)e^z + \frac{z^2}{2} = \sum_{n=1}^{\infty} \frac{z^{n+4}}{(n+2)(n+4)n!} + C_2 \tag{3.30}$$

With $C_2 = 3$, found by setting $z = 0$, we have

$$\frac{(z^2 - 3z + 3)e^z + \frac{z^2}{2} - 3}{z^4} = \sum_{n=1}^{\infty} \frac{z^n}{(n+2)(n+4)n!} \tag{3.31}$$

Then, with $z = 2$, we obtain

$$S = \sum_{n=1}^{\infty} \frac{2^n}{(n+2)(n+4)n!} = \frac{e^2 - 1}{16} \tag{3.32}\square$$

Summing a series using differentiation

Since derivatives of a power series yield factors of the summation index in the numerator, differentiation is an appropriate manipulation when there are such factors in the numerator of the power series we are trying to sum.

Example 3.8: Differentiation to sum a series

Since the series

$$S = \sum_{n=0}^{\infty} (-1)^n \frac{(n+2)(2n+3)\pi^{2n}}{4^n(2n+1)!} \tag{3.33a}$$

contains the factors $1/(2n+1)!$ and $(-1)^n$, we look to manipulating the series for $\sin z$ given in eq. 3.10. The factors of $(n+2)$ and $(2n+3)$ in the numerator suggest that we manipulate a known series using differentiation.

The terms in the series for $\sin z$ contain only odd powers of z. Differentiating an odd power of z cannot yield the factor $(n+2)$. But we note that the factor $(n+2)$ can be written as $(2n+4)/2$ and that $4^n = 2^{2n}$. Therefore, we write the series as

$$S = \sum_{n=0}^{\infty} (-1)^n \frac{(2n+4)(2n+3)}{(2n+1)!} \left(\frac{\pi}{2}\right)^{2n} \tag{3.33b}$$

From eq. 3.10, we have

$$z^3 \sin z = \sum_{n=0}^{\infty} (-1)^n \frac{z^{2n+4}}{(2n+1)!} \tag{3.34}$$

Therefore,

$$\frac{d^2(z^3 \sin z)}{dz^2} = \left(6z - z^3\right)\sin z + 6z^2 \cos z$$

$$= \sum_{n=0}^{\infty} (-1)^n \frac{(2n+4)(2n+3)z^{2n+2}}{(2n+1)!} = z^2 \sum_{n=0}^{\infty} (-1)^n \frac{(2n+4)(2n+3)z^{2n}}{(2n+1)!} \tag{3.35}$$

Setting $z = \pi/2$, this becomes

$$S = \sum_{n=0}^{\infty} (-1)^n \frac{(2n+4)(2n+3)}{(2n+1)!} \left(\frac{\pi}{2}\right)^{2n} = \left(3\pi - \frac{\pi^3}{8}\right)\frac{4}{\pi^2} = \frac{12}{\pi} - \frac{\pi}{2} \tag{3.36}\square$$

Example 3.9: Differentiation to sum a series

The terms in the series

$$S = \sum_{n=0}^{\infty} \frac{(n+1)(n+3)}{3^n} \tag{3.37}$$

do not contain factors involving n in the denominator. The geometric series given in eq. 3.8, which converges for $|z| < 1$, has this property. Therefore, with the factor $1/3^n$, this suggests that we manipulate the geometric series in z, then set $z = 1/3$.

To obtain the factors $(n + 1)$ and $(n + 3)$ in the numerator of terms in the geometric series, we must differentiate z^{n+1} and z^{n+3}. Therefore, we multiply the geometric series by z^3 and take one derivative to obtain the factor $(n + 3)$. We then obtain $(n + 1)$ by dividing by z then taking a second derivative. The result is

$$\frac{d}{dz}\left[\frac{1}{z}\frac{d}{dz}\left(\frac{z^3}{1-z}\right)\right] = \frac{3-z}{(1-z)^3} = \sum_{n=0}^{\infty}(n+3)(n+1)z^n \qquad (3.38)$$

With $z = 1/3$, we find

$$S = \sum_{n=0}^{\infty}\frac{(n+1)(n+3)}{3^n} = 9 \qquad (3.39)\square$$

Summing a series using integration and differentiation

When the terms of a series contain factors involving the summation index in both the numerator and denominator, we may be able to sum such a series using integration and differentiation.

Example 3.10: Integration and differentiation to sum a series

For the series

$$S = \sum_{n=0}^{\infty}\frac{(n+2)}{(n+3)}\frac{2^n}{n!} \qquad (3.40)$$

the factor $1/n!$ suggests that we manipulate the exponential series of eq. 3.6. The factor $(n + 2)$ in the numerator is obtained from a derivative of z^{n+2} and $(n + 3)$ in the denominator is obtained by integrating z^{n+2}. Therefore, we consider

$$\int\left[z\frac{d}{dz}\left(z^2 e^z\right)\right]dz = e^z\left(z^3 - z^2 + 2z - 2\right) = \sum_{n=0}^{\infty}\frac{(n+2)}{(n+3)}\frac{z^{n+3}}{n!} + C \qquad (3.41)$$

By setting $z = 0$, we find the constant of integration to be $C = -2$. Therefore,

$$\sum_{n=0}^{\infty}\frac{(n+2)}{(n+3)}\frac{z^n}{n!} = \frac{e^z(z^3 - z^2 + 2z - 2) + 2}{z^3} \qquad (3.42)$$

With $z = 2$,

$$S = \sum_{n=0}^{\infty} \frac{(n+2)}{(n+3)} \frac{2^n}{n!} = \frac{3e^2 + 1}{4} \tag{3.43}\square$$

Example 3.11: Differentiation and integration to sum a series

The factors $(-1)^n$ and $1/(2n)!$ in

$$S = \sum_{n=0}^{\infty} (-1)^n \frac{(2n+3)}{(n+1)(2n)!} \frac{\pi^{2n}}{4^n} \tag{3.44}$$

suggest manipulations of $\cos z$ given in eq. 3.11. The factor $(2n + 3)$ in the numerator is obtained by differentiating z^{2n+3}. To obtain $(n + 1)$ in the denominator suggests that we integrate z^n. But the series for $\cos z$ only contains terms in z^{2n}. Therefore, we write

$$(n + 1) = \tfrac{1}{2}(2n + 2) \tag{3.45a}$$

and

$$4^n = 2^{2n} \tag{3.45b}$$

Then, the series of eq. 3.44 can be written as

$$S = 2\sum_{n=0}^{\infty} (-1)^n \frac{(2n+3)}{(2n+2)(2n)!} \left(\frac{\pi}{2}\right)^{2n} \tag{3.46}$$

We then consider

$$\int \frac{1}{z} \frac{d}{dz} \left[z^3 \cos z\right] dz = z\sin z + \left(z^2 + 1\right)\cos z$$
$$= \sum_{n=0}^{\infty} (-1)^n \frac{(2n+3)}{(2n+2)} \frac{z^{2n+1}}{(2n)!} + C \tag{3.47}$$

Taking $z = 0$, we find $C = 1$. Then, setting $z = \pi/2$, we have

$$\frac{\pi}{2} - 1 = \sum_{n=0}^{\infty} (-1)^n \frac{(2n+3)}{(2n+2)} \frac{1}{(2n)!} \left(\frac{\pi}{2}\right)^{2n+1} \tag{3.48a}$$

from which

$$S = \sum_{n=0}^{\infty} (-1)^n \frac{(2n+3)}{(2n+2)} \frac{1}{(2n)!} \left(\frac{\pi}{2}\right)^{2n} = 2 - \frac{4}{\pi} \tag{3.48b}\square$$

Bernoulli numbers and the Riemann zeta series

Bernoulli numbers B_n are defined by the series

$$F_B(z) = \frac{z}{e^z - 1} \equiv \sum_{k=0}^{\infty} B_n \frac{z^k}{k!} \qquad (3.49)$$

It is straightforward to see that

$$\lim_{z \to 0} F_B(z) = \lim_{z \to 0} \frac{z}{e^z - 1} = 1 \qquad (3.50)$$

Thus, F_B is analytic at $z = 0$, so it can be expanded in the MacLaurin series

$$F_B(z) = \sum_{k=0}^{\infty} \left(\frac{d^k F_B}{dz^k} \right)_{x=0} \frac{z^k}{k!} \qquad (3.51a)$$

Comparing this to the series of eq. 3.49, we have

$$B_k = \frac{d^k}{dz^k} \left(\frac{z}{e^z - 1} \right) \Bigg|_{z=0} \qquad (3.51b)$$

The Bernoulli numbers can be determined using a recurrence relation. From the binomial expansion, we have

$$(B + 1)^n - B^n = 0 = \sum_{k=0}^{n-1} \frac{n!}{k!(n-k)!} B^k \qquad (3.52)$$

We then make the replacement

$$B^k \to B_k \qquad (3.53a)$$

to obtain an expression for B_{n-1} in terms of Bernoulli numbers of lower index

$$B_{n-1} = -\sum_{k=0}^{n-2} \frac{(n-1)!}{k!(n-k)!} B_k \qquad (3.53b)$$

From eq. 3.51b [or equivalently eq. 3.50] we see that

$$B_0 = 1 \qquad (3.54a)$$

n	B_n
0	1
1	−1/2
2	1/6
4	−1/30
6	1/42
8	−1/30
10	5/66

Table 3.1 Short table of non-zero Bernoulli numbers

and from eq. 3.51b

$$B_1 = \frac{d}{dz}\left(\frac{z}{e^z - 1}\right)\bigg|_{z=0} = \lim_{z \to 0}\left[\frac{e^z - 1 - ze^z}{(e^z - 1)^2}\right] =$$

$$\lim_{z \to 0}\left[\frac{z + \frac{1}{2}z^2 + \ldots - z - z^2}{(z + \ldots)^2}\right] = -\frac{1}{2} \tag{3.54b}$$

Using eq. 3.51b or eq. 3.53b, it is straightforward to show that for integers $m \geq 1$,

$$B_{2m+1} = 0 \tag{3.55}$$

That is, all odd index Bernoulli numbers except B_1 are zero.

It can be shown (see Cohen, H., 1992, pp. 153–155) that the Bernoulli numbers of even index are related to the Riemann zeta function of even integers by

$$B_{2n} = (-1)^{n+1}2\frac{(2n)!}{(2\pi)^{2n}}\varsigma(2n) = (-1)^{n+1}2\frac{(2n)!}{(2\pi)^{2n}}\sum_{k=1}^{\infty}\frac{1}{k^{2n}} \tag{3.56}$$

Therefore, once one determines the value of the Bernoulli numbers from eq. 3.51b or eq. 3.53b, one can sum the Riemann zeta series $\zeta(2n)$ exactly.

Table 3.1 below is a short list of non-zero Bernoulli numbers of the index n for $0 \leq n \leq 10$ generated from either eq. 3.51b or eq. 3.53b.

Example 3.12: The Riemann zeta series and Bernoulli numbers

We see from Table 3.1 that

$$B_2 = \frac{1}{6} \tag{3.57a}$$

and

$$B_4 = -\frac{1}{30} \tag{3.57b}$$

Therefore, from eq. 3.56, we have

$$\varsigma(2) = \sum_{k=1}^{\infty} \frac{1}{k^2} = \frac{\pi^2}{6} \tag{3.58a}$$

and

$$\varsigma(4) = \sum_{k=1}^{\infty} \frac{1}{k^4} = \frac{\pi^4}{90} \tag{3.58b}$$

In Problem 9, part (a), the reader will determine the exact value of the Riemann series $\zeta(6)$. $\square$

Euler numbers and summing an alternating series

Euler numbers are defined by the MacLaurin series for

$$F_E(z) = \sec(z) = \sum_{k=0}^{\infty} (-1)^k E_{2k} \frac{z^{2k}}{(2k)!} \tag{3.59}$$

Clearly, $F_E(z)$ is singular for z an odd multiple of $\pi/2$. We note that eq. 3.59 defines Euler numbers of even index only.

From the MacLaurin series for $F_E(z)$,

$$F_E(z) = \sum_{k=0}^{\infty} \left(\frac{d^k F_E}{dz^k} \right) \bigg|_{x=0} \frac{z^k}{k!} \tag{3.60a}$$

we have

$$E_k = (-1)^k \left(\frac{d^k \sec(z)}{dz^k} \right) \bigg|_{z=0} \tag{3.60b}$$

The Euler numbers can also be obtained from a recurrence relation similar to that satisfied by the Bernoulli numbers. From the binomial expansion we have

$$(E+1)^n + (E-1)^n = 0 = \sum_{k=0}^{n} \frac{n!}{k!(n-k)!} \left[1 + (-1)^{n-k} \right] E^k \tag{3.61}$$

and making the replacement

$$E^k \to E_k \tag{3.62a}$$

we obtain

$$E_n = -\frac{1}{2}\sum_{k=0}^{n-2}\frac{n!}{k!(n-k)!}\left[1+(-1)^{n-k}\right]E_k \qquad (3.62b)$$

If n is odd, eq. 3.62b becomes

$$E_n = -\frac{1}{2}\sum_{k=0}^{n-2}\frac{n!}{k!(n-k)!}\left[1-(-1)^k\right]E_k = -\sum_{\substack{k=0 \\ k\ odd}}^{n-2}\frac{n!}{k!(n-k)!}E_k \qquad (3.63)$$

Referring to eq. 3.60b, it is straightforward that

$$E_1 = 0 \qquad (3.64a)$$

Therefore, from eq. 3.63 we have

$$E_3 = 3E_1 = 0 \qquad (3.64b)$$

from which eq. 3.63 yields

$$E_5 = 5E_1 + 10E_3 = 0 \qquad (3.64c)$$

and so on. In this way, it is straightforward to show that all odd index Euler numbers are zero.

Euler numbers of even index are found from either eq. 3.60b or eq. 3.62b. Table 3.2 below is a short list of values of even index Euler numbers.

As shown by Cohen, H., 1992, pp. 160–162, the Euler numbers are related to the alternating series

$$\sum_{k=0}^{\infty}(-1)^k\frac{1}{(2k+1)^{2n+1}} = (-1)^n\frac{\pi^{2n+1}}{4(2^{2n})(2n)!}E_{2n} \qquad (3.65)$$

Therefore, this alternating series can be summed exactly in terms of known values of Euler numbers.

$2n$	E_{2n}
0	1
2	−1
4	5
6	−61
8	1,385
10	−50,521

Table 3.2 Short table of even index Euler numbers

Example 3.13: Summing an alternating series and Euler numbers

Using the values of the Euler numbers given in Table 3.2, we have

$$\sum_{k=0}^{\infty} (-1)^k \frac{1}{(2k+1)} = \frac{\pi}{4} \tag{3.66a}$$

and

$$\sum_{k=0}^{\infty} (-1)^k \frac{1}{(2k+1)^3} = \frac{\pi^3}{32} \tag{3.66b}$$

In Problem 9, part (b), the reader will determine the exact value of the alternating series of eq. 3.65 for $n = 2$. □

3.3 Approximating the Sum of a Convergent Series

If a convergent series cannot be summed exactly, there are methods by which the value of a series can be approximated. In Appendix 2, it was noted that a series can be written as a finite sum plus a remainder series as

$$S(z) = \sum_{n=n_0}^{N-1} \sigma_n(z) + \sum_{n=N}^{\infty} \sigma_n(z) \equiv S_1(z) + S_2(z) \tag{A2.1}$$

Since $S_1(z)$ is a finite sum, its exact value can be determined. Therefore, we approximate $S(z)$ by estimating the value of $S_2(z)$.

Truncating a series

Since $S(z)$ converges, for some value(s) of z, we can choose $S_2(z)$ (by choosing N) so that the magnitudes of the terms in $S_2(z)$ decrease with increasing index n so that ignoring the remainder series results in a small error. Then, $S(z)$ is approximated by the finite sum $S_1(z)$. This approach is called *truncating the series*.

Example 3.14: Approximating a series by truncation

From eq. 3.6, the series for $e^{1/2}$ is

$$e^{\frac{1}{2}} = \sum_{n=0}^{\infty} \frac{1}{2^n n!} = 1.64872 \tag{3.67}$$

The finite sum approximation

$$e^{\frac{1}{2}} \simeq \sum_{n=0}^{3} \frac{1}{2^n n!} = 1 + \frac{1}{2} + \frac{1}{8} + \frac{1}{48} = 1.64583 \qquad (3.68a)$$

compares reasonably well with the exact value.

We can improve the accuracy, we take more terms in the finite sum. For example,

$$e^{\frac{1}{2}} \simeq \sum_{n=0}^{5} \frac{1}{2^n n!} = 1 + \frac{1}{2} + \frac{1}{8} + \frac{1}{48} + \frac{1}{384} + \frac{1}{3840} = 1.64870 \qquad (3.68b)$$

Thus, for the exponential series, for relatively small values of z, truncation yields fairly accurate results with a finite sum of just a few terms.

However, if z is large, a large number of terms are required to achieve accurate results. For example, for $z = 5$, the finite sum

$$e^5 \simeq \sum_{n=0}^{3} \frac{5^n}{n!} = 1 + 5 + \frac{25}{2} + \frac{125}{6} = 39.33333 \qquad (3.69a)$$

is an extremely poor approximation to

$$e^5 = 148.41316 \qquad (3.69b)$$

To achieve the value given in eq. 3.69b to five decimals requires a finite sum of 23 terms. $\square$

Approximation to the value of an absolute series

It is shown in Appendix 2 that for an absolute series, the bounds on the remainder series $S_2(z)$ are given in terms of the envelope integral by

$$\int_N^\infty |\sigma(n,z)| dn < S_2(z) < |\sigma_N(z)| + \int_N^\infty |\sigma(n,z)| dn \qquad (A2.9)$$

Therefore, if the envelope integral can be evaluated in closed form, or accurately approximated by some method such as numerical integration, then $S_2(z)$ can be approximated by some average of the upper and lower bounds as

$$S_2(z) = \alpha |\sigma_n(z)| + \alpha \int_N^\infty |\sigma(n,z)| dn + (1 - \alpha) \int_N^\infty |\sigma(n,z)| dn$$

$$= \int_N^\infty |\sigma(n,z)| dn + \alpha |\sigma_n(z)| \qquad (3.70)$$

with $0 < \alpha < 1$. Unless there is information specifying the value of α, it is reasonable to take it to be $1/2$. Then, $S_2(z)$ is approximated by

$$S_2(z) \simeq \tfrac{1}{2}|\sigma_N(z)| + \int_N^\infty |\sigma(n,z)|\,dn \qquad (3.71)$$

Example 3.15: Estimate of an absolute series

We note that

$$S = \sum_{n=0}^\infty \frac{1}{(2n+1)(2n+2)} = \sum_{n=0}^\infty \left[\frac{1}{(2n+1)} - \frac{1}{(2n+2)}\right]$$

$$= 1 - \frac{1}{2} + \frac{1}{3} - \frac{1}{4} + \ldots = \ell n(1+z)|_{z=1} = \ell n(2) = 0.69315 \qquad (3.72)$$

Writing this as

$$S = \sum_{n=0}^4 \frac{1}{(2n+1)(2n+2)} + S_2 \qquad (3.73)$$

we have

$$S_1 = \sum_{n=0}^4 \frac{1}{(2n+1)(2n+2)} = 1 - \frac{1}{2} + \frac{1}{3} - \frac{1}{4} + \ldots + \frac{1}{9} - \frac{1}{10} = 0.64563 \quad (3.74a)$$

which is not very accurate. The approximation to S_2 given in eq. 3.71 is

$$S_2 \simeq \frac{1}{2} \frac{1}{(2 \times 5 + 1)(2 \times 5 + 2)} + \int_5^\infty \frac{1}{(2n+1)(2n+2)}\,dn$$

$$= \frac{1}{264} + \frac{1}{2}\ell n\left(\frac{12}{11}\right) = 0.04729 \qquad (3.74b)$$

Adding this to S_1 we obtain

$$S \simeq 0.69292 \qquad (3.75)$$

which differs from the exact value by about 0.03%. □

Example 3.16: Estimate of an absolute series

In example 3.14, it is shown that for large positive values of z ($z = 5$), the exponential series is poorly approximated by truncating the series with a small number of terms for $S_1(z)$. It is possible to estimate the exponential series by a finite sum with a small number of terms added to the approximation given in eq. 3.71.

As such, we consider

$$S = e^5 = \sum_{n=0}^{N-1} \frac{5^n}{n!} + \sum_{n=N}^{\infty} \frac{5^n}{n!} \simeq \sum_{n=0}^{N-1} \frac{5^n}{n!} + S_2 \qquad (3.76)$$

with

$$S_2 \simeq \frac{5^N}{2 \times N!} + \int_N^{\infty} \frac{5^n}{n!} \, dn \qquad (3.77)$$

Clearly, this integral cannot be evaluated in closed form. However, it can be evaluated numerically.

As shown in Appendix 3, for "large" n, the *Stirling* approximation for $n!$ is

$$n! \simeq n^n e^{-n} \sqrt{2\pi n} \qquad (A3.54)$$

An improved approximation given by Namias, V., (1986), pp. 25–29, is

$$n! \simeq n^n e^{-n} \sqrt{2\pi n} \, e^{\left(\frac{1}{12n} - \frac{1}{360n^3} + \frac{1}{1260n^5} - \cdots\right)} \qquad (A3.58)$$

For example, the exact value of 5! is 120. Using eq. A3.54, we have

$$5! \simeq 5^5 e^{-5} \sqrt{10\pi} = 118.01917 \qquad (3.78a)$$

The approximation to 5! using the Namias expression yields

$$5! \simeq 5^5 e^{-5} \sqrt{10\pi} e^{(1/60 - 1/45000 + 1/3937500)} = 120.00000 \qquad (3.78b)$$

Thus, eq. A3.58 provides an accurate approximation to $n!$ for $n \geq 5$.

In ch. 4, we demonstrate that an integral can be approximated numerically by a sum over discrete points. One such method, called an N *point Gauss Legendre quadrature sum,* approximates an integral as

$$\int_{-1}^{1} F(x) dx \simeq \sum_{k=1}^{N} w_k F(x_k) \qquad (3.79)$$

where the numbers w_k and x_k are given in ch. 4, Table 4.1.

To use this method to evaluate the integral in eq. 3.77, we use the transformation

$$n(x) = \frac{2N}{(1+x)} \qquad (3.80)$$

Taking $N = 5$, we obtain

$$S_1 = \sum_{n=0}^{4} \frac{5^n}{n!} = 65.37500 \tag{3.81}$$

Then

$$S \simeq S_1 + \frac{1}{2} \frac{5^5}{5!} + 10 \int_{-1}^{1} \frac{5^{n(x)}}{n(x)!} \frac{1}{(1+x)^2} dx$$

$$\simeq 65.37500 + 13.02083 + 10 \sum_{k=1}^{10} w_k \frac{5^{n(x_k)}}{n(x_k)!} \frac{1}{(1+x_k)^2} \simeq 148.19159 \tag{3.82}$$

which differs from the exact value e^5 by 0.15%. $\square$

Approximation to the value of an alternating series

In Appendix 2, it is shown that when $S(z)$ is an alternating series, the remainder series is bounded by

$$|\sigma_N(z)| - |\sigma_{N+1}(z)| < S_2(z) < |\sigma_N(z)| \tag{A2.14}$$

where N must be even.

As with an absolute series, the value of $S_2(z)$ is somewhere between the bounds, and unless there is some information to suggest otherwise, it is reasonable to take $S_2(z)$ to be the average of the upper and lower bounds;

$$S_2(z) \simeq |\sigma_N(z)| - \tfrac{1}{2}|\sigma_{N+1}(z)| \tag{3.83}$$

Example 3.17: Estimate of an alternating series

Referring to eq. 3.10, we recognize that

$$S = \sum_{n=0}^{\infty} (-1)^n \frac{2^{2n}}{(2n+1)!} = \frac{1}{2} \sum_{n=0}^{\infty} (-1)^n \frac{2^{2n+1}}{(2n+1)!} = \frac{\sin(2)}{2} = 0.45465 \tag{3.84}$$

Taking

$$S_1 = \sum_{n=0}^{1} (-1)^n \frac{2^{2n}}{(2n+1)!} = 1 - \frac{4}{3!} = 0.33333 \tag{3.85a}$$

This is a poor approximation to the exact value of S. Thus, truncation at the $n = 1$ term is not a good estimate of this series. The remainder series is

$$S_2 \simeq |\sigma_2| - \tfrac{1}{2}|\sigma_3| = \frac{2^4}{5!} - \frac{1}{2}\frac{2^6}{7!} = 0.12698 \tag{3.85b}$$

Adding this to S_1 we obtain

$$S \simeq 0.46032 \tag{3.86}$$

which is fairly accurate, particularly since S_1 is approximated by only two terms. This result differs from the exact value by about 1.25%. $\square$

Pade Approximant

A Pade Approximant introduced in Appendix 1 can be used to approximate a function represented by an infinite power series. The general form of a Pade Approximant is the ratio of polynomials;

$$S^{[N,M]}(z) = \frac{P_N(z)}{Q_M(z)} \tag{3.87}$$

where $P_N(z)$ and $Q_M(z)$ are polynomials in z of order N and M, respectively.

Since $Q_M(z)$ is an Mth order polynomial, it has M roots (some of which may be repeated roots). Therefore, there are up to M possible points at which the Pade Approximant may be singular. Since $S(z)$ is not singular, the Pade Approximant can approximate $S(z)$ only at points that are not roots of $Q_M(z)$.

It is discussed in Appendix 1 that the Pade Approximant is most accurate for $M = N$ or $M = N-1$. Taking $M = N$, we consider approximating

$$S(z) = \sum_{n=0}^{\infty} c_n z^n \tag{3.88}$$

by

$$S^{[N,N]}(z) \simeq \frac{p_0 + p_1 z + p_2 z^2 + ... + p_N z^N}{1 + q_1 z + q_2 z^2 + ... + q_N z^N} \tag{3.89}$$

where the sets of coefficients $\{p_k\}$ and $\{q_k\}$ are found by requiring the Pade Approximant and the infinite series to be identical for all powers of z up to z^{2N}.

Thus, $S(z) - S^{[N, \, N]}(z)$ is a power series in which the lowest power of z is $2N + 1$. Such a power series is denoted by $O(z^{2N+1})$. This requirement can be written as

$$\left(1 + q_1 z + q_2 z^2 + \ldots + q_N z^N\right) \sum_{n=0}^{\infty} c_n z^n - \left(p_0 + p_1 z + p_2 z^2 + \ldots + p_N z^N\right)$$

$$= \sum_{n=2N+1}^{\infty} r_n z^n \equiv O\left(z^{2N+1}\right) \tag{3.90}$$

Example 3.18: Estimate of a series by Pade Approximant

Let us approximate the logarithm series

$$S(z) = -\ell n(1 - z) = z + \frac{z^2}{2} + \frac{z^3}{3} + \frac{z^4}{4} + \frac{z^5}{5} + \ldots \tag{3.91}$$

by

$$S^{[2,2]}(z) = \frac{p_0 + p_1 z + p_2 z^2}{1 + q_1 z + q_2 z^2} \tag{3.92}$$

We find the coefficients by requiring that

$$\left(1 + q_1 z + q_2 z^2\right)\left(z + \frac{z^2}{2} + \frac{z^3}{3} + \frac{z^4}{4} + \frac{z^5}{5} + \ldots\right) - \left(p_0 + p_1 z + p_2 z^2\right) = O(z^5) \tag{3.93}$$

Thus, the coefficients of z^0, z, z^2, z^3, and z^4 must be zero. From this we have

$$p_0 = 0 \tag{3.94a}$$

$$1 - p_1 = 0 \tag{3.94b}$$

$$\tfrac{1}{2} + q_1 - p_2 = 0 \tag{3.94c}$$

$$\tfrac{1}{3} + \tfrac{1}{2} q_1 - q_2 = 0 \tag{3.94d}$$

and

$$\tfrac{1}{4} + \tfrac{1}{3} q_1 + \tfrac{1}{2} q_2 = 0 \tag{3.94e}$$

from which we obtain

$$(-\ell n(1 - z))^{[2,2]} = \frac{z - \tfrac{1}{2} z^2}{1 - z + \tfrac{1}{6} z^2} \tag{3.95}$$

This Pade Approximant has singularities at $z \cong 4.73205$ and 1.26795, both of which are outside the range of convergence of the series. Thus, eq. 3.95 should yield reasonably accurate approximations to the logarithm function at all $z \, \varepsilon \, [-1, 1)$.

Substituting $z = 1/2$ in eq. 3.95, the [2,2] Pade Approximant yields

$$\left(-\ell n(1 - \tfrac{1}{2})\right)^{[2,2]} = 0.69231 \qquad (3.96a)$$

The 4th order finite sum

$$S_1\left(\tfrac{1}{2}\right) = \sum_{n=1}^{4} \frac{1}{2^n n} = 0.68229 \qquad (3.96b)$$

is somewhat less accurate than the [2,2] Pade Approximant in approximating

$$-\ell n\left(1 - \tfrac{1}{2}\right) = 0.69315 \qquad (3.97)\square$$

3.4 Fourier Series

A function that is periodic with period T has the property

$$S(z + T) = S(z) \qquad (3.98)$$

A *Fourier series* is an infinite series representing a periodic function. The *Fourier exponential series* is of the form

$$F(z) = \sum_{n=-\infty}^{\infty} f_n e^{in\omega z} \qquad (3.99)$$

where

$$\omega \equiv \frac{2\pi}{T} \qquad (3.100)$$

For any integer N,

$$F(z + NT) = \sum_{n=-\infty}^{\infty} f_n e^{in\omega(z+NT)} = \sum_{n=-\infty}^{\infty} f_n e^{in\omega z} e^{i2\pi nN} \qquad (3.101)$$

Since n and N are integers, their product is an integer, and

$$e^{i2\pi nN} = \cos(2\pi nN) + i\sin(2\pi nN) = 1 \qquad (3.102)$$

Therefore,

$$F(z + NT) = \sum_{n=-\infty}^{\infty} f_n e^{in\omega z} = F(z) \qquad (3.103)$$

Let m and n be integers, and let the beginning of an interval be z_0. When $n = m$,

$$\int_{z_0}^{z_0+T} e^{inwz} e^{-imwz} dz = \int_{z_0}^{z_0+T} dz = T \tag{3.104a}$$

Referring to eqs. 3.100 and 3.102, if $n \neq m$,

$$\int_{z_0}^{z_0+T} e^{inwz} e^{-imwz} dz =$$

$$\frac{e^{i(n-m)\omega z_0} e^{i(n-m)\omega T} - e^{i(n-m)\omega z_0}}{i(n-m)\omega} = \frac{e^{i(n-m)\omega z_0} - e^{i(n-m)\omega z_0}}{i(n-m)\omega} = 0$$

$$\tag{3.104b}$$

To find the coefficients f_n of the Fourier exponential series, we consider

$$\frac{1}{T}\int_{z_0}^{z_0+T} F(z) e^{-im\omega z} dz = \frac{1}{T}\sum_{n=-\infty}^{\infty} f_n \int_{z_0}^{z_0+T} e^{i(n-m)\omega z} dz = f_m \tag{3.105}$$

Another form of the Fourier series for a periodic function, the *Fourier sine–cosine series*, is given by

$$F(z) = \sum_{n=0}^{\infty} [A_n\cos(n\omega z) + B_n\sin(n\omega z)]$$

$$= A_0 + \sum_{n=1}^{\infty} [A_n\cos(n\omega z) + B_n\sin(n\omega z)] \tag{3.106}$$

Since the two forms of the Fourier series are equivalent, it is straightforward to show that

$$A_0 = f_0 \tag{3.107a}$$

$$A_n = f_n + f_{-n} \quad n > 0 \tag{3.107b}$$

and

$$B_n = i(f_n - f_{-n}) \quad n > 0 \tag{3.107c}$$

or

$$f_0 = A_0 \tag{3.108a}$$

$$f_n = \tfrac{1}{2}(A_n - iB_n) \quad n > 0 \tag{3.108b}$$

and

$$f_{-n} = \tfrac{1}{2}(A_n + iB_n) \quad n>0 \tag{3.108c}$$

Convergence of a Fourier series

The Fourier series for a periodic function $F(z)$ will converge if:

- $F(z)$ does not have an infinite number of discontinuities over one period. This means that $F(z)$ can have discontinuities at a finite number of points. It cannot be discontinuous over an extended range of z.
- $F(z)$ does not have an infinite number of maxima and minima over one period.
- $\int_{z_0}^{z_0+T} |F(z)| dz$ must be finite.

For a detailed discussion of these conditions, the reader is referred to Spiegel, M., (1973), pp. 35–37.

Example 3.19. Fourier series for a square wave and summing a series.

The square wave of period T shown in Fig. 3.1 is defined by

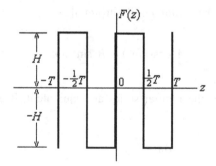

Fig. 3.1 Square wave of period T

$$F(z) \equiv \begin{cases} +H & nT<z<\left(n+\tfrac{1}{2}\right)T \\ -H & \left(n+\tfrac{1}{2}\right)T<z<(n+1)T \end{cases} \tag{3.109}$$

With $z_0 = 0$, the coefficients of the Fourier exponential series is given by

$$f_n = \frac{1}{T} \int_0^T F(z)e^{-in\omega z} dz = \frac{1}{T} \left[\int_0^{T/2} He^{-in\omega z} dz + \int_{T/2}^T -He^{-in\omega z} dz \right] \tag{3.110a}$$

from which we obtain

$$f_n = \begin{cases} 0 & n = 0, \ even \\ \dfrac{2H}{in\pi} & n \ odd \end{cases} \tag{3.110b}$$

Therefore, the Fourier exponential series is

$$
\begin{aligned}
F(z) &= \frac{2H}{i\pi} \left[\left(\frac{e^{i\omega z}}{1} + \frac{e^{-i\omega z}}{-1} \right) + \left(\frac{e^{3i\omega z}}{3} + \frac{e^{-3i\omega z}}{-3} \right) + \left(\frac{e^{5i\omega z}}{5} + \frac{e^{-5i\omega z}}{-5} \right) + \ldots \right] \\
&= \frac{4H}{\pi} \left[\frac{\sin(\omega z)}{1} + \frac{\sin(3\omega z)}{3} + \frac{\sin(5\omega z)}{5} + \ldots \right] \\
&= \frac{4H}{\pi} \sum_{n=0}^{\infty} \frac{\sin[(2n+1)\omega z]}{(2n+1)}
\end{aligned} \tag{3.111}
$$

If this Fourier series is approximated by a finite sum,

$$F(z) = \frac{4H}{\pi} \sum_{n=0}^{N} \frac{\sin[(2n+1)\omega z]}{(2n+1)} \tag{3.112}$$

we obtain the graphs shown in Fig. 3.2 for $N = 3$, 10, and 100.

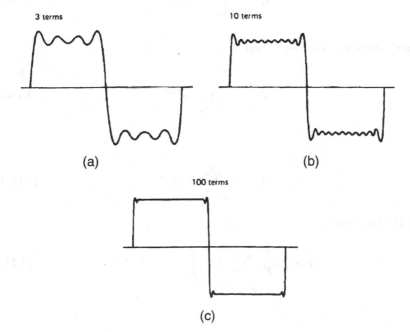

Fig. 3.2 Partial sum approximations to the Fourier series for a square wave

We notice that, for a large enough number of terms, the partial sum is a reasonable approximation to the square wave. However, there is a spike in the curve at the points of discontinuity. This is known at the *Gibbs phenomenon*. As noted by Mathews, J., and Walker, R. L., (1970), pp. 98, 120, the spikes are no larger than about 18% of the height of the square wave. These spikes result from the fact that the finite sum, which is a continuous function, is approximating a function with discontinuities. □

As an example of a practical application of a Fourier sum, there exist electronic devices that can generate wave forms in various shapes. Commercial AC electricity is a sine wave of some fundamental frequency. Circuits can be designed that create sinusoidal AC electricity of frequencies that are integer multiples of the fundamental. These wave forms are combined by finite sums to create approximations to square waves, triangular (saw tooth) waves, etc. (See problem 14 of Mathews, J., 1992, p. 306.)

Parseval's theorem

Let $F(z)$ and $G(z)$ be two periodic functions that have the same period T. The average of their product over one period is defined by

$$\langle FG \rangle \equiv \frac{1}{T} \int_{z_0}^{z_0+T} F(z)G(z)dz \tag{3.113}$$

Substituting the two Fourier series

$$F(z) = \sum_{n=-\infty}^{\infty} f_n e^{in\omega z} \tag{3.114a}$$

and

$$G(z) = \sum_{m=-\infty}^{\infty} g_n e^{im\omega z} \tag{3.114b}$$

eq. 3.113 becomes

$$\langle FG \rangle = \frac{1}{T} \sum_{\substack{m=-\infty \\ n=-\infty}}^{\infty} f_n g_m \int_{z_0}^{z_0+T} e^{i(n+m)\omega z}dz \tag{3.115}$$

With

$$\omega = \frac{2\pi}{T} \tag{3.100}$$

we obtain

$$\int_{z_0}^{z_0+T} e^{i(n+m)\omega z} dz = e^{i(n+m)\omega z_0} \frac{e^{i(n+m)2\pi} - 1}{i(n+m)\omega} = \begin{cases} T & m = -n \\ 0 & m \neq -n \end{cases} \tag{3.116}$$

Thus, only one term in the sum over m is non-zero, and eq. 3.115 becomes

$$\langle FG \rangle = \sum_{n=-\infty}^{\infty} f_n g_{-n} = f_0 g_0 + \sum_{n=-\infty}^{-1} f_n g_{-n} + \sum_{n=1}^{\infty} f_n g_{-n}$$

$$= f_0 g_0 + \sum_{n=1}^{\infty} (f_n g_{-n} + f_{-n} g_n) \tag{3.117}$$

This is the *generalized Parseval theorem*.
   If $G(z) = F(z)$, this becomes

$$\langle F^2 \rangle = f_0^2 + 2 \sum_{n=1}^{\infty} f_n f_{-n} \tag{3.118a}$$

Referring to eqs. 3.108, Parseval's theorem expressed in terms of the coefficients of the sinc–cosine series is

$$\langle F^2 \rangle = A_0^2 + \frac{1}{2} \sum_{n=1}^{\infty} (A_n^2 + B_n^2) \tag{3.118b}$$

Example 3.20: Summing a series using Parseval's theorem

Applying this to the square wave of example 3.19, we see from eq. 3.111 that the coefficients of the cosine series are

$$A_n = 0 \tag{3.119a}$$

and the coefficients of the sine series are

$$B_n = \frac{4H}{\pi} \frac{1}{(2n+1)} \tag{3.119b}$$

The average of $[F(z)]^2$ over one period is

$$\langle F^2 \rangle = \frac{1}{T} \int_0^T [F(z)]^2 dz = \frac{1}{T} \left[\int_0^{T/2} H^2 dz + \int_{T/2}^T H^2 dz \right] = H^2 \qquad (3.120a)$$

and from Parseval's theorem we have

$$\langle F^2 \rangle = \frac{8H^2}{\pi^2} \sum_{n=0}^{\infty} \frac{1}{(2n+1)^2} \qquad (3.120b)$$

Equating these, we can sum the series

$$S = \sum_{n=0}^{\infty} \frac{1}{(2n+1)^2} = \frac{\pi^2}{8} \qquad (3.121)\square$$

Problems

1. Using the Cauchy ratio test, determine the convergence or divergence of the series $S = \sum_{n=2}^{\infty} \frac{1}{n \times \ell n^p(n)}$ for $p > 0, p = 0$, and $p < 0$.

2. Using the Cauchy integral test, determine the values of p for which the series $S = \sum_{n=2}^{\infty} \frac{1}{n \times \ell n^p(n)}$ converges.

3. For what values of p does the series $S = \sum_{n=2}^{\infty} (-1)^n \frac{1}{n \times \ell n^p(n)}$ converge?

4. Use algebraic manipulation to sum the series $S = \sum_{n=1}^{\infty} \frac{1}{(n+2)(n+3)(n+4)}$

5. Use calculus manipulation to sum the series $S = \sum_{n=1}^{\infty} \frac{1}{(n+2)(n+3)(n+4)}$

6. Use calculus manipulation to sum the series $S(z) = \sum_{n=1}^{\infty} n^2 e^{-nz}$

7. Use calculus manipulation to sum the series $S = \sum_{n=1}^{\infty} (-1)^n \frac{n}{(2n+1)!} \pi^n$

8. Use calculus manipulation to sum the series $S = \sum_{n=0}^{\infty} (-1)^n \frac{(n+1)}{(2n+3)!} \left(\frac{\pi^2}{16}\right)^n$

9. (a) Sum the Riemann zeta series $\zeta(6)$.
   (b) Sum the alternating series of eq. 3.65 for $n = 2$.

10. For the absolute series $S = \sum_{n=0}^{\infty} n^2 e^{-n}$
    (a) Approximate this series by the finite sum $S_1 = \sum_{n=0}^{2} n^2 e^{-n}$
    (b) Approximate this series by the finite sum of part (a) plus an estimate of the remainder series.

    Compare each of the results of parts (a) and (b) to the exact value obtained in problem 6.

11. For the alternating series $S = \sum\limits_{n=0}^{\infty} (-1)^n \dfrac{2^n}{n!}$

  (a) Approximate this series by the finite sum $S_1 = \sum\limits_{n=0}^{5} (-1)^n \dfrac{2^n}{n!}$

  (b) Approximate this series by the finite sum of part (a) plus an estimate of the remainder series.

      Compare each of the results of parts (a) and (b) to the exact value of S.

12. (a) Find the [2,2] Pade Approximants for each of the series

    (1) $S(z) = \sum\limits_{n=0}^{\infty} (-1)^n \dfrac{z^n}{n!}$ and

    (2) $S(z) = \sum\limits_{n=0}^{\infty} n^2 e^{-nz}$

  (b) For each series, compare the value of the [2,2] Pade Approximant and the 4th order finite sum at $z = 1/2$ to their exact values [see eq. 3.6 and Problem 6].

13. For the square wave defined in example 3.19, take $z_0 = -T/4$

  (a) Find the Fourier series for $F(z)$.

  (b) Determine the sum of a series obtained by applying Parseval's theorem for this Fourier series.

14. (a) Find the Fourier series for the periodic linear function defined by
    $$F(z) = z \quad -\pi < z < \pi$$

  (b) Graph the finite sum approximation consisting of the first 15 terms of this Fourier series

  (c) Use Parseval's theorem to sum the series $S = \sum\limits_{n=0}^{\infty} (-1)^n \dfrac{1}{(2n+1)}$

Chapter 4
INTEGRATION

In this chapter, we present methods for approximating an integral that cannot be evaluated exactly.

4.1   Expanding the Integrand in a Series

One approach is to expand the integrand in a series, each term of which can be integrated exactly. One then integrates that series term by term.

Example 4.1: Integration by expanding the integrand in a series

The integral

$$I = \int_0^1 e^{-x^2} dx \tag{4.1}$$

cannot be evaluated exactly. Referring to eq. 3.6, we write

$$e^{-x^2} = \sum_{k=0}^{\infty} (-1)^k \frac{x^{2k}}{k!} \tag{4.2}$$

Then,

$$I = \sum_{k=0}^{\infty} (-1)^k \frac{1}{(2k+1)k!} \tag{4.3}$$

This series can then be approximated using one of the techniques presented in ch. 3. □

H. Cohen, *Numerical Approximation Methods*, DOI 10.1007/978-1-4419-9837-8_4,
© Springer Science+Business Media, LLC 2011

Example 4.2: Integration by expanding the integrand in a series

Let $f(x)$ be a periodic function of period $T = 2\pi/\omega$, and let us evaluate

$$\int_0^{2T/3} f(x)dx$$

As shown in eq. 3.99, $f(x)$ can be expanded in a Fourier exponential series

$$f(x) = \sum_{n=-\infty}^{\infty} c_n e^{in\omega x} \tag{4.4}$$

where, from eq. 3.105,

$$c_n = \frac{1}{T} \int_{x_0}^{x_0+T} f(x)e^{-in\omega x}dx \tag{4.5}$$

Then

$$I = \int_0^{2T/3} f(x)dx = \sum_{n=-\infty}^{\infty} c_n \int_0^{2T/3} e^{in\omega x}dx = \sum_{-\infty}^{\infty} c_n \frac{e^{in4\pi/3} - 1}{in\omega}$$

$$= \frac{2T}{3} + \sum_{n=1}^{\infty} \left[c_n \frac{e^{in4\pi/3} - 1}{in\omega} - c_{-n} \frac{e^{-in4\pi/3} - 1}{in\omega} \right] \tag{4.6}\square$$

4.2   Euler–MacLaurin Approximation

The *Euler–MacLaurin approximation* is designed to estimate an integral of the form

$$I = \int_0^1 f(x)dx \tag{4.7}$$

When the limits of the integral are not 0 and 1, a transformation of integration variable can be made to obtain an integral from 0 to 1.

Let us consider a general integral of the form

$$J \equiv \int_a^b F(y)dy \tag{4.8}$$

with a and b constants.

1. If a and b are finite numbers, we make the substitution

$$y = (b - a)x + a \qquad (4.9)$$

Then

$$J = \int_0^1 (b - a)F[(b - a)x + a]dx = \int_0^1 f(x)dx \qquad (4.10)$$

2. Let a be finite and non-zero and $b = \infty$. The substitution

$$y = \frac{a}{x} \qquad (4.11)$$

yields

$$J = \int_0^1 \frac{a}{x^2}F\left(\frac{a}{x}\right)dx = \int_0^1 f(x)dx \qquad (4.12)$$

3. For $a = 0$ and $b = \infty$, the substitution

$$y = \tan\left(x\frac{\pi}{2}\right) \qquad (4.13)$$

results in

$$J = \int_0^1 \frac{\pi}{2}\sec^2\left(x\frac{\pi}{2}\right)F\left[\tan\left(x\frac{\pi}{2}\right)\right]dx = \int_0^1 f(x)\,dx \qquad (4.14)$$

4. When $a = -\infty$ and $b = \infty$, we substitute

$$y = \tan\left((2x - 1)\frac{\pi}{2}\right) \qquad (4.15)$$

Then

$$J = \int_0^1 \pi\sec^2\left((2x - 1)\frac{\pi}{2}\right)F\left[\tan\left((2x - 1)\frac{\pi}{2}\right)\right]dx = \int_0^1 f(x)dx \qquad (4.16)$$

Thus, we see that any integral can be cast into the form

$$I = \int_0^1 f(x)dx \qquad (4.7)$$

To develop the Euler–MacLaurin method, we develop a set of properties of the *Bernoulli functions* denoted by $\beta_k(s)$. Like the Bernoulli numbers introduced in ch. 3, the Bernoulli functions are defined from the coefficients of z^k in the series

$$\frac{ze^{sz}}{(e^z - 1)} \equiv \sum_{k=0}^{\infty} \beta_k(s) \frac{z^k}{k!} \tag{4.17}$$

One of the properties is obtained by setting $s = 0$ to obtain

$$\frac{z}{(e^z - 1)} = \sum_{k=0}^{\infty} \beta_k(0) \frac{z^k}{k!} \tag{4.18}$$

Referring to eq. 3.49, we see that $e^z/(e^z-1)$ is the generator of the Bernoulli numbers. Therefore,

$$\beta_k(0) = B_k \tag{4.19}$$

Setting $s = 1$ in eq. 4.17, we obtain

$$\frac{ze^z}{e^z - 1} = \frac{-z}{e^{-z} - 1} = \sum_{k=0}^{\infty} \beta_k(1) \frac{z^k}{k!} \tag{4.20}$$

Again, referring to eq. 3.49, we see that

$$\frac{-z}{e^{-z} - 1} = \sum_{k=0}^{\infty} B_k \frac{(-z)^k}{k!} = \sum_{k=0}^{\infty} (-1)^k B_k \frac{z^k}{k!} \tag{4.21}$$

Equating these results, we have

$$\sum_{k=0}^{\infty} \beta_k(1) \frac{z^k}{k!} = \sum_{k=0}^{\infty} (-1)^k B_k \frac{z^k}{k!} \tag{4.22}$$

from which

$$\beta_k(1) = (-1)^k B_k \tag{4.23}$$

From

$$\frac{\partial}{\partial s}\left[\frac{ze^{sz}}{e^z - 1}\right] = \sum_{k=0}^{\infty} \beta'_k(s) \frac{z^k}{k!} = z\left[\frac{ze^{sz}}{e^z - 1}\right] = \sum_{k=0}^{\infty} \beta_k(s) \frac{z^{k+1}}{k!} \tag{4.24}$$

we obtain

$$\sum_{k=0}^{\infty} \beta_k(s)\frac{z^{k+1}}{k!} = \sum_{k=1}^{\infty} \beta_{k-1}(s)\frac{z^k}{(k-1)!} = \beta_0'(s) + \sum_{k=1}^{\infty} \beta_k'(s)\frac{z^k}{k!} \qquad (4.25)$$

Equating terms in like powers of z, the term containing z^0 yields

$$\beta_0'(s) = 0 \Rightarrow \beta_0(s) = constant \qquad (4.26a)$$

Setting $s = 0$, and referring to eq. 4.19 that

$$\beta_0(s) = constant = \beta_0(0) = B_0 = 1 \qquad (4.26b)$$

With $\beta'_0 = 0$, eq. 4.25 becomes

$$\sum_{k=1}^{\infty} \beta_{k-1}(s)\frac{z^k}{(k-1)!} = \sum_{k=1}^{\infty} \beta_k'(s)\frac{z^k}{k!} \qquad (4.27)$$

from which

$$\beta_k'(s) = k\beta_{k-1}(s) \quad k \geq 1 \qquad (4.28)$$

Therefore, with eq. 4.26b, we have, for example,

$$\beta_1'(s) = \beta_0(s) = B_0 = 1 \qquad (4.29a)$$

from which

$$\beta_1(s) = s + C \qquad (4.29b)$$

To evaluate the constant of integration C, we set $s = 0$ to obtain

$$\beta_1(0) = B_1 = -\frac{1}{2} = C \qquad (4.30)$$

so that

$$\beta_1(s) = s - \frac{1}{2} \qquad (4.31a)$$

By the same process, we obtain

$$\beta_2(s) = s^2 - s + \frac{1}{6} \qquad (4.31b)$$

and so on. Table 4.1 contains a short list of Bernoulli functions.

n	$\beta_n(s)$
0	1
1	$s-1/2$
2	$s^2-s+1/6$
3	$s^3-3s^2/2+s/2$
4	$s^4-2s^3+s^2-1/30$
5	$s^5-5s^4/2+5s^3/3-s/6$
6	$s^6-3s^5+5s^4/2-s^2/2+1/42$
7	$s^7-7s^6/2+7s^5/2-7s^3/6+s/6$
8	$s^8-4s^7+14s^6/3-7s^4/3+2s^2/3-1/30$
9	$s^9-9s^8/2+6s^7-21s^5/5+2s^3-3s/10$
10	$s^{10}-5s^9+15s^8/2-7s^6+5s^4-3s^2/2+5/66$

Table 4.1 A short list of Bernoulli functions

We begin the development of the Euler–MacLaurin approximation by noting from eq. 4.29a that

$$\beta'_1(x) = 1 \tag{4.32}$$

Therefore, we can write eq. 4.7 as

$$I = \int_0^1 f(x)\beta'_1(x)dx \tag{4.33}$$

With $u = f(x)$ and $dv = \beta'_1(x)dx$, integration by parts yields

$$I = [f(1)\beta_1(1) - f(0)\beta_1(0)] - \int_0^1 f'(x)\beta_1(x)dx \tag{4.34}$$

Referring to eqs. 4.19, 4.23, and 4.28, we have

$$\beta_1(0) = B_1 = -\frac{1}{2} \tag{4.35a}$$

$$\beta_1(1) = (-1)^1 B_1 = \frac{1}{2} \tag{4.35b}$$

and

$$\beta_1(x) = \frac{1}{2}\beta'_2(x) \tag{4.35c}$$

Therefore, eq. 4.34 becomes

$$I = \frac{1}{2}[f(1) + f(0)] - \frac{1}{2}\int_0^1 f'(x)\beta'_2(x)dx \tag{4.36}$$

We again integrate by parts, taking $u = f'(x)$ and $dv = \beta'_2(x)dx$. With $\beta_2(1) = \beta_2(0) = B_2 = 1/6$, this becomes

$$I = \frac{1}{2}[f(1) + f(0)] - \frac{1}{12}[f'(1) - f'(0)] - \frac{1}{3!}\int_0^1 f''(x)\beta'_3(x)dx \qquad (4.37)$$

After integrating by parts N times, the result is

$$I = \frac{1}{2}[f(1) + f(0)] - \sum_{n=1}^N \frac{B_{2n}}{(2n)!}\left[f^{(2n-1)}(1) - f^{(2n-1)}(0)\right]$$

$$-\frac{1}{(2N)!}\int_0^1 f^{(2N)}(x)\beta_{2N}(x)dx \qquad (4.38)$$

where $f^{(m)}(x)$ represents the mth derivative of $f(x)$.

Table 4.2 is a short list of non-zero Bernoulli numbers divided by the factorial of the index.

n	$B_n/n!$
0	1.00000
1	−0.50000
2	0.08333
4	−1.38889 × 10^{-3}
6	3.30688 × 10^{-5}
8	−8.62720 × 10^{-7}
10	2.08768 × 10^{-8}

Table 4.2 A short list of the non-zero Bernoulli numbers of index $0 \leq n \leq 10$ divided by $n!$

We note that $B_{2n}/(2n)!$ decreases with increasing n. Therefore, for N large enough, it is expected that the remainder integral in eq. 4.38 is a small correction and can be ignored. Then, the Euler–MacLaurin approximation to the integral of eq. 4.7 is

$$I \simeq \frac{1}{2}[f(1) + f(0)] - \sum_{k=1}^N \frac{B_{2k}}{(2k)!}\left[f^{(2k-1)}(1) - f^{(2k-1)}(0)\right] \qquad (4.39)$$

Example 4.3: The Euler–MacLaurin approximation to an integral

(a) We consider

$$I = \int_0^1 e^x dx = e = 1.71828 \qquad (4.40)$$

Using the first three non-zero terms in the Euler–MacLaurin approximation, we have

$$I \simeq \frac{1}{2}(e+1) - \frac{B_2}{2!}(e-1) - \frac{B_4}{4!}(e-1)$$
$$= 1.85914 - 0.14318 + 0.00239 = 1.71835 \tag{4.41}$$

(b) For

$$I = \int_2^\infty \frac{1}{(1+y)^2} dy = \frac{1}{3} \tag{4.42}$$

we substitute

$$y = \frac{2}{x} \tag{4.43}$$

Then

$$I = \int_0^1 \frac{2}{(x+2)^2} dx \simeq \frac{1}{2}\left(\frac{2}{3^2} + \frac{2}{2^2}\right) - \frac{B_2}{2!}\left(\frac{-4}{3^3} + \frac{4}{2^3}\right) - \frac{B_4}{4!}\left(\frac{-48}{3^5} + \frac{48}{2^5}\right)$$
$$= 0.36111 - 0.02932 + 0.00181 = 0.33360 \tag{4.44}$$

We see that both of these integrals are well approximated by the Euler–MacLaurin technique. □

The Euler–MacLaurin scheme approximates an integral by a sum that involves the derivatives of the integrand. *Quadrature methods* do not involve derivatives of the integrand. They approximate an integral as

$$\int_a^b f(x)dx \simeq \sum_{k=1}^N w_k f(x_k) \tag{4.45}$$

The numbers w_k and x_k are the *weights* and *abscissae* (or *abscissas*) for that N-point quadrature rule, respectively.

4.3   Newton-Cotes Quadratures

Newton-Cotes quadrature rules are based on the interpretation that the integral of a function over the limits $[a, b]$ is the area under the functional curve between these limits. With the range of integration divided into segments, made up of intervals each of width h as shown in Fig. 4.1. If the range of integration is divided by N equally spaced points $\{x_1, \ldots, x_N\}$, this common width is given by

$$h = x_{k+1} - x_k = \frac{(b-a)}{(N-1)} = constant \tag{4.46}$$

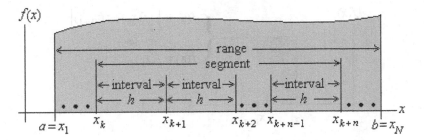

Fig. 4.1 The range $[a, b]$ is divided into segments comprised of equally spaced intervals

Each Newton-Cotes quadrature rule is developed by approximating the integrand $f(x)$ by a particular spline interpolation over each segment. Then

$$I \simeq \sum_{segments} \int_{x_k}^{x_{k+n}} f_{\substack{interpolated \\ polynomial}} (x)dx$$

$$= \sum_{segments} \int_{x_k}^{x_k+n\Delta x} f_{\substack{interpolated \\ polynomial}} (x)dx \tag{4.47}$$

where n is the number of intervals that make up a segment for that spline interpolation. The abscissae are the points x_k in each segment and the weights are obtained from integration of the spline interpolated polynomial over a segment.

Rectangular quadrature rule

As discussed in ch. 1, the simplest spline interpolation is the cardinal spline, in which the function is approximated by a constant over an interval. One choice for this constant is

$$f_{interpolated} = f\left(\tfrac{1}{2}(x_k + x_{k+1})\right)^0 \equiv f(\bar{x}_k) \tag{1.22a}$$

A second option is

$$f_{interpolated}(x) = \tfrac{1}{2}[f(x_k) + f(x_{k+1})] \quad x_k \le x \le x_{k+1} \tag{1.22b}$$

Thus, the cardinal spline forms rectangles between adjacent points as shown in Fig. 4.2. For this reason, the Newton-Cotes quadrature for this spline interpolation is called the *rectangular quadrature rule* (or simply the rectangular rule).

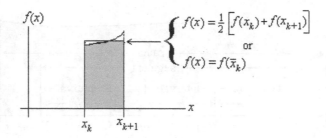

Fig. 4.2 Rectangles formed by approximating $f(x)$ by a cardinal spline

We note that since the cardinal spline is defined between two consecutive points, the intervals are the same as the segments. Therefore, there are $N - 1$ segments for the N-point rectangular Newton-Cotes rule.

Using the approximation of eq. 1.22a, the integral given by the sum of areas of the rectangles is

$$I \simeq h[f(\bar{x}_1) + f(\bar{x}_2) + \ldots + f(x_{N-2}) + f(x_{N-2})] \qquad (4.48a)$$

Thus, all the weights of this rectangular rule are 1 and the abscissae are the midpoints of the segments. Approximating $f(x)$ by the cardinal spline given in eq. 1.22b, we obtain

$$\begin{aligned}
I &\simeq \tfrac{1}{2}[f(x_1) + f(x_2)]h + \tfrac{1}{2}[f(x_2) + f(x_3)]h + \ldots \\
&\ldots + \tfrac{1}{2}[f(x_{N-2}) + f(x_{N-1})]h + \tfrac{1}{2}[f(x_{N-1}) + f(x_N)]h \\
&= [\tfrac{1}{2}f(x_1) + f(x_2) + f(x_3) + \ldots + f(x_{N-2}) + f(x_{N-1}) + \tfrac{1}{2}f(x_N)]h \qquad (4.48b)
\end{aligned}$$

We see that the weights of this rectangular rule are given by

$$w_k = \begin{cases} \tfrac{1}{2}h & k = 1, N \\ h & 2 \le k \le N - 1 \end{cases} \qquad (4.49)$$

and the abscissae are the points that define the rectangular segments7.

Trapezoidal rule

If the function is approximated by a linear spline between two adjacent points, then, for $x_k \le x \le x_{k+1}$,

$$\begin{aligned}
f_{interpolated}(x) &= \frac{(x - x_{k+1})}{(x_k - x_{k+1})}f(x_k) + \frac{(x - x_k)}{(x_{k+1} - x_k)}f(x_{k+1}) \\
&= \frac{[f(x_{k+1}) - f(x_k)]}{h}x + \frac{[x_k f(x_{k+1}) - x_{k+1} f(x_k)]}{h} \qquad (4.50)
\end{aligned}$$

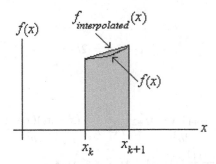

Fig. 4.3 Trapezoid formed by approximating $f(x)$ by a linear spline

which describes the trapezoid shown in Fig. 4.3.

The area of such a trapezoid is given by

$$A_k = \int_{x_k}^{x_{k+1}} f_{interpolated}(x)dx = \int_{x_k}^{x_k + \Delta x} f_{interpolated}(x)dx$$

$$= \tfrac{1}{2}[f(x_k) + f(x_{k+1})]h \tag{4.51}$$

and the sum of these areas over all segments results in

$$I \simeq \tfrac{1}{2}[f(x_1) + f(x_2)]h + \tfrac{1}{2}[f(x_2) + f(x_3)]h + \ldots$$
$$\ldots + \tfrac{1}{2}[f(x_{N-2}) + f(x_{N-1})]h + \tfrac{1}{2}[f(x_{N-1}) + f(x_N)]h$$
$$= [\tfrac{1}{2}f(x_1) + f(x_2) + f(x_3) + \ldots + f(x_{N-2}) + f(x_{N-1}) + \tfrac{1}{2}f(x_N)]h \tag{\textit{4.48b}}$$

That is, the *trapezoidal rule* is identical to the rectangular rule generated by taking the average of $f(x)$ over one segment. As such, the rectangular rule will mean that generated by approximating $f(x)$ by its value at the midpoint of a segment.

We note that for this trapezoidal rule, each segment is comprised of one interval. Therefore, the range $[a, b]$ can be divided into any number of segments. This is not true for higher order quadrature rules.

Simpson's 1/3 rule

Simpson's 1/3 rule is obtained by approximating $f(x)$ by a quadratic spline, which requires fitting the function to a parabola over three consecutive points. That is, for $x_k \leq x \leq x_{k+2}$, we have

$$f_{interpolated}(x) = \frac{(x - x_{k+1})(x - x_{k+2})}{(x_k - x_{k+1})(x_k - x_{k+2})} f(x_k) +$$

$$\frac{(x - x_k)(x - x_{k+2})}{(x_{k+1} - x_k)(x_{k+1} - x_{k+2})} f(x_{k+1}) + \frac{(x - x_k)(x - x_{k+1})}{(x_{k+2} - x_k)(x_{k+2} - x_{k+1})} f(x_{k+2}) \tag{4.52}$$

With

$$x_{k+2} - x_k = 2h \tag{4.53}$$

eq. 4.52 becomes

$$f_{interpolated}(x) = \frac{(x - x_{k+1})(x - x_{k+2})}{2h^2} f(x_k) - \frac{(x - x_k)(x - x_{k+2})}{h^2} f(x_{k+1})$$

$$+ \frac{(x - x_k)(x - x_{k+1})}{2h^2} f(x_{k+2}) \quad x_k \le x \le x_{k+2} \tag{4.54}$$

The area under such a parabola is given by

$$A_k = \int_{x_k}^{x_{k+2}} f_{interpolated}(x)dx = \int_{x_k}^{x_k+2h} f_{interpolated}(x)dx$$

$$= \tfrac{1}{3}h[f(x_k) + 4f(x_{k+1}) + f(x_{k+2})] \tag{4.55}$$

Summing these areas, we obtain

$$I \simeq \frac{h}{3}[(f(x_1) + 4f(x_2) + f(x_3)) + (f(x_3) + 4f(x_4) + f(x_5)) + \ldots]$$

$$= \frac{h}{3}[f(x_1) + 4f(x_2) + 2f(x_3) + 4f(x_4) + 2f(x_5) + \ldots$$

$$+ \ldots + 2f(x_{N-2}) + 4f(x_{N-1}) + f(x_N)] \tag{4.56}$$

which is Simpson's 1/3 rule.

Since segments are defined by three points containing two intervals, each segment begins and ends at a point with an odd index and is two intervals wide. Therefore, to apply this quadrature rule, one must divide $[a, b]$ using an odd number of points, or equivalently an even number of intervals.

Note also that for each x_k (k odd) except for x_1 and x_N, there are two terms containing $f(x_k)$ coming from the end of a segment which is the beginning of the next segment. Therefore, all odd index $f(x_k)$ except for $f(x_1)$ and $f(x_N)$ are multiplied by 2, and all even index $f(x_k)$ are multiplied by 4. From this, we see that the weights for Simpson's 1/3 rule are given by

$$w_k = \begin{cases} \tfrac{1}{3}h & k = 1, N \\ \tfrac{4}{3}h & 2 \le k \le N - 1 \quad k \; even \\ \tfrac{2}{3}h & 3 \le k \le N - 2 \quad k \; odd \end{cases} \tag{4.57}$$

Simpson's 3/8 rule

When $f(x)$ is interpolated by a cubic spline, then for $x_k \leq x \leq x_{k+3}$,

$$
\begin{aligned}
f_{interpolated}(x) &= \frac{(x - x_{k+1})(x - x_{k+2})(x - x_{k+3})}{(x_k - x_{k+1})(x_k - x_{k+2})(x_k - x_{k+3})} f(x_k) \\
&+ \frac{(x - x_k)(x - x_{k+2})(x - x_{k+3})}{(x_{k+1} - x_k)(x_{k+1} - x_{k+2})(x_{k+1} - x_{k+3})} f(x_{k+1}) \\
&+ \frac{(x - x_k)(x - x_{k+1})(x - x_{k+3})}{(x_{k+2} - x_k)(x_{k+2} - x_{k+1})(x_{k+2} - x_{k+3})} f(x_{k+2}) \\
&+ \frac{(x - x_k)(x - x_{k+1})(x - x_{k+2})}{(x_{k+3} - x_k)(x_{k+3} - x_{k+1})(x_{k+3} - x_{k+2})} f(x_{k+3}) \\
&= -\frac{(x - x_{k+1})(x - x_{k+2})(x - x_{k+3})}{6h^3} f(x_k) \\
&+ \frac{(x - x_k)(x - x_{k+2})(x - x_{k+3})}{2h^3} f(x_{k+1}) \\
&- \frac{(x - x_k)(x - x_{k+1})(x - x_{k+3})}{2h^3} f(x_{k+2}) \\
&+ \frac{(x - x_k)(x - x_{k+1})(x - x_{k+2})}{6h^3} f(x_{k+3})
\end{aligned}
\tag{4.58}
$$

The area under this cubic polynomial is given by

$$
\begin{aligned}
A_{cubic} &= \int_{x_k}^{x_k+3h} f_{interpolated}(x)\,dx \\
&= \frac{3}{8} h[f(x_k) + 3f(x_{k+1}) + 3f(x_{k+2}) + f(x_{k+3})]
\end{aligned}
\tag{4.59}
$$

This results in the *Simpson's 3/8 rule*

$$
\begin{aligned}
I = \frac{3}{8} h[&f(x_1) + 3f(x_2) + 3f(x_3) + 2f(x_4) + \cdots \\
&\cdots + 2f(x_{N-3}) + 3f(x_{N-2}) + 3f(x_{N-1}) + f(x_N)]
\end{aligned}
\tag{4.60}
$$

The reader can easily deduce the weights of this quadrature rule.

Higher order Newton-Cotes quadrature rules

A fourth order spline interpolation of $f(x)$ over segments defined by five points defining four intervals results in *Hardy's quadrature rule*

$$I = \frac{h}{5}\left[\frac{7}{15}f(x_1) + \frac{27}{10}f(x_2) + \frac{11}{3}f(x_3) + \frac{27}{10}f(x_4) + \frac{14}{15}f(x_5) + \cdots \right.$$

$$\left. \cdots + \frac{14}{15}f(x_{N-4}) + \frac{27}{10}f(x_{N-3}) + \frac{11}{3}f(x_{N-2}) + \frac{27}{10}f(x_{N-1}) + \frac{7}{15}f(x_N)\right]$$

$$(4.61a)$$

and *Weddle's quadrature rule*

$$I = \frac{3}{10}h[f(x_1) + 5f(x_2) + f(x_3) + 6f(x_4)$$
$$+ f(x_5) + 5f(x_6) + 2f(x_7) + \cdots$$
$$\cdots + 2f(x_{N-6}) + 5f(x_{N-5}) + f(x_{N-4}) + 6f(x_{N-3})$$
$$+ f(x_{N-2}) + 5f(x_{N-1}) + f(x_N)]$$

$$(4.61b)$$

is obtained from a sixth order spline interpolation of $f(x)$ over segments defined by seven points divided into six intervals.

These higher order quadrature rules are rarely used. If one does not obtain acceptable accuracy for a given number of segments using a low order quadrature rule (Simpson's 1/3 rule is often satisfactory), then rather than use a higher order quadrature it is more common to use one of these lower order rules and divide $[a, b]$ into a larger number of segments.

Example 4.4: Evaluation of an integral by Newton-Cotes quadratures

We evaluate

$$\int_1^3 \frac{1}{(1+x)}\,dx = \ell n(2) = 0.69315 \qquad (4.62)$$

using the rectangular, trapezoidal, and Simpson's 1/3 rule.

Since Simpson's 1/3 rule requires the range $[1, 3]$ to be divided into an even number of intervals using an odd number of points, we will first consider using five abscissa points $\{1.0, 1.5, 2.0, 2.5, 3.0\}$. This will produce four segments for the rectangular and trapezoidal rules and two segments for the Simpson's 1/3 rule.

(a) The midpoints of the intervals are $\{1.25, 1.75, 2.25, 2.75\}$. Therefore, with $h = 0.5$, the rectangular rule of eq. 4.48a yields

$$I \simeq 0.5\left[\frac{1}{(1+1.25)} + \frac{1}{(1+1.75)} + \frac{1}{(1+2.25)} + \frac{1}{(1+2.75)}\right]$$

$$= 0.69122 \qquad (4.63a)$$

(b) The trapezoidal rule yields

$$\int_1^3 \frac{1}{(1+x)}dx \simeq 0.5\left(\frac{1}{2}\frac{1}{(1+1)}+\frac{1}{(1+1.5)}+\frac{1}{(1+2)}+\frac{1}{(1+2.5)}+\frac{1}{2}\frac{1}{(1+3)}\right)$$
$$= 0.69702 \tag{4.63b}$$

(c) For the Simpson's 1/3 rule, we obtain

$$\int_1^3 \frac{1}{(1+x)}dx \simeq \frac{0.5}{3}\left(\frac{1}{(1+1)}+\frac{4}{(1+1.5)}+\frac{2}{(1+2)}+\frac{4}{(1+2.5)}+\frac{1}{(1+3)}\right)$$
$$= 0.69325 \tag{4.63c}$$

As expected, the Simpson's 1/3 rule yields a more accurate approximation than rectangular or trapezoidal quadrature rules.

(d) For the Simpson's 3/8 rule, we would have to use $3N - 1$ points to divide $[1, 3]$ into N segments. With Weddle's rule, we would have to use $6N - 1$ points to produce N segments.

These higher order rules are rarely used. As mentioned earlier, most often one would use Simpson's 1/3 rule and divide the range $[1, 3]$ into a larger number of segments. By increasing the number of segments (and intervals), each segment (interval) is narrowed, and $f(x)$ is better approximated by $f_{interpolated}(x)$.

For example, taking $h = 0.25$, using nine abscissae $\{1, 1.25, \ldots, 2.75, 3\}$ to divide $[1, 3]$ into three segments, the Simpson's 1/3 rule yields

$$I = \frac{0.25}{3}\left(\frac{1}{(1+1)}+\frac{4}{(1+1.25)}+\frac{2}{(1+1.5)}+\frac{4}{(1+1.75)}+\frac{2}{(1+2)}\right.$$
$$\left. +\frac{4}{(1+2.25)}+\frac{2}{(1+2.5)}+\frac{4}{(1+2.75)}+\frac{1}{(1+3)}\right) = 0.69316 \tag{4.64}$$

Thus, the Simpson's 1/3 rule using two segments is accurate to 0.015%. With three segments the accuracy is improved to 0.001%. □

Accuracy of Newton-Cotes quadrature rules

Let a Newton-Cotes quadrature rule be developed by approximating $f(x)$ by a polynomial of order $N - 1$ over each segment divided into $N - 1$ intervals. Then

$$I = \int_a^b f(x)dx \simeq \sum_{segments}\int_{x_k}^{x_k+(N-1)h} f_{interpolated}(x)dx = \sum_{k=1}^N w_k f(x_k) \tag{4.65a}$$

The exact value of this integral can also be written as

$$I = \int_a^b f(x)dx = \sum_{segments} \int_{x_k}^{x_k+(N-1)h} f(x)dx \qquad (4.65b)$$

Clearly, if $f(x)$ is a polynomial of order $N-1$,

$$f_{interpolated}(x) = f(x) \qquad (4.66)$$

at all points within each segment. Therefore, a Newton-Cotes quadrature rule yields an exact value of the integral of a polynomial, if the order of that polynomial is less than or equal to the order of the spline interpolation polynomial used to obtain the quadrature rule. Thus, for example, Simpson's 1/3 rule yields the exact value of

$$\int_a^b f(x)dx$$

if $f(x)$ is a constant, a linear function or a second order polynomial.

Example 4.5: Accuracy of the integral of a polynomial using Simpson's 1/3 rule

The segment for Simpson's 1/3 rule is made up of two intervals and is defined by a second order spline polynomial interpolation over each segment. Therefore,

$$h = x_{k+1} - x_k = x_{k+2} - x_{k+1} = \tfrac{1}{2}(x_{k+2} - x_k) \qquad (4.67a)$$

Thus,

$$x_{k+1} = x_k + \tfrac{1}{2}(x_{k+2} - x_k) = \tfrac{1}{2}(x_{k+2} + x_k) \qquad (4.67b)$$

For

$$f(x) = 1 \qquad (4.68)$$

the integral over one segment is

$$\int_{x_k}^{x_{k+2}} f(x)dx = \int_{x_k}^{x_{k+2}} dx = x_{k+2} - x_k \qquad (4.69a)$$

and the quadrature rule over one segment yields this exact result:

$$\tfrac{1}{3}[f(x_k) + 4f(x_{k+1}) + f(x_{k+2})]h$$
$$= \tfrac{1}{3}[1 + 4 + 1]\tfrac{1}{2}(x_{k+2} - x_k) = x_{k+2} - x_k \qquad (4.69b)$$

With

$$f(x) = x \tag{4.70}$$

the integral is given by

$$\int_{x_k}^{x_{k+2}} x\,dx = \tfrac{1}{2}(x_{k+2}^2 - x_k^2) \tag{4.71a}$$

and, with eq. 4.67b, the quadrature sum is

$$\tfrac{1}{3}[x_k + 4x_{k+1} + x_{k+2}]h = \tfrac{1}{3}[x_k + 2(x_{k+2} + x_k) + x_{k+2}]\tfrac{1}{2}(x_{k+2} - x_k)$$
$$= \tfrac{1}{2}(x_{k+2}^2 - x_k^2) \tag{4.71b}$$

which is the exact result given in eq. 4.71a.

If

$$f(x) = x^2 \tag{4.72}$$

the integral is

$$\int_{x_k}^{x_{k+2}} x^2\,dx = \tfrac{1}{3}(x_{k+2}^3 - x_k^3) \tag{4.73a}$$

which is identical to the quadrature sum

$$\tfrac{1}{3}[x_k^2 + 4x_{k+1}^2 + x_{k+2}^2]h = \tfrac{1}{3}[x_k^2 + x_k x_{k+2} + x_{k+2}^2](x_{k+2} - x_k)$$
$$= \tfrac{1}{3}(x_{k+2}^3 - x_k^3) \tag{4.73b}$$

Thus we see that the Simpson's 1/3 rule yields exact results for integrands that are zeroth order, first order, and second order polynomials.

We consider the integral of

$$\int_{x_k}^{x_{k+2}} x^3\,dx = \tfrac{1}{4}(x_{k+2}^4 - x_k^4) \tag{4.74}$$

over one segment. If we were to use Simpson's 3/8 rule, that quadrature sum would yield this result.

The quadrature sum for Simpson's 1/3 rule yields

$$\frac{1}{3}\left[x_k^3 + 4\left(\frac{x_{k+2} + x_k}{2}\right)^3 + x_{k+2}^3\right]h = \tfrac{1}{2}(x_{k+2}^3 + x_{k+2}^2 x_k + x_{k+2} x_k^2 + x_k^3)h$$
$$= \tfrac{1}{4}(x_{k+2}^3 + x_{k+2}^2 x_k + x_{k+2} x_k^2 + x_k^3)(x_{k+2} - x_k) = \tfrac{1}{4}(x_{k+2}^4 - x_k^4) \tag{4.75}$$

That is, because of the arithmetic involved, Simpson's 1/3 rule also yields an exact value of the integral of a third order polynomial, the same level of accuracy as the Simpson's 3/8 rule. For this reason, the 3/8 rule is rarely if ever used.

In Problem 3, the reader will demonstrate that Simpson's 1/3 rule does not yield an exact value for an integral of x^4. □

From the general statement about their accuracy, we can get a sense of the error involved in using Newton-Cotes quadrature rules.

Let a quadrature rule yield an exact value for the integral of a polynomial of order M (or lower). We expand the integrand in a Taylor series about some point within the range of integration. Expanding about $a \leq x_0 \leq b$, we have

$$
\begin{aligned}
f(x) &= \sum_{k=0}^{\infty} \frac{f^{(k)}(x_0)}{k!}(x-x_0)^k \\
&= \sum_{k=0}^{M} \frac{f^{(k)}(x_0)}{k!}(x-x_0)^k + \sum_{k=M+1}^{\infty} \frac{f^{(k)}(x_0)}{k!}(x-x_0)^k
\end{aligned}
\tag{4.76}
$$

The error in approximating the integral by a Newton-Cotes quadrature sum is defined as

$$
\begin{aligned}
E &\equiv \sum_{k=1}^{N} w_k f(x_k) - \int_a^b f(x)dx \\
&= \sum_{k=1}^{N} w_k f(x_k) - \int_a^b \sum_{k=0}^{M} \frac{f^{(k)}(x_0)}{k!}(x-x_0)^k dx \\
&\quad - \int_a^b \sum_{k=M+1}^{\infty} \frac{f^{(k)}(x_0)}{k!}(x-x_0)^k dx
\end{aligned}
\tag{4.77}
$$

Since the quadrature sum is the exact value of the integral of the polynomial of order M (or lower), the terms in the second line of eq. 4.77 are identical. Therefore,

$$
\begin{aligned}
E &= -\int_a^b \sum_{k=M+1}^{\infty} \frac{f^{(k)}(x_0)}{k!}(x-x_0)^k dx \\
&= -\sum_{k=M+1}^{\infty} \frac{f^{(k)}(x_0)}{k!(k+1)} \left[(b-x_0)^{k+1} - (a-x_0)^{k+1} \right]
\end{aligned}
\tag{4.78}
$$

If $f^{(k)}(x_0)/k!$ decreases rapidly enough with increasing k, a reasonable estimate of the error can be obtained by taking the first few terms in the sum in eq. 4.78.

Example 4.6: Error in using Simpson's 1/3 rule

Since Simpson's 1/3 rule is accurate up to a third order polynomial, the leading term in the error is proportional to the fourth derivative of the integrand. That is, $M = 4$ in eq. 4.78.

We consider the integral

$$
I = \int_0^1 e^{x/2} dx = 2(\sqrt{e} - 1) = 1.2974425
\tag{4.79}
$$

The Simpson's 1/3 quadrature rule, using two segments, is

$$I \simeq \frac{1}{12} \left[e^{0/2} + 4e^{0.25/2} + 2e^{0.5/2} + 4e^{0.75/2} + e^{1/2} \right] = 1.2974443 \qquad (4.80)$$

Therefore, the error involved using the 1/3 quadrature rule is

$$E \simeq 1.8 \times 10^{-6} = 1.4 \times 10^{-4}\% \qquad (4.81)$$

Taking $x_0 = 0$, the first term in the error as given in eq. 4.78, is

$$E \simeq \frac{e^{0/2}}{2^5 \, 5!} \left\{ \frac{1}{12} \left[0 + 4(.25)^5 + 2(.5)^5 + 4(.75)^5 + 1 \right] - \frac{1}{6} \right\}$$
$$= 3.3 \times 10^{-7} = 1.6 \times 10^{-5}\% \qquad (4.82)$$

Taking additional terms in the sum in eq. 4.78 would bring the error closer to $1.8 \times 10^{-6}\%$. $\square$

By increasing the number of segments within [a, b], the smaller each segment is and the better $f(x)$ is approximated by the polynomial $f_{interpolated}(x)$. Therefore, by increasing the number of segments, one increases the accuracy of a Newton-Cotes quadrature approximation of an integral.

4.4 Gaussian Quadratures

It was shown above that if $f(x)$ is a polynomial, the order of which is less than or equal to the interpolation polynomial used to approximate $f(x)$ over one segment, then the Newton-Cotes quadrature sum is the exact value of the integral. *Gaussian quadrature rules* are defined by requiring that if $f(x)$ is a polynomial of some order, then

$$\int_a^b \rho(x) f(x) dx = \sum_{k=1}^{N} w_k f(x_k) \qquad (4.83)$$

is exact. Based on the values of a and b, that polynomial $f(x)$ satisfying eq. 4.83 is a specific orthogonal polynomial, sometimes referred to as a *special function*. The function $\rho(x)$, which is unique to each special orthogonal polynomial, is called the *weighting function*. We denote these polynomials of orders M and N as $Q_M(x)$ and $Q_N(x)$. They satisfy the property

$$\int_a^b \rho(x) Q_N(x) Q_M(x) dx = \Gamma_N \delta_{NM} = \begin{cases} \Gamma_N & N = M \\ 0 & N \neq M \end{cases} \qquad (4.84)$$

When $N = M$, the integral is a non-zero constant Γ_N called the *normalization constant*. When $N \neq M$ the integral is zero. This property is called *orthogonality* of the polynomials. Together, the property of eq. 4.84 is called the *orthonormality* condition.

In this presentation, we will develop the three most commonly used Gaussian quadrature rules, based on three sets of orthogonal polynomials, *Legendre* polynomials, *Laguerre* polynomials, and *Hermite* polynomials. As will be demonstrated, these three quadrature rules can be used to evaluate any type of integral.

Weights and abscissae of a Gaussian quadrature rule

To get a sense of how the weights and abscissae of a Gaussian quadrature rule are found, let us consider approximating an integral by a two point rule given by

$$\int_a^b \rho(x)f(x)dx = w_1 f(x_1) + w_2 f(x_2) \tag{4.85}$$

Since there are four quantities to be determined, w_1, x_1, w_2, and x_2, four equations are required. Therefore, these weights and abscissae are found by requiring that eq. 4.85 be exact for the four lowest powers of x,

$$f(x) = \left\{1, x, x^2, x^3\right\} \tag{4.86}$$

This results in the equations

$$w_1 + w_2 = \int_a^b \rho(x)dx = F_0(a,b) \tag{4.87a}$$

$$w_1 x_1 + w_2 x_2 = \int_a^b x\rho(x)dx = F_1(a,b) \tag{4.87b}$$

$$w_1 x_1^2 + w_2 x_2^2 = \int_a^b x^2\rho(x)dx = F_2(a,b) \tag{4.87c}$$

and

$$w_1 x_1^3 + w_2 x_1^3 = \int_a^b x^3\rho(x)dx = F_3(a,b) \tag{4.87d}$$

where the quantities $F_k(a,b)$ are known constants. Clearly, it is very difficult to solve this set of highly nonlinear equations for $\{w_1, w_2\}$ and $\{x_1, x_2\}$. Thus, some method other than solving a system of equations is needed to determine the sets of weights and abscissae.

Using the above example as a guide, we see that since an N-point Gaussian quadrature rule is defined by N abscissae and N weights, we must determine $2N$ quantities by requiring that the integrals of the lowest $2N - 1$ powers of x be given exactly by the N-point Gaussian quadrature sum.

To find these quantities, let

$$Q_N(x) \equiv (x - x_1)(x - x_2) \dots (x - x_N) \tag{4.88}$$

be a polynomial of order N formed from $\{x_1, x_2, \dots, x_N\}$, the abscissae of the quadrature rule. Let $R_M(x)$ be a polynomial of order M, with $0 \leq M \leq N - 1$. Then $R_M(x)Q_N(x)$ is a polynomial, the order of which is between N and $2N - 1$. Since the quadrature sum is an exact value of the integral,

$$\int_a^b \rho(x)R_M(x)Q_N(x)dx = \sum_{k=1}^N w_k R_M(x_k)Q_N(x_k)dx \tag{4.89}$$

But each x_k is a root of $Q_N(x)$. Therefore,

$$Q_N(x_k) = 0 \quad 1 \leq k \leq N \tag{4.90}$$

from which

$$\int_a^b \rho(x)R_M(x)Q_N(x)dx = 0 \tag{4.91}$$

Since $R_M(x)$ and $Q_N(x)$ are polynomials of different order, eq. 4.91 is an orthogonality relation as that given in eq. 4.84. Thus, $R_M(x)$ must be a member of the same set of orthogonal polynomials as $Q_N(x)$. That is,

$$R_M(x) = Q_M(x) \tag{4.92}$$

with $M < N$.

Since $Q_M(x)$ is a different polynomial from $Q_N(x)$, the roots of $Q_M(x)$ are different from those of $Q_N(x)$. Therefore,

$$Q_M(x_k) \neq 0 \tag{4.93}$$

To find the N weights of the Gaussian quadrature rule, we define the set of $N - 1$ order polynomials

$$\mu_k(x) \equiv \frac{Q_N(x)}{(x - x_k)Q_N'(x_k)} \tag{4.94}$$

where x_k is one of the roots of $Q_N(x)$. It is straightforward to see that for any abscissa point x_m,

$$\lim_{x \to x_m} \mu_k(x) = \mu_k(x_m) = \frac{1}{Q'_N(x_k)} \lim_{x \to x_m} \frac{Q_N(x)}{(x - x_k)} = \delta_{km} = \begin{cases} 1 & k = m \\ 0 & k \neq m \end{cases} \tag{4.95}$$

Let $S_{N-1}(x)$ to be a polynomial of order $N - 1$. Since $S_{N-1}(x)$ and $\mu_k(x)$ are polynomials of order $N - 1$, the interpolation

$$S_{N-1}(x) = \sum_{k=1}^{N} \mu_k(x) S_{N-1}(x_k) \tag{4.96}$$

is an exact representation of $S_{N-1}(x)$. Because $S_{N-1}(x)$ is a polynomial of order less than $2N - 1$,

$$\int_a^b \rho(x) S_{N-1}(x) dx = \sum_{k=1}^{N} w_k S_{N-1}(x_k) \tag{4.97}$$

is exact. Substituting eq. 4.96 into the integral of eq. 4.97, we obtain

$$\sum_{k=1}^{N} w_k S_{N-1}(x_k) = \sum_{k=1}^{N} S_{N-1}(x_k) \int_a^b \rho(x) \mu_k(x) dx \tag{4.98}$$

Equating the coefficients of $S_{N-1}(x_k)$, the weights of the quadrature rule are given by

$$w_k = \int_a^b \rho(x) \mu_k(x) dx = \frac{1}{Q'_N(x_k)} \int_a^b \rho(x) \frac{Q_N(x)}{(x - x_k)} dx \tag{4.99}$$

Gauss–Legendre quadratures

Properties of Legendre polynomials needed for the development of Gauss–Legendre quadratures are:

Weighting Function:

$$\rho(x) = 1 \tag{4.100a}$$

Orthonormality:

$$\int_{-1}^{1} P_n(x) P_m(x) dx = \frac{2}{2n + 1} \delta_{nm} \tag{4.100b}$$

Summation representation of Legendre polynomials:

$$P_\ell(x) = \frac{1}{2^\ell} \begin{cases} \displaystyle\sum_{k=0}^{(\ell-1)/2} (-1)^k \frac{(2\ell-2k)!}{k!\,(\ell-k)!\,(\ell-2k)!} x^{\ell-2k} & \ell \text{ odd} \\[3ex] \displaystyle\sum_{k=0}^{\ell/2} (-1)^k \frac{(2\ell-2k)!}{k!\,(\ell-k)!\,(\ell-2k)!} x^{\ell-2k} & \ell \text{ even} \end{cases} \qquad (4.100c)$$

Five lowest order polynomials:

$$P_0(x) = 1 \qquad P_1(x) = x \qquad P_2(x) = \tfrac{1}{2}(3x^2 - 1)$$
$$P_3(x) = \tfrac{1}{2}(5x^3 - 3x) \qquad P_4(x) = \tfrac{1}{8}(35x^4 - 30x^2 + 3) \qquad (4.100d)$$

The derivation of these and other properties of Legendre polynomials can be found in the literature (see, for example, Cohen, H., 1992, pp. 288–299).

Since $\rho(x) = 1$ and $x \,\varepsilon\, [-1, 1]$ for Legendre polynomials, the N-point Gauss–Legendre quadrature rule is

$$\int_{-1}^{1} f(x)dx \simeq \sum_{k=1}^{N} w_k f(x_k) \qquad (4.101)$$

where $\{x_k\}$ are the zeros of the Legendre polynomial $P_N(x)$ and the weights $\{w_k\}$, found using eq. 4.99, are given by

$$w_k = \frac{1}{P'_N(x_k)} \int_{-1}^{1} \frac{P_N(x)}{(x - x_k)} dx \qquad (4.102)$$

Example 4.7: Abscissae and weights for a two point Gauss–Legendre quadrature rule

From

$$P_2(x) = \tfrac{1}{2}(3x^2 - 1) \qquad \textbf{(4.100d)}$$

the abscissae of the two point Gauss–Legendre quadrature rule are

$$x_{1,2} = \pm\frac{1}{\sqrt{3}} = \pm 0.57735 \qquad (4.103a)$$

and the weights are given by

$$w_1 = w_2 = \pm\frac{1}{3 \times (1/\sqrt{3})} \int_{-1}^{1} \frac{\tfrac{1}{2}(3x^2 - 1)}{(x \mp 1/\sqrt{3})} dx = 1.00000 \qquad (4.103b)\square$$

In Table 4.3, we present abscissae and weights for 5-, 10-, and 20-point Gauss–Legendre quadrature rules.

x	w
$N = 5$	
0.00000	0.56889
±0.53847	0.47863
±0.90618	0.23693
$N = 10$	
±0.14887	0.29552
±0.43340	0.26927
±0.67941	0.21909
±0.86506	0.14945
±0.97391	0.06667
$N = 20$	
±0.07653	0.15275
±0.22779	0.14917
±0.37371	0.14210
±0.51087	0.13169
±0.63605	0.11819
±0.74633	0.10193
±0.83912	0.08328
±0.91223	0.06267
±0.96397	0.04060
±0.99313	0.01761

Tables 4.3 Gauss–Legendre quadrature points

For a more complete compilation of Legendre quadrature data, the reader is referred to Stroud, A.H., and Secrest, D., (1966), pp. 100–151.

We note that integrals that are approximated by Gauss–Legendre quadratures are integrated between symmetric limits [–1, 1]. If the integrand of such an integral is an odd function $f_{odd}(x)$,

$$\int_{-1}^{1} f_{odd}(x)dx = 0 \tag{4.104a}$$

We see from Table 4.3 that the abscissae of every even order Gauss–Legendre quadrature rule occur in pairs, one positive, one negative, and for each pair, the weights of the positive and negative points are the same. Each odd order rule also contains positive and negative pairs of non-zero abscissae with a common weight plus the point $x = 0$. Since an odd function is zero for zero argument, every Gauss–Legendre quadrature rule satisfies

$$\sum_{k=1}^{N} w_k f_{odd}(x_k) = w_0 f_{odd}(0) + w_1[f_{odd}(x_1) + f_{odd}(-x)] + \dots = 0 \tag{4.104b}$$

That is, the integral of an odd function over [–1, 1] is given exactly by every Gauss–Legendre quadrature rule; it is zero.

Example 4.8: Approximating an integral by a Gauss–Legendre quadrature sum

To approximate

$$\int_1^2 \cos^2 y \, dy = \tfrac{1}{4}(2 + \sin 4 - \sin 2) = 0.08348 \tag{4.105}$$

by a Gauss–Legendre quadrature rule, the limits must be transformed to [–1, 1].
    To do this, we take

$$y = \tfrac{1}{2}(x + 3) \tag{4.106}$$

Then

$$\int_1^2 \cos^2 y \, dy = \frac{1}{2} \int_{-1}^1 \cos^2\left(\tfrac{1}{2}(x + 3)\right) dx \simeq \frac{1}{2} \sum_{k=1}^{N} w_k \cos^2\left(\tfrac{1}{2}(x_k + 3)\right) \tag{4.107}$$

Referring to example 4.7 and Table 4.3, we approximate this integral using 2, 5, 10, and 20 point Gauss–Legendre rules. The results are given in Table 4.4. □

N	Quadrature sum of eq. 4.107	%Error
2	0.08524	2.1×10^0
5	0.08347	2.8×10^{-4}
10	0.08348	4.2×10^{-5}
20	0.08348	4.2×10^{-5}

Table 4.4 Results for the approximation of the integral of eq. 4.105 using Gauss–Legendre quadrature rules

From Table 4.4, we conclude that an integral between finite limits of an integrand that is analytic over the range of integration is well approximated by transforming the range of integration to [–1, 1] and using a Gauss–Legendre quadrature sum of reasonable size.

Gauss–Laguerre quadratures

Properties of Laguerre polynomials needed for the development of Gauss–Laguerre quadratures are:

Weighting Function:

$$\rho(x) = e^{-x} \tag{4.108a}$$

Orthonormality:

$$\int_0^\infty e^{-x} L_n(x) L_m(x) dx = \delta_{nm} \qquad (4.108b)$$

Summation representation of Laguerre polynomials:

$$L_n(x) = \sum_{k=0}^n (-1)^k \frac{n!}{(k!)^2 (n-k)!} x^k \qquad (4.108c)$$

Five lowest order polynomials:

$$
\begin{aligned}
L_0(x) &= 1 \qquad L_1(x) = 1 - x \qquad L_2(x) = \tfrac{1}{2}x^2 - 2x + 1 \\
L_3(x) &= -\tfrac{1}{6}x^3 + \tfrac{3}{2}x^2 - 3x + 1 \qquad L_4(x) = \tfrac{1}{24}x^4 - \tfrac{2}{3}x^3 + 3x^2 - 4x + 1
\end{aligned} \qquad (4.108d)
$$

(For the derivation of these and other properties of Laguerre polynomials see Cohen, H., 1992, pp. 320–324.)

The N-point Gauss–Laguerre quadrature rule is

$$\int_0^\infty e^{-x} f(x) dx \simeq \sum_{k=1}^N w_k f(x_k) \qquad (4.109)$$

where $\{x_k\}$ are the zeros of the Nth order Laguerre polynomial. The weights $\{w_k\}$ are found from eq. 4.99 to be

$$w_k = \frac{1}{L_n'(x_k)} \int_0^\infty e^{-x} \frac{L_n(x)}{(x - x_k)} dx \qquad (4.110)$$

Example 4.9: Abscissae and weights for a two-point Gauss–Laguerre quadrature rule

The zeros of

$$L_2(x) = \tfrac{1}{2}x^2 - 2x + 1 \qquad \textbf{\textit{(4.108d)}}$$

are

$$x_{1,2} = 2 \pm \sqrt{2} = 3.41421, \; 0.58579 \qquad (4.111a)$$

and the weights are given by

$$
\begin{aligned}
w_{1,2} &= \frac{1}{L_2'(x_{1,2})} \int_0^\infty e^{-x} \frac{L_2(x)}{(x - x_{1,2})} dx = \frac{1}{\pm\sqrt{2}} \int_0^\infty e^{-x} \frac{(\tfrac{1}{2}x^2 - 2x + 1)}{(x - 2 \mp \sqrt{2})} dx \\
&= \frac{1}{\pm 2\sqrt{2}} \int_0^\infty e^{-x} \left(x - 2 \pm \sqrt{2}\right) dx = \begin{cases} 0.14645 \\ 0.85355 \end{cases}
\end{aligned} \qquad (4.111b)\square
$$

x	w	we^{-x}
	$N = 5$	
0.26356	0.52176×10^{0}	0.67909
1.41340	0.39867×10^{0}	1.63849
3.59643	0.75942×10^{-1}	2.76944
7.08581	0.36118×10^{-2}	4.31566
12.64080	0.23370×10^{-4}	7.21919
	$N = 10$	
0.13779	0.30844×10^{0}	0.35401
0.72945	0.40112×10^{0}	0.83190
1.80834	0.21807×10^{0}	1.33029
3.40143	0.62087×10^{-1}	1.86306
5.55250	0.95015×10^{-2}	2.45026
8.33015	0.75301×10^{-3}	3.12276
11.84379	0.28259×10^{-4}	3.93415
16.27926	0.42493×10^{-6}	4.99241
21.99659	0.18396×10^{-8}	6.57220
29.92070	0.99118×10^{-12}	9.78470
	$N = 20$	
0.07054	0.16875×10^{0}	0.18108
0.37212	0.29125×10^{0}	0.42256
0.91658	0.26669×10^{0}	0.66691
1.70731	0.16600×10^{0}	0.91535
2.74920	0.74826×10^{-1}	1.16954
4.04893	0.24964×10^{-1}	1.43135
5.61517	0.62026×10^{-2}	1.70298
7.45902	0.11450×10^{-2}	1.98702
9.59439	0.15574×10^{-3}	2.28664
12.03880	0.15401×10^{-4}	2.60583
14.81429	0.10865×10^{-5}	2.94978
17.94890	0.53301×10^{-7}	3.32540
21.47879	0.17580×10^{-8}	3.74226
25.45170	0.37255×10^{-10}	4.21424
29.93255	0.47675×10^{-12}	4.76252
35.01343	0.33728×10^{-14}	5.42173
40.83306	0.11550×10^{-16}	6.25401
47.61999	0.15395×10^{-19}	7.38731
55.81080	0.52864×10^{-23}	9.15133
66.52442	0.16565×10^{-27}	12.89339

Table 4.5 Gauss–Laguerre quadrature points

The third column in Table 4.5 is a list of the values of $\{(we^x)_k\}$. These data are used when the integrand does not explicitly contain the weighting factor e^{-x}. Then the Gauss–Laguerre rule can be written as

$$\int_0^\infty f(x)dx = \int_0^\infty e^{-x}(e^x f(x))dx \simeq \sum_{k=1}^N w_k e^{x_k} f(x_k) \qquad (4.112)$$

For a more complete compilation of Gauss–Laguerre quadrature data, the reader is referred to Stroud, A.H., and Secrest, D., (1966), pp. 254–274, or Abramowitz, M., and Stegun, I., (1964), p. 923.

Example 4.10: Approximating various integrals by Gauss–Laguerre quadratures

(a) We consider the integral

$$\int_0^\infty \frac{e^{-x}}{(e^{-x}+1)^3}dx = -\int_0^\infty \frac{d(e^{-x})}{(e^{-x}+1)^3} = \frac{1}{2}\left[\frac{1}{(e^{-x}+1)^2}\right]_0^\infty = \frac{3}{8} \qquad (4.113)$$

Since this integrand contains e^{-x} explicitly, the integral is in the form of eq. 4.109. Therefore, the Laguerre quadrature approximation to this integral is given by

$$\int_0^\infty \frac{e^{-x}}{(e^{-x}+1)^3}dx \simeq \sum_{k=1}^N w_k \frac{1}{(e^{-x_k}+1)^3} \qquad (4.114)$$

Results obtained using 2, 5, 10, and 20 point Gauss–Laguerre quadrature rules is given in Table 4.6.

N	Quadrature sum of eq. 4.114	%Error
2	0.35917	4.2×10^0
5	0.37544	1.2×10^{-1}
10	0.37501	1.5×10^{-3}
20	0.37500	5.3×10^{-6}

Table 4.6 Results for the approximation of the integral of eq. 4.113 using Gauss–Laguerre quadrature rules

(b) The integrand of

$$\int_0^\infty xe^{-x^2}dx = \frac{1}{2}\int_0^\infty e^{-x^2}d(x^2) = \frac{1}{2} \qquad (4.115)$$

does not contain the Gauss–Laguerre weighting factor, e^{-x} explicitly. To evaluate this by the Gauss–Laguerre rule, we write

$$\int_0^\infty xe^{-x^2}dx = \int_0^\infty e^{-x}\left(e^x xe^{-x^2}\right)dx \simeq \sum_{k=1}^{N}(w_k e^{x_k})x_k e^{-x_k^2} \qquad (4.116)$$

In Table 4.7, we present results for 2-, 5-, 10-, and 20-point Gauss–Laguerre approximations to this integral.

N	Quadrature sum of eq. 4.116	%Error
2	0.63744	2.8×10^1
5	0.48113	3.8×10^0
10	0.49578	8.4×10^{-1}
20	0.49988	2.4×10^{-2}

Table 4.7 Results for the approximation of the integral of eq. 4.115 using Gauss–Laguerre quadrature rules

(c) We consider

$$\int_0^\infty \frac{1}{(1+x)^2}dx = 1.00000 \qquad (4.117)$$

As in part (b), we can approximate this by a Gauss–Laguerre rule by writing

$$\int_0^\infty \frac{1}{(1+x)^2}dx \simeq \sum_{k=1}^{N}(w_k e^{x_k})\frac{1}{(1+x_k)^2} \qquad (4.118)$$

The results for this approximation are shown in Table 4.8. □

N	Quadrature sum of eq. 4.118	%Error
2	0.83817	1.6×10^1
5	0.94254	5.7×10^0
10	0.97261	2.7×10^0
20	0.98663	1.3×10^0

Table 4.8 Results for the approximation of the integral of eq. 4.117 using Gauss–Laguerre quadrature rules

Referring to the discussion of sect. 4.1, it is possible to transform an integral between 0 to ∞ to an integral from -1 to 1. This would allow us to use the Gauss–Legendre rule to evaluate an integral of the form

$$\int_0^\infty F(y)dy$$

This transformation is achieved by defining x' as

$$x = \frac{4}{\pi}\tan^{-1}y - 1 \qquad (4.119)$$

The integral can then be approximated by

$$\int_0^\infty F(y)\,dy = \frac{\pi}{4}\int_{-1}^1 F\left[\tfrac{\pi}{4}(1+x)\right]\sec^2\left[\tfrac{\pi}{4}(1+x)\right]dx$$
$$\simeq \frac{\pi}{4}\sum_{k=1}^N w_k F\left[\tfrac{\pi}{4}(1+x_k)\right]\sec^2\left[\tfrac{\pi}{4}(1+x_k)\right] \qquad (4.120)$$

where the weights $\{w_k\}$ and abscissae $\{x_k\}$ in eq. 4.120 are those of a Gauss–Legendre rule.

Example 4.11: Approximating the integrals of Example 4.10 by Gauss–Legendre quadratures

(a) Making the substitution given in eq. 4.119, the integral of eq. 4.113 can be approximated by

$$\int_0^\infty \frac{e^{-y}}{(e^{-y}+1)^3}\,dy =$$
$$\frac{\pi}{4}\int_{-1}^1 e^{-\tan\left[\tfrac{\pi}{4}(1+x)\right]}\left[e^{-\tan\left[\tfrac{\pi}{4}(1+x)\right]}+1\right]^{-3}\sec^2\left[\tfrac{\pi}{4}(1+x)\right]dx \qquad (4.121)$$
$$\simeq \frac{\pi}{4}\sum_{k=1}^N w_k e^{-\tan\left[\tfrac{\pi}{4}(1+x_k)\right]}\left[e^{-\tan\left[\tfrac{\pi}{4}(1+x_k)\right]}+1\right]^{-3}\sec^2\left[\tfrac{\pi}{4}(1+x_k)\right]$$

Results using 2-, 5-, 10-, and 20-point Gauss–Legendre quadrature rules are given in Table 4.9.

N	Quadrature sum of eq. 4.121	%Error
2	0.47105	2.6×10^1
5	0.38821	3.5×10^0
10	0.37539	1.0×10^{-1}
20	0.37500	4.8×10^{-4}

Table 4.9 Results for the approximation of the integral of eq. 4.113 using Gauss–Legendre quadrature rules

(b) With the transformation of eq. 4.119, the integral of eq. 4.117 becomes

$$\int_0^\infty y e^{-y^2} dy =$$

$$= \frac{\pi}{4} \int_{-1}^{1} \tan\left[\tfrac{\pi}{4}(1+x)\right] e^{-\tan^2\left[\tfrac{\pi}{4}(1+x)\right]} \sec^2\left[\tfrac{\pi}{4}(1+x)\right] dx \qquad (4.122)$$

$$\simeq \frac{\pi}{4} \sum_{k=1}^{N} w_k \tan\left[\tfrac{\pi}{4}(1+x_k)\right] e^{-\tan^2\left[\tfrac{\pi}{4}(1+x_k)\right]} \sec^2\left[\tfrac{\pi}{4}(1+x_k)\right]$$

Table 4.10 lists the approximation of this integral by various Gauss–Legendre quadrature rules.

N	Quadrature sum of eq. 4.122	%Error
2	0.27371	4.5×10^1
5	0.49123	1.8×10^0
10	0.49931	1.4×10^{-1}
20	1.00000	2.7×10^{-4}

Table 4.10 Results for the approximation of the integral of eq. 4.115 using Gauss–Legendre quadrature rules

(c) Transformating the integral of eq. 4.117, we obtain

$$\int_0^\infty \frac{1}{(1+y)^2} dy = \frac{\pi}{4} \int_{-1}^{1} \frac{\sec^2\left(\tfrac{\pi}{4}(1+x)\right)}{\left[1 + \tan\left(\tfrac{\pi}{4}(1+x)\right)\right]^2} dx$$

$$\simeq \frac{\pi}{4} \sum_{k=1}^{N} w_k \frac{\sec^2\left(\tfrac{\pi}{4}(1+x_k)\right)}{\left[1 + \tan\left(\tfrac{\pi}{4}(1+x_k)\right)\right]^2} \qquad (4.123)$$

The results of these Gauss–Legendre quadrature rules are given in Table 4.11. □

N	Quadrature sum of eq. 4.123	%Error
2	0.97191	2.8×10^0
5	0.99997	2.6×10^{-3}
10	1.00000	8.5×10^{-6}
20	1.00000	8.5×10^{-6}

Table 4.11 Results for the approximation of the integral of eq. 4.115 using Gauss–Legendre quadrature rules

The results given in Tables 4.6–4.11 suggest the following conclusions about Gauss–Laguerre quadrature rules:

- Comparing the errors in Tables 4.6 and 4.9, we conclude that when the integrand explicitly contains e^{-x}, the integral over $[0, \infty]$ converges to the correct result quickly (i.e., with relatively small quadrature rules yielding accurate results) when approximated by a Gauss–Laguerre quadrature rule. Although it would not make sense to do so, one could also transform the integral to the range $[-1, 1]$ and use a comparable sized Gauss–Legendre quadratures to obtain a result that is about as accurate as that obtained with a Gauss–Laguerre rule.
- From Tables 4.7 and 4.10, we conclude that when the integrand does not explicitly contain e^{-x}, but when the integrand satisfies

$$\lim_{x \to \infty} f(x)e^x = 0 \tag{4.124}$$

the integral over $[0, \infty]$ converges somewhat slowly to the correct result (i.e., requires a relatively large Gauss–Laguerre quadrature sum to obtain accurate results). If the integral is transformed to the range $[-1, 1]$ and Gauss–Legendre quadrature rule is used, the integral converges to the correct value more quickly.
- Comparing the results given in Tables 4.8 and 4.11, they suggest that when the integrand does not explicitly contain e^{-x}, and when the integrand satisfies

$$\lim_{x \to \infty} f(x)e^x = \infty \tag{4.125}$$

approximating the integral over $[0, \infty]$ by a Gauss–Laguerre quadrature sum converges very slowly to the correct result. If the integral is transformed to the range $[-1, 1]$ and a Gauss–Legendre quadrature rule is used, the integral converges to the correct value much more quickly.

Gauss–Hermite quadrature rules

Properties of Hermite polynomials used in the development of Gauss–Hermite quadratures are:

Weighting function:

$$\rho(x) = e^{-x^2} \tag{4.126a}$$

Orthonormality:

$$\int_{-\infty}^{\infty} e^{-x^2} H_n(x) H_m(x) dx = 2^n n! \sqrt{\pi} \delta_{nm} \tag{4.126b}$$

Summation representation of Hermite polynomials:

$$
H_n(x) = \begin{cases} \displaystyle\sum_{k=0}^{n/2} (-1)^k \frac{n!}{(k!)\,(n-2k)!}\,(2x)^{n-2k} & n \text{ even} \\[2em] \displaystyle\sum_{k=0}^{(n-1)/2} (-1)^k \frac{n!}{(k!)\,(n-2k)!}\,(2x)^{n-2k} & n \text{ odd} \end{cases} \qquad (4.126c)
$$

Five lowest order polynomials:

$$
\begin{aligned}
&H_0(x) = 1 \qquad H_1(x) = 2x \qquad H_2(x) = 4x^2 - 2 \\
&H_3(x) = 8x^3 - 12x \qquad H_4(x) = 16x^4 - 48x^2 + 12
\end{aligned} \qquad (4.126d)
$$

(The derivation of these and other properties of Hermite polynomials are given by Cohen, H., (1992), pp. 330–336.)

The N-point Gauss–Hermite quadrature rule is given by

$$
\int_{-\infty}^{\infty} e^{-x^2} f(x)\,dx \simeq \sum_{k=1}^{N} w_k f(x_k) \qquad (4.127)
$$

The abscissae $\{x_k\}$ are the zeros of the Hermite polynomials and the weights $\{w_k\}$ are given by

$$
w_k = \frac{1}{H_N'(x_k)} \int_{-\infty}^{\infty} e^{-x^2} \frac{H_N(x)}{(x - x_k)}\,dx \qquad (4.128)
$$

Example 4.12: Abscissae and weights for a two point Gauss–Hermite quadrature rule

The roots of

$$
H_2(x) = 4x^2 - 2 \qquad \textbf{(4.126d)}
$$

which are the abscissae of the two-point Gauss–Hermite quadrature rule, are given by

$$
x_{1,2} = \pm \frac{1}{\sqrt{2}} = \pm 0.70711 \qquad (4.129a)
$$

The corresponding weights are

$$
w_1 = w_2 = \frac{\pm\sqrt{2}}{8} \int_{-\infty}^{\infty} e^{-x^2} \frac{(4x^2 - 2)}{(x \mp 1/\sqrt{2})}\,dx = \frac{\sqrt{\pi}}{2} = 0.88623 \qquad (4.129b)\square
$$

Table 4.12 contains quadrature data for 5-, 10-, and 20-point Gauss–Hermite quadratures.

x	w	we^{x^2}
	$N = 5$	
0.00000	0.94531×10^{0}	0.94531
± 0.95857	0.39362×10^{0}	0.98658
± 2.02018	0.19953×10^{-1}	1.18149
	$N = 10$	
± 0.34290	0.61086	0.68708
± 1.03661	0.24014	0.70330
± 1.75668	0.33874×10^{-1}	0.74144
± 2.53273	0.13436×10^{-2}	0.82067
± 3.43616	0.76404×10^{-5}	1.02545
	$N = 20$	
± 0.24534	0.46224×10^{0}	0.49092
± 0.73747	0.28668×10^{0}	0.49834
± 1.23408	0.10902×10^{0}	0.49992
± 1.73854	0.24811×10^{-1}	0.50968
± 2.25497	0.32438×10^{-2}	0.52408
± 2.78881	0.22834×10^{-3}	0.54485
± 3.34785	0.78026×10^{-5}	0.57526
± 3.94476	0.10861×10^{-6}	0.62228
± 4.60368	0.43993×10^{-9}	0.70433
± 5.38748	0.22294×10^{-12}	0.89859

Table 4.12 Abscissae and weights for 5, 10, and 20 point Gauss–Hermite quadrature points

A more complete set of Gauss–Hermite quadratures can be found by Stroud, A.H., and Secrest, D., (1966), pp. 218–252, and Abramowitz, M., and Stegun, I., (1964), p. 924.

As with the Gauss–Laguerre quadratures, the table for the Gauss–Hermite quadratures contains a third column of quantities,

$$we^{x^2}$$

If the integrand does not explicitly contain the weighting factor

$$e^{-x^2}$$

the integral can be written and approximated by a Gauss–Hermite quadrature as

$$\int_{-\infty}^{\infty} f(x)dx = \int_{-\infty}^{\infty} e^{-x^2}\left(e^{x^2}f(x)\right)dx \simeq \sum_{k=1}^{N} w_k e^{x_k^2} f(x_k) \qquad (4.130)$$

Above we discussed the fact that the integral

$$\int_{-1}^{1} f_{odd}(x)dx = 0 \qquad (\textbf{\textit{4.104a}})$$

is given exactly by any Gauss–Legendre rule. Likewise, because Gauss–Hermite quadratures contain pairs of points, one positive, one negative, with a common weight (and one point that is zero for odd order quadratures), the integrals

$$\int_{-\infty}^{\infty} e^{-x^2} f_{odd}(x)dx = 0 \qquad (4.131a)$$

and

$$\int_{-\infty}^{\infty} f_{odd}(x)dx = 0 \qquad (4.131b)$$

are given exactly by

$$\sum_{k=1}^{N} w_k f_{odd}(x_k) = 0 \qquad (4.132a)$$

and

$$\sum_{k=1}^{N} w_k e^{x_k^2} f_{odd}(x_k) = 0 \qquad (4.132b)$$

respectively.

Example 4.13: Approximating an integral using Gauss–Hermite quadratures

(a) We consider

$$\int_{-\infty}^{\infty} e^{-x} e^{-x^2} dx = e^{\frac{1}{4}} \int_{-\infty}^{\infty} e^{-\left(x+\frac{1}{2}\right)^2} dx = \sqrt{\pi} e^{\frac{1}{4}} \simeq 2.27588 \qquad (4.133)$$

Since the weighting function appears explicitly in the integrand, the Gauss–Hermite quadrature rule for this integral is given by

$$\int_{-\infty}^{\infty} e^{-x} e^{-x^2} dx \simeq \sum_{k=1}^{N} w_k e^{-x_k} \qquad (4.134)$$

The results of approximating this integral by various Gauss–Hermite quadrature sums is given in Table 4.13.

N	Quadrature sum of eq. 4.134	%Error
2	2.23434	1.8×10^0
5	2.27587	9.5×10^{-5}
10	2.27588	4.2×10^{-6}
20	2.27588	4.2×10^{-6}

Table 4.13 Approximations to the integral of eq. 4.133 by various Gauss–Hermite quadrature sums

(b) To evaluate the integral

$$I = \int_{-\infty}^{\infty} x^4 e^{-x^4} dx = 2 \int_0^{\infty} x^4 e^{-x^4} dx \qquad (4.135)$$

we substitute

$$y = x^4 \qquad (4.136)$$

Then, referring to eq. A3.1 of Appendix 3,

$$I = \frac{1}{2} \int_0^{\infty} y^{1/4} e^{-y} dy = \frac{1}{2}\Gamma(1.25) \qquad (4.137)$$

Although this gamma function cannot be evaluated exactly, we have established methods for obtaining approximate values of it. Using the Stirling approximation for the Γ function given in Appendix 3, eq. A3.58 for N large, we have

$$\Gamma(N + p + 1) = (N + p)! \simeq$$

$$(N + p)^{(N+p)} e^{-(N+p)} \sqrt{2\pi(N + p)}\, e^{\left(\frac{1}{12(N+p)} - \frac{1}{360(N+p)^3} + \frac{1}{1260(N+p)^5}\right)} \qquad (4.138a)$$

Then, from the iteration property of the gamma function, we have

$$\Gamma(1 + p) \simeq \frac{(N + p)^{N+p} e^{-(N+p)} \sqrt{2\pi(N + p)}}{(N + p)(N + p - 1)...(1 + p)}\, e^{\left(\frac{1}{12(N+p)} - \frac{1}{360(N+p)^3} + \frac{1}{1260(N+p)^5}\right)} \qquad (4.138b)$$

With $p = 0.25$, we find that for $N \geq 5$, we obtain

$$\Gamma(1.25) \simeq 0.90640 \qquad (4.138c)$$

(This result can also be found in tables of gamma functions such as Abramowitz, M., and Stegun, I., (1964), p. 267–270.) Thus, the integral of eq. 4.137 is

$$I = \int_{-\infty}^{\infty} x^4 e^{-x^4} dx \simeq 0.45320 \tag{4.139}$$

Approximating this integral by a Gauss–Hermite quadrature rule, we have

$$I \simeq \sum_{k=1}^{N} w_k e^{x_k^2} x_k^4 e^{-x_k^4} \tag{4.140}$$

the values of which, for various values of N, are given in Table 4.14.

N	Quadrature sum of eq. 4.140	%Error
2	0.56897	2.6×10^1
5	0.71612	5.8×10^1
10	0.53164	1.7×10^1
20	0.44993	7.2×10^{-1}

Table 4.14 Approximations to the integral of eq. 4.139 by various Gauss–Hermite quadrature sums

(c) By substituting

$$x = \tan\phi \tag{4.141}$$

it is straightforward to show that

$$\int_{-\infty}^{\infty} \frac{1}{(x^2 + 1)^2} dx = \frac{\pi}{2} = 1.57080 \tag{4.142}$$

To use a Gauss–Hermite quadrature rule, we write this integral as

$$\int_{-\infty}^{\infty} e^{-x^2} \frac{e^{x^2}}{(x^2 + 1)^2} dx \simeq \sum_{k=1}^{N} w_k e^{x_k^2} \frac{1}{(x_k^2 + 1)^2} \tag{4.143}$$

Results using various Gauss–Hermite quadrature rules are given in Table 4.15. $\square$

N	Quadrature sum of eq. 4.143	%Error
2	1.29879	1.7×10^1
5	1.57272	1.2×10^{-1}
10	1.55823	8.0×10^{-1}
20	1.56754	2.1×10^{-1}

Table 4.15 Approximations to the integral of eq. 4.142 by various Gauss–Hermite quadrature sums

It is also possible to evaluate integrals over $[-\infty, \infty]$ by Gauss–Legendre quadratures. To do so, we make the substitution

$$x = \frac{2}{\pi}\tan^{-1}y \tag{4.144}$$

Then

$$\int_{-\infty}^{\infty} f(y)dy = \frac{\pi}{2}\int_{-1}^{1} f\left[\tan\left(\frac{\pi}{2}x\right)\right]\sec^2\left(\frac{\pi}{2}x\right)dx$$
$$\simeq \sum_{k=1}^{N} w_k f\left[\tan\left(\frac{\pi}{2}x_k\right)\right]\sec^2\left(\frac{\pi}{2}x_k\right) \tag{4.145}$$

where $\{w_k\}$ and $\{x_k\}$ are the weights and abscissae of a Gauss–Legendre quadrature rule.

Example 4.14: Evaluation of the integrals of Example 4.13 by Gauss–Legendre quadratures

(a) To approximate

$$\int_{-\infty}^{\infty} e^{-y}e^{-y^2}dy = \sqrt{\pi}e^{\frac{1}{4}} \simeq 2.27588 \tag{4.133}$$

using a Gauss–Legendre rule, we make the substitution given in eq. 4.144 to obtain

$$\int_{-\infty}^{\infty} e^{-y}e^{-y^2}dy = \frac{\pi}{2}\int_{-1}^{1}\sec^2\left(\frac{\pi}{2}x\right)e^{-\left[\tan\left(\frac{\pi}{2}x\right)+\tan^2\left(\frac{\pi}{2}x\right)\right]}dx$$
$$\simeq \frac{\pi}{2}\sum_{k=1}^{N} w_k\sec^2\left(\frac{\pi}{2}x_k\right)e^{-\left[\tan\left(\frac{\pi}{2}x_k\right)+\tan^2\left(\frac{\pi}{2}x_k\right)\right]} \tag{4.146}$$

In Table 4.16, we present the results obtained for approximating this integral by various Legendre quadrature sums.

N	Quadrature sum of eq. 4.146	%Error
2	3.12413	3.7×10^1
5	2.52670	1.1×10^1
10	2.27769	8.0×10^{-2}
20	2.27549	1.7×10^{-2}

Table 4.16 Approximations to the integral of eq. 4.133 by various Gauss–Legendre quadrature sums

(b) With the transformation of eq. 4.144, the integral of eq. 4.135 becomes

$$\int_{-\infty}^{\infty} y^4 e^{-y^4} dy = \frac{\pi}{2} \int_{-1}^{1} \sec^2\left(\frac{\pi}{2}x\right) \tan^4\left(\frac{\pi}{2}x\right) e^{-\tan^4\left(\frac{\pi}{2}x\right)} dx$$

$$= \frac{1}{2}\Gamma(1.25) = 0.45320$$

(4.147)

Approximating the integral over $[-1, 1]$ by a Gauss–Legendre quadrature rule, this becomes

$$\int_{-\infty}^{\infty} y^4 e^{-y^4} dy \simeq \frac{\pi}{2} \sum_{k=1}^{N} w_k \sec^2\left(\frac{\pi}{2}x_k\right) \tan^4\left(\frac{\pi}{2}x_k\right) e^{-\tan^4\left(\frac{\pi}{2}x_k\right)}$$

(4.148)

The results of this approximation by various Gauss–Legendre quadrature sums is given in Table 4.17.

N	Quadrature sum of eq. 4.148	%Error
2	1.53082	2.4×10^2
5	1.09482	1.4×10^2
10	0.39580	1.3×10^1
20	0.44771	1.2×10^0

Table 4.17 Approximations to the integral of eq. 4.135 by various Gauss–Legendre quadrature sums

(c) Transforming

$$\int_{-\infty}^{\infty} \frac{1}{(y^2 + 1)^2} dy \simeq 1.57085$$

(*4.142*)

to the interval [–1, 1] and approximating the resulting integral by a Gauss–Legendre quadrature rule, we have

$$\int_{-\infty}^{\infty} \frac{1}{(y^2 + 1)^2} dy = \frac{\pi}{2} \int_{-1}^{1} \cos^2\left(\frac{\pi}{2}x\right) dx \simeq \frac{\pi}{2} \sum_{k=1}^{N} w_k \cos^2\left(\frac{\pi}{2}x_k\right) \qquad (4.149)$$

Table 4.18 shows the results of this approximation for various Gauss–Legendre quadrature rules. □

N	Quadrature sum of eq. 4.149	%Error
2	1.19283	2.4×10^1
5	1.57084	3.1×10^{-3}
10	1.57080	2.9×10^{-9}
20	1.57080	6.9×10^{-9}

Table 4.18   Approximations to the integral of eq. 4.142 by various Gauss–Legendre quadrature sums

The results presented in Tables 4.13–4.18 suggest the following conclusions about Gauss–Hermite quadrature rules:

- Comparing the results given in Tables 4.13 and 4.16, we see that when the integrand contains the weighting factor $\exp(-x^2)$ explicitly, convergence to the correct value with Hermite quadrature rules is quite rapid (errors become quite small for low order quadrature rules) but converge to the exact value more slowly with Legendre quadratures.
- From the results given in Tables 4.14 and 4.17, we see that when

$$\lim_{x\to\infty} e^{x^2} f(x) = 0 \qquad\qquad (4.124)$$

is satisfied, the value of the integral converges to the correct result faster with Gauss–Hermite quadratures than with Gauss–Legendre rules, even though the integrand does not contain $\exp(-x^2)$ explicitly.
- A comparison of the results given in Tables 4.15 and 4.18 indicate that when

$$\lim_{x\to\infty} e^{x^2} f(x) = \infty \qquad\qquad (4.125)$$

transforming the limits to [–1, 1] and using Gauss–Legendre quadratures yields more rapid convergence to the correct value than when Gauss–Hermite quadratures are used.

4.5   **Integral of a Function with Singularities**

When $f(x)$ is infinite at $x_0 \, \varepsilon \, [a, \, b]$, care must be exercised in numerically approximating

$$\int_a^b f(x)dx$$

Weak singularities

If

$$\lim_{x \to x_0} f(x) = \infty \qquad (4.150a)$$

and

$$\lim_{x \to x_0} (x - x_0)f(x) = 0 \qquad (4.150b)$$

the singularity of $f(x)$ at x_0 is called a *weak singularity*. Examples of such functions that arise in applied problems are

$$f(x) = \frac{L(x)}{|x - x_0|^p} \qquad 0 < p < 1 \qquad (4.151a)$$

and

$$f(x) = L(x)\ell n(|x - x_0|) \qquad (4.151b)$$

where $L(x)$ is analytic at all x in the range of integration.
    To evaluate the integral of such a function, we write

$$\int_a^b \frac{L(x)}{|x - x_0|^p}\, dx = \int_a^b \frac{L(x) - L(x_0)}{|x - x_0|^p}\, dx + L(x_0)\int_a^b \frac{1}{|x - x_0|^p}\, dx \qquad (4.152a)$$

and

$$\int_a^b L(x)\ell n|x - x_0|dx =$$

$$\int_a^b [L(x) - L(x_0)]\ell n|x - x_0|dx + L(x_0)\int_a^b \ell n|x - x_0|dx \qquad (4.152b)$$

with $a < x_0 < b$. Because $L(x)$ is analytic at all $x \ \varepsilon \ [a, b]$, each term in the Taylor series of $L(x) - L(x_0)$ contains a factor $(x - x_0)^n$ with $n \geq 1$. Therefore, referring to eq. 4.150b,

$$\lim_{x \to x_0} \frac{L(x) - L(x_0)}{|x - x_0|^p} = \lim_{x \to x_0} [L(x) - L(x_0)]\ell n|x - x_0| = 0 \qquad (4.153)$$

Thus, the first integrals on the right hand sides of eq. 4.152b can be approximated by quadrature sums

$$\int_a^b \frac{L(x) - L(x_0)}{|x - x_0|^p} dx \simeq \sum_{k=1}^{N} w_k \frac{L(x_k) - L(x_0)}{|x_k - x_0|^p} \qquad (4.154a)$$

and

$$\int_a^b [L(x) - L(x_0)]\ell n|x - x_0| dx \simeq \sum_{k=1}^{N} w_k [L(x_k) - L(x_0)]\ell n|x_k - x_0| \qquad (4.154b)$$

where the kth term in the sum is zero if $x_0 = x_k$.

The second integrals on the right hand sides of eqs. 4.152 can be evaluated exactly as

$$\int_a^b \frac{1}{|x - x_0|^p} dx = \int_a^{x_0} \frac{1}{|x - x_0|^p} dx + \int_{x_0}^b \frac{1}{|x - x_0|^p} dx =$$
$$\int_a^{x_0} \frac{1}{(x_0 - x)^p} dx + \int_{x_0}^b \frac{1}{(x - x_0)^p} dx = \frac{(x_0 - a)^{1-p} + (b - x_0)^{1-p}}{(1 - p)} \qquad (4.155a)$$

and

$$\int_a^b \ell n|x - x_0| dx = \int_a^{x_0} \ell n(x_0 - x) dx + \int_{x_0}^b \ell n(x - x_0) dx \qquad (4.155b)$$
$$= (x_0 - a)[\ell n(x_0 - a) - 1] + (b - x_0)[\ell n(b - x_0) - 1]$$

Example 4.15: Evaluation of the integral of a function with a weak singularity

It is straightforward to show that

$$\int_{-1}^{1} \frac{x}{\sqrt{|x - 0.2|}} dx = \int_{-1}^{0.2} \frac{x}{\sqrt{0.2 - x}} dx + \int_{0.2}^{1} \frac{x}{\sqrt{x - 0.2}} dx \qquad (4.156)$$
$$= -0.4\sqrt{1.2} + \frac{2.8}{3}\sqrt{0.8} = 0.39662$$

To evaluate this integral by the method described above, we write

$$\int_{-1}^{1} \frac{x}{\sqrt{|x-0.2|}}\, dx = \int_{-1}^{1} \frac{x-0.2}{\sqrt{|x-0.2|}}\, dx + 0.2 \int_{-1}^{1} \frac{1}{\sqrt{|x-0.2|}}\, dx \qquad (4.157)$$

Then, using a ten-point Gauss–Legendre quadrature rule to approximate

$$\int_{-1}^{1} \frac{x-0.2}{\sqrt{|x-0.2|}}\, dx \simeq \sum_{k=1}^{4} w_k \frac{x_k-0.2}{\sqrt{|x_k-0.2|}} = -0.42529 \qquad (4.158a)$$

and with

$$0.2 \int_{-1}^{1} \frac{1}{\sqrt{|x-0.2|}}\, dx = 0.2 \left[\int_{-1}^{0.2} \frac{1}{\sqrt{0.2-x}}\, dx + \int_{0.2}^{1} \frac{1}{\sqrt{x-0.2}}\, dx \right]$$

$$= 0.4 \left[\sqrt{1.2} + \sqrt{0.8} \right] = 0.79595 \qquad (4.158b)$$

we obtain

$$\int_{-1}^{1} \frac{x}{\sqrt{|x-0.2|}}\, dx \simeq 0.37066 \qquad (4.159)$$

which is in reasonably good agreement with the exact value. □

As the reader will show in Problem 9, this approach also yields a reasonable result when the integrand contains a logarithmic infinity.

Cauchy or pole singularity

When $f(x)$ can be written as

$$f(x) = \frac{L(x)}{(x-x_0)} \qquad (4.160)$$

with $L(x)$ analytic at x_0, the singularity of $f(x)$ at x_0 is called a *simple pole*. It is occasionally referred to as a *Cauchy singularity*. The integral of $f(x)$ that is well defined is the *Cauchy principal value integral*, which is given by

$$P \int_{a}^{b} \frac{L(x)}{(x-x_0)}\, dx \equiv \lim_{\varepsilon \to 0} \left[\int_{a}^{x_0-\varepsilon} \frac{L(x)}{(x-x_0)}\, dx + \int_{x_0+\varepsilon}^{b} \frac{L(x)}{(x-x_0)}\, dx \right] \qquad (4.161)$$

with $\varepsilon > 0$. From this definition, we see that the integration accesses all x from a to b except x_0.

As we did with weakly singular integrands, we write this integral as

$$P \int_a^b \frac{L(x)}{(x-x_0)} dx = \int_a^b \frac{[L(x)-L(x_0)]}{(x-x_0)} dx + L(x_0) P \int_a^b \frac{1}{(x-x_0)} dx \qquad (4.162)$$

Since

$$\lim_{x \to x_0} \frac{L(x)-L(x_0)}{(x-x_0)} = L'(x_0) \qquad (4.163)$$

the integrand of the first integral on the right side of eq. 4.162 is analytic at x_0. Therefore, this integral is not a principal value integral and it can be approximated using a Gauss quadrature rule as

$$\int_a^b \frac{[L(x)-L(x_0)]}{(x-x_0)} dx \simeq \sum_{k=1}^N w_k \left[\frac{[L(x_k)-L(x_0)]}{(x_k-x_0)} \right] \qquad (4.164a)$$

with

$$\frac{[L(x_k)-L(x_0)]}{(x_k-x_0)} \simeq L'(x_0) \qquad (4.164b)$$

when $|x_k - x_0|$ is less than some small value ε defined by the user.
    The second integral on the right side of eq. 4.162 is evaluated as

$$P \int_a^b \frac{1}{(x-x_0)} dx = \lim_{\varepsilon \to 0} \left[\int_a^{x_0-\varepsilon} \frac{1}{(x-x_0)} dx + \int_{x_0+\varepsilon}^b \frac{1}{(x-x_0)} dx \right]$$

$$= \lim_{\varepsilon \to 0} \left[\ell n |x-x_0| \Big|_a^{x_0-\varepsilon} + \ell n |x-x_0| \Big|_{x_0+\varepsilon}^b \right]$$

$$= \lim_{\varepsilon \to 0} \left[\ell n \left(\frac{\varepsilon}{x_0-a} \right) + \ell n \left(\frac{b-x_0}{\varepsilon} \right) \right] = \ell n \left(\frac{b-x_0}{x_0-a} \right) \qquad (4.165)$$

Therefore,

$$P \int_a^b \frac{L(x)}{(x-x_0)} dx = \int_a^b \frac{[L(x)-L(x_0)]}{(x-x_0)} dx + L(x_0) \ell n \left(\frac{b-x_0}{x_0-a} \right)$$

$$\simeq \sum_{k=1}^N w_k \left[\frac{[L(x_k)-L(x_0)]}{(x_k-x_0)} \right] + L(x_0) \ell n \left(\frac{b-x_0}{x_0-a} \right) \qquad (4.166)$$

Example 4.16: Evaluation of a Cauchy principal value integral

Using partial fractions, it is straightforward to show that

$$P \int_{-1}^1 \frac{1}{(x^2+4)(x-\frac{1}{2})} dx = \frac{(2\tan^{-1}2 - \pi - 4\ell n(3))}{17} = -0.31304 \qquad (4.167)$$

Clearly, the pole singularity of this integrand is at

$$x = \tfrac{1}{2} \tag{4.168}$$

Therefore, we write

$$P \int_{-1}^{1} \frac{1}{(x^2 + 4)(x - \tfrac{1}{2})} dx =$$

$$\int_{-1}^{1} \left[\frac{1}{(x^2 + 4)} - \frac{1}{(\tfrac{1}{4} + 4)} \right] \frac{1}{(x - \tfrac{1}{2})} dx + \ell n \left(\frac{1 - \tfrac{1}{2}}{\tfrac{1}{2} + 1} \right) \times \left[\frac{1}{(\tfrac{1}{4} + 4)} \right] \tag{4.169}$$

$$= \int_{-1}^{1} \left[\frac{1}{(x^2 + 4)} - \frac{4}{17} \right] \frac{1}{(x - \tfrac{1}{2})} dx - \frac{4}{17} \ell n(3)$$

Since the remaining integral contains no singularities, it can be approximated by a Gauss–Legendre quadrature rule as

$$P \int_{-1}^{1} \frac{1}{(x^2 + 4)(x - \tfrac{1}{2})} dx \simeq \sum_{k=1}^{N} w_k \left[\frac{1}{(x_k^2 + 4)} - \frac{4}{17} \right] \frac{1}{(x_k - \tfrac{1}{2})} - \frac{4}{17} \ell n(3) \tag{4.170a}$$

Again, if one of the abscissa points is equal to or is very close to 1/2, the integrand is

$$L'\left(\frac{1}{2}\right) = \frac{d}{dx} \left[\frac{1}{(x^2 + 4)} - \frac{4}{17} \right]_{x=1/2} = -\left(\frac{4}{17}\right)^2 \tag{4.170b}$$

The results for various Gauss–Legendre quadrature rules are given in Table 4.19. □

N	Approximate value of the integral	%Error
2	−0.31280	7.9×10^{-2}
5	−0.31304	1.2×10^{-6}
10	−0.31304	6.2×10^{-11}
20	−0.31304	9.3×10^{-10}

Table 4.19 Approximation to the principal value integral of eq. 4.167 using various Legendre quadrature rules

Another technique for evaluating principal value integrals relies on two properties of the Gauss–Legendre quadratures:

- The abscissae, which range over [−1, 1], are symmetric about $x = 0$. For each abscissa $+x_k$ with weight w_k, there is an abscissa $−x_k$ with the same weight w_k.
- For any Legendre polynomial of order N, $P_N(\pm 1) \neq 0$. Therefore, +1 and −1 are never abscissae of any Legendre quadrature rule.

Since $L(x)$ is analytic at x_0, let x be a point close enough to x_0 that we can approximate $L(x) \cong L(x_0)$. When x approaches x_0 from $x > x_0$, the sign of the integrand is opposite the sign when x approaches x_0 from $x < x_0$. With

$$|x - x_0| \equiv \varepsilon \tag{4.171}$$

we see that with ε small

$$\frac{L(x)}{(x - x_0)} \simeq \frac{L(x_0)}{\varepsilon} \quad x > x_0 \tag{4.172a}$$

and

$$\frac{L(x)}{(x - x_0)} \simeq -\frac{L(x_0)}{\varepsilon} \quad x < x_0 \tag{4.172b}$$

Taking $L(x_0) > 0$ for the sake of discussion, the behavior of $L(x_0)/(x-x_0)$ at points near x_0 is shown in Fig. 4.4.

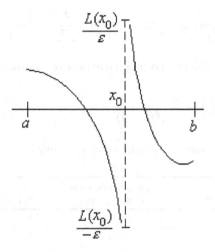

Fig. 4.4 $f(x)/(x-x_0)$ near x_0

To evaluate

$$P \int_a^b \frac{L(x)}{(x - x_0)} dx$$

we divide the range $[a, b]$ into three segments. If x_0 is closer to a than to b, we divide $[a, b]$ as shown in Fig. 4.5a. If x_0 is closer to b than to a, $[a, b]$ is segmented as shown in Fig. 4.5b.

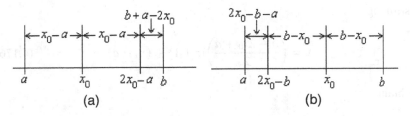

Fig. 4.5 (**a, b**) Dividing the interval $[a, b]$ into three segments

Such divisions yield segments on either side of x_0 with equal widths.

The integrals over the two segments separated by x_0 are then transformed to integrals over $[-1, 1]$ and approximated by the same Gauss–Legendre quadrature rule. The third segment can be approximated by any numerical integration method the user chooses. However, it is often convenient to use one quadrature rule for all three segments.

Referring to Fig. 4.5a, with x_0 closer to a, we write

$$P \int_a^b \frac{L(x)}{(x - x_0)} dx =$$

$$\lim_{\varepsilon \to 0} \int_a^{x_0 - \varepsilon} \frac{L(x)}{(x - x_0)} dx + \lim_{\varepsilon \to 0} \int_{x_0 + \varepsilon}^{2x_0 - a} \frac{L(x)}{(x - x_0)} dx + \int_{2x_0 - a}^b \frac{L(x)}{(x - x_0)} dx \qquad (4.173)$$

Substituting

$$x = \frac{(x_0 - a)}{2}(z - 1) + x_0 \qquad (4.174a)$$

we obtain

$$\lim_{\varepsilon \to 0} \int_a^{x_0 - \varepsilon} \frac{L(x)}{(x - x_0)} dx = \lim_{\varepsilon \to 0} \int_{-1}^{1 - \varepsilon} \frac{L\left[\frac{1}{2}(x_0 - a)(z - 1) + x_0\right]}{(z - 1)} dz \qquad (4.174b)$$

With

$$x = \frac{(x_0 - a)}{2}(z + 1) + x_0 \qquad (4.175a)$$

we have

$$\lim_{\varepsilon \to 0} \int_{x_0 + \varepsilon}^{2x_0 - a} \frac{L(x)}{(x - x_0)} dx = \lim_{\varepsilon \to 0} \int_{-1 + \varepsilon}^{1} \frac{L\left[\frac{1}{2}(x_0 - a)(z + 1) + x_0\right]}{(z + 1)} dz \qquad (4.175b)$$

Substituting

$$x = \left(\frac{b - 2x_0 + a}{2}\right)(z + 1) + (x_0 - a) \qquad (4.176a)$$

we obtain

$$\int_{2x_0-a}^{b} \frac{L(x)}{(x - x_0)} dx =$$

$$\tfrac{1}{2}(b - 2x_0 + a) \int_{-1}^{1} \frac{L\left[\tfrac{1}{2}(b - 2x_0 + a)(z + 1) + (2x_0 - a)\right]}{\tfrac{1}{2}(b - 2x_0 + a)(z + 1) + (x_0 - a)} dz \qquad (4.176b)$$

We see that the singularity at $x = x_0$ in the integral of eq. 4.174a has been transformed to the singularity at $z = 1$ in eq. 4.174b. Likewise the singularity at $x = x_0$ in the integral of eq. 4.175a has been transformed to the singularity at $z = -1$ in eq. 4.175a.

Approximating these integrals by the same N-point Gauss–Legendre quadrature sum, we see from Fig. 4.6 that for the approximation

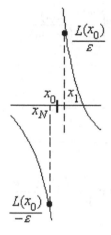

Fig. 4.6 Points on two sides of x_0 for which the integrands have opposite signs

$$\int_{-1}^{1-\varepsilon} \frac{L\left[\tfrac{1}{2}(x_0 - a)(z - 1) + x_0\right]}{(z - 1)} dz \simeq \sum_{k=1}^{N} w_k \frac{L\left[\tfrac{1}{2}(x_0 - a)(x_k - 1) + x_0\right]}{(x_k - 1)} \qquad (4.177a)$$

the quadrature point nearest to $z = 1$ is x_N. This $k = N$ term in the quadrature sum is

$$w_N \frac{L\left[\tfrac{1}{2}(x_0 - a)(x_N - 1) + x_0\right]}{(x_N - 1)} \simeq w_N \frac{L(x_0)}{(x_N - 1)} \qquad (4.177b)$$

Likewise, for

$$\lim_{\varepsilon \to 0} \int_{-1+\varepsilon}^{1} \frac{L\left[\frac{1}{2}(x_0 - a)(z + 1) + x_0\right]}{(z + 1)} dz$$

$$\simeq \sum_{k=1}^{N} w_k \frac{L\left[\frac{1}{2}(x_0 - a)(x_k + 1) + x_0\right]}{(x_k + 1)} \tag{4.178a}$$

the point in this quadrature sum closest to -1 is x_1. Since $x_1 = -x_N$ and $w_1 = w_N$, the $k = 1$ term in this sum can be written as

$$w_1 \frac{L\left[\frac{1}{2}(x_0 - a)(x_1 + 1) + x_0\right]}{(x_1 + 1)} \simeq w_N \frac{L(x_0)}{(-x_N + 1)} = -w_N \frac{L(x_0)}{(x_N - 1)} \tag{4.178b}$$

Therefore, the $k = N$ term in the quadrature sum of eq. 4.177a and the $k = 1$ term in the sum of eq. 4.178a add to zero, which mimics the elimination of the point $x = x_0$ implied by the principal value. Thus, the principle value integral can be approximated by

$$P \int_{a}^{b} \frac{L(x)}{(x - x_0)} dx \simeq \sum_{k=1}^{N} w_k \frac{L\left[\frac{1}{2}(x_0 - a)(x_k - 1) + x_0\right]}{(x_k - 1)} +$$

$$\sum_{k=1}^{N} w_k \frac{L\left[\frac{1}{2}(x_0 - a)(x_k + 1) + x_0\right]}{(x_k + 1)} +$$

$$\frac{1}{2}(b - 2x_0 + a) \sum_{k=1}^{N} w_k \frac{L\left[\frac{1}{2}(b - 2x_0 + a)(x_k + 1) + (2x_0 - a)\right]}{\frac{1}{2}(b - 2x_0 + a)(x_k + 1) + (x_0 - a)} \tag{4.179}$$

Example 4.17: Evaluating a principal value integral by dividing the range into three segments and using a Gauss–Legendre quadrature

Using the integral of example 4.16, we write

$$P \int_{-1}^{1} \frac{1}{(x^2 + 4)(x - \frac{1}{2})} dx = \int_{-1}^{0} \frac{1}{(x^2 + 4)(x - \frac{1}{2})} dx +$$

$$P \int_{0}^{1/2} \frac{1}{(x^2 + 4)(x - \frac{1}{2})} dx + P \int_{1/2}^{1} \frac{1}{(x^2 + 4)(x - \frac{1}{2})} dx \tag{4.180}$$

We transform each of these three integrals to integrals over $[-1, 1]$ using the transformation

$$x = \frac{1}{2}(z - 1) \tag{4.181a}$$

for the first integral,

$$x = \frac{1}{4}(z + 1) \tag{4.181b}$$

for the second integral, and

$$x = \tfrac{1}{4}(z+3) \tag{4.181c}$$

for the third. Then

$$P \int_{-1}^{1} \frac{1}{(x^2+4)(x-\tfrac{1}{2})} dx = \int_{-1}^{1} \frac{1}{\left(\tfrac{1}{4}(z-1)^2+4\right)(z-2)} dz$$

$$P \int_{-1}^{1} \frac{1}{\left(\tfrac{1}{16}(z+1)^2+4\right)(z-1)} dz + P \int_{-1}^{1} \frac{1}{\left(\tfrac{1}{16}(z+3)^2+4\right)(z+1)} dz \simeq$$

$$\sum_{k=1}^{N} w_k \left[\frac{1}{\left(\tfrac{1}{4}(x_k-1)^2+4\right)(x_k-2)} + \right.$$

$$\left. \frac{1}{\left(\tfrac{1}{16}(x_k+1)^2+4\right)(x_k-1)} + \frac{1}{\left(\tfrac{1}{16}(x_k+3)^2+4\right)(x_k+1)} \right] \tag{4.182}$$

Results using various Gauss–Legendre rules are given in Table 4.20. □

N	Approximation to the integral of eq. 4.180	%Error
2	−0.31126	5.7×10^{0}
5	−0.31304	3.4×10^{-4}
10	−0.31304	8.1×10^{-11}
20	−0.31304	4.5×10^{-9}

Table 4.20 Approximation to the principal value integral of eq. 4.167 using various Legendre quadrature rules

4.6   Monte Carlo Integration

The Monte Carlo method of integration is based on the interpretation of an integral as the area under a functional curve. The method requires that the user have computer software that contains a random number generator.

Integral of a function of one variable

Let $f(x)$ be integrated over $[a, b]$ and let f_{max} be the maximum value of $f(x)$ in that range. We construct a rectangle of width $b-a$ and height $H \geq f_{max}$ so that the area of the rectangle is larger than the area under the curve.

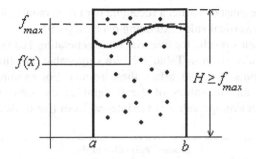

Fig. 4.7 A rectangle enclosing the area under the curve $f(x)$

To integrate by the Monte Carlo method, one generates a large sample of pairs of random numbers. For each pair, one number is used to define $x \, \varepsilon \, [a, b]$ the other to define $y \, \varepsilon \, [0, H]$.

As noted in Fig. 4.7, some of the points (x, y) are on or under the curve $f(x)$ and some are above the curve. The integral is found by using the statistical identity that for a large sample of random pairs (x, y)

$$\frac{area \ under f(x)}{area \ of \ the \ rectangle} \simeq \frac{number \ of \ points \ on/under f(x)}{total \ number \ of \ points \ generated} \tag{4.183a}$$

Since the area under $f(x)$ is the integral being evaluated, and the area of the rectangle is $(b - a) \times H$, we have

$$\int_a^b f(x) dx = \left[\frac{number \ of \ points \ on/under f(x)}{total \ number \ of \ points \ generated} \right] \times (b - a)H \tag{4.183b}$$

An efficient method of determining the maximum of $f(x)$ in the interval $[a, b]$ is to find the zeros of df/dx (perhaps numerically using one of the methods described in ch. 2). If none of these maxima are at the end points of $[a, b]$, one must compare these values to $f(a)$ and $f(b)$.

Example 4.18: Monte Carlo integration

Let us evaluate

$$\int_0^2 \left(1 + x^3 \right) dx = 6.00000 \tag{4.184}$$

by the Monte Carlo method outlined above.

Since $f(x) = (1 + x^3)$ increases with increasing x, we determine that the maximum of $f(x)$ in the interval $[0, 2]$ is given by

$$\left(1 + x^3 \right)_{max} = 9 \tag{4.185}$$

This means that we must construct a rectangle with horizontal sides extending over $x \varepsilon [0, 2]$ and with vertical sides extending over $y \varepsilon [0, 9]$.

Table 4.21 presents results for five trial runs, generating 10,000 random pairs in each trial. The results given in Table 4.21 are very stable. We find that increasing the number of points does not affect these results. For example, using 40,000 random points results in values of the integral of the same size as those in Table 4.21, with an average value of the integral from five trials of 5.97940. □

Trial	Monte Carlo value of the integral of eq. 4.184	%Error
1	5.97241	4.6×10^{-1}
2	5.98047	3.3×10^{-1}
3	6.02102	3.5×10^{-1}
4	5.96609	5.7×10^{-1}
5	5.97064	4.9×10^{-1}
Average	5.98213	4.4×10^{-1}

Table 4.21 Average of five trials of a program to evaluate the integral of eq. 4.184 by the Monte Carlo method

Integral of a function of two or more variables

The Monte Carlo method can easily be extended to multiple integrals. For example, to evaluate

$$I = \int_{x=a}^{x=b} \int_{y=c}^{y=d} f(x,y)dxdy \qquad (4.186)$$

one must define a three dimensional rectangular box in the space defined by x, y, and $z = f(x,y)$ with sides defined by $x \varepsilon [a, b]$, $y \varepsilon [c, d]$, and $z \varepsilon [0, H]$, where H is as large or larger than the maximum of $f(x,y)$. One then generates a large sample of triplets of random numbers to define x, y, and z. Then

$$\frac{volume \ under \ the \ surface \ z = f(x,y)}{volume \ of \ the \ rectangular \ box} \simeq$$
$$\frac{number \ of \ points \ under \ the \ surface \ z = f(x,y)}{total \ number \ of \ points \ (x,y,z) \ generated} \qquad (4.187)$$

from which

$$\int_{x=a}^{x=b} \int_{y=c}^{y=d} f(x,y)dxdy \simeq$$
$$\frac{number \ of \ points \ under \ the \ surface \ z = f(x,y)}{total \ number \ of \ points \ (x,y,z) \ generated} \times (b-a)(d-c)H \qquad (4.188)$$

4.7   Evaluating Integrals Using the Gamma and Beta Functions

As shown in Appendix 3, the *gamma* and *beta* functions are defined by integrals. Therefore, an integral that can be cast into a form of one of these functions can be evaluated or approximated using the properties of that function.

Integrals telated to the gamma function

The gamma function has an integral representation given by

$$\Gamma(p+1) \equiv \int_0^\infty z^p e^{-z} dz \quad \mathrm{Re}(p) > 0 \qquad (A3.1)$$

Therefore, an integral that can be cast into this form can be described in terms of the gamma function, and, if it cannot be evaluated exactly, its value can be estimated by approximating the gamma function.

Example 4.19: An integral related to a gamma function

Generalizing the integral discussed in example 4.13(b), we consider

$$I(m,n) \equiv \int_0^\infty x^m e^{-x^n} dx \quad m > 0, \ n > 0 \qquad (4.189)$$

Substituting

$$x = z^{\frac{1}{n}} \qquad (4.190)$$

this becomes

$$I(m,n) = \frac{1}{n} \int_0^\infty z^{\frac{m+1}{n}-1} e^{-z} dz = \frac{1}{n} \Gamma\left(\frac{m+1}{n}\right) \qquad (4.191)$$

If $(m+1)/n$ is a positive integer or a half integer, this can be evaluated exactly. Otherwise, one must use an approximate value of the Γ function found by the Stirling approximation, MacLaurin series approximation, a Gauss–Laguerre quadrature sum, or values presented in the literature. □

Integrals related to the beta function

Three integral representations of $\beta(p,q)$ are given in Appendix 3;

$$\beta(p,q) = 2 \int_0^{\pi/2} \cos^{2p-1}\theta \sin^{2q-1}\theta d\theta \quad p > 0, \ q > 0 \qquad (A3.20)$$

$$\beta(p,q) = \int_0^1 u^{p-1}(1-u)^{q-1}du \quad p>0,\ q>0 \qquad (A3.21)$$

and

$$\beta(p,q) = \int_0^\infty \frac{x^{p-1}}{(1+x)^{p+q}}dx \quad p>0,\ q>0 \qquad (A3.26)$$

Integrals that can be cast into one of these forms can be evaluated exactly or can be approximated using

$$\beta(p,q) = \beta(q,p) = \frac{\Gamma(p)\Gamma(q)}{\Gamma(p+q)} \qquad (A3.21)$$

Example 4.20: Integrals related to a beta function

(a) We consider

$$I_a \equiv \int_0^1 \frac{1}{x^{1/8}(1-x)^{5/12}}dx \qquad (4.192)$$

Comparing this to the definition of the β function given in eq. A3.26, we have

$$p - 1 = -\frac{1}{8} \Rightarrow p = \frac{7}{8} > 0 \qquad (4.193a)$$

and

$$q - 1 = -\frac{5}{12} \Rightarrow q = \frac{7}{12} > 0 \qquad (4.193b)$$

Therefore,

$$I_a = \beta\left(\tfrac{7}{8},\tfrac{7}{12}\right) = \frac{\Gamma\left(\tfrac{7}{8}\right)\Gamma\left(\tfrac{7}{12}\right)}{\Gamma\left(\tfrac{35}{24}\right)} \qquad (4.194)$$

Referring to the tables in Abramowitz, M., and Stegun, I., (1964), pp. 268 and 269, we find

$$\Gamma\left(\frac{7}{8}\right) = \frac{8}{7}\Gamma\left(1\tfrac{7}{8}\right) \simeq 1.08965 \qquad (4.195a)$$

By a linear interpolation between $\Gamma(1.580)$ and $\Gamma(1.585)$, which are given in the tables, we estimate that

$$\Gamma\left(\frac{7}{12}\right) = \frac{12}{7}\Gamma\left(1\frac{7}{12}\right) \simeq 1.52871 \tag{4.195b}$$

and a linear interpolation between $\Gamma(1.455)$ and $\Gamma(1.460)$ yields an estimate

$$\Gamma\left(\frac{35}{24}\right) = \Gamma\left(1\frac{11}{24}\right) \simeq 0.88561 \tag{4.195c}$$

From these values, we have

$$\int_0^1 \frac{1}{x^{1/8}(1-x)^{5/12}}\,dx \simeq 1.88092 \tag{4.196}$$

(b) We consider

$$I_b \equiv \int_0^{\pi/2} \sqrt{\tan x}\,dx \tag{4.197}$$

We see from eq. A3.20 that I_b is in the form of $\beta(p, q)$ with

$$2p - 1 = -\frac{1}{2} \Rightarrow p = \frac{1}{4} > 0 \tag{4.198a}$$

and

$$2q - 1 = \frac{1}{2} \Rightarrow q = \frac{3}{4} > 0 \tag{4.198b}$$

Therefore,

$$I_b = \tfrac{1}{2}\beta\left(\tfrac{1}{4}, \tfrac{3}{4}\right) = \frac{1}{2}\frac{\Gamma\left(\frac{1}{4}\right)\Gamma\left(\frac{3}{4}\right)}{\Gamma(1)} = \frac{1}{2}\Gamma\left(\tfrac{1}{4}\right)\Gamma\left(\tfrac{3}{4}\right) \tag{4.199}$$

This product of gamma functions can be determined in two equivalent ways. From the identity

$$\Gamma(N - p)\Gamma(p - N + 1) = (-1)^{N+1}\frac{\pi}{\sin(\pi p)} \tag{A3.29}$$

we set $N = 1$ and $p = 1/4$. Then, this becomes

$$\Gamma\left(\tfrac{1}{4}\right)\Gamma\left(\tfrac{3}{4}\right) = \frac{\pi}{\sin(\pi/4)} = \pi\sqrt{2} \tag{4.200}$$

Equivalently, from the Legendre duplication formula

$$\frac{\Gamma(q+1)\Gamma\left(q+\frac{1}{2}\right)}{\Gamma(2q+1)} = \frac{\sqrt{\pi}}{2^{2q}} \qquad (A3.36b)$$

we set $q = 1/4$. Then

$$\frac{\Gamma\left(\frac{5}{4}\right)\Gamma\left(\frac{3}{4}\right)}{\Gamma\left(\frac{3}{2}\right)} = \frac{\frac{1}{4}\Gamma\left(\frac{1}{4}\right)\Gamma\left(\frac{3}{4}\right)}{\frac{1}{2}\sqrt{\pi}} = \sqrt{\frac{\pi}{2}} \qquad (4.201)$$

from which

$$\Gamma\left(\tfrac{1}{4}\right)\Gamma\left(\tfrac{3}{4}\right) = \pi\sqrt{2} \qquad (4.202)$$

From this result, we have

$$\int_0^{\pi/2} \sqrt{\tan x}\, dx = \frac{\pi}{\sqrt{2}} \qquad (4.203)\square$$

As in the above example, we see that for certain values of p and q,

$$\beta(p,q) = \beta(q,p) = \frac{\Gamma(p)\Gamma(q)}{\Gamma(p+q)} \qquad (A3.21)$$

can determined exactly. This is possible when

• the exact value of each gamma function can be determined. This occurs when the arguments of the gamma function are positive integers and/or half integers.
• $p + q$ is an integer N with $N \geq 1$. Then $\Gamma(p+q) = (N-1)!$ and the numerator is $\Gamma(p)\Gamma(N-p)$. The exact value of $\Gamma(p)\Gamma(N-p)$ can be determined from

$$\Gamma(N-p)\Gamma(p-N+1) = (-1)^{N+1}\frac{\pi}{\sin(\pi p)} \qquad (A3.29)$$

where,

$$\Gamma(p) = (p-1)(p-2)...(p-N+1)\Gamma(p-N+1) \qquad (4.204)$$

Therefore,

$$\beta(p, N-p) = (-1)^{N+1}\frac{(p-1)(p-2)...(p-N+1)}{(N-1)!}\frac{\pi}{\sin(\pi p)} \qquad (4.205)$$

- p and q can be written in terms of an integer $N \geq 1$ as

$$p = \frac{N+2}{4} \qquad (4.206a)$$

and

$$q = \frac{N}{4} \qquad (4.206b)$$

Then

$$\beta\left(\frac{N+2}{4}, \frac{N}{4}\right) = \frac{\Gamma\left(\frac{N+2}{4}\right)\Gamma\left(\frac{N}{4}\right)}{\Gamma\left(\frac{N+1}{2}\right)} \qquad (4.207)$$

If N is even, then $(N+2)/4$ and $N/4$ are half integers and $(N+1)/2$ is a half integer. The exact values of these three Γ functions can be determined. If N is odd, it can be expressed in terms of an integer M as $2M + 1$. Then,

$$\beta\left(\frac{N+2}{4}, \frac{N}{4}\right) = \frac{\Gamma\left(\frac{2M+3}{4}\right)\Gamma\left(\frac{2M+1}{4}\right)}{M!} \qquad (4.208)$$

With

$$q = \frac{2M-1}{4} \qquad (4.209)$$

the Legendre duplication formula

$$\frac{\Gamma(q+1)\Gamma(q+\frac{1}{2})}{\Gamma(2q+1)} = \frac{\sqrt{\pi}}{2^{2q}} \qquad (A3.36b)$$

becomes

$$\Gamma\left(\frac{2M+3}{4}\right)\Gamma\left(\frac{2M+1}{4}\right) = \frac{\sqrt{\pi}}{2^{\left(M-\frac{1}{2}\right)}}\Gamma\left(M+\frac{1}{2}\right) \qquad (4.210)$$

Then eq. 4.208 becomes

$$\beta\left(\frac{N+2}{4}, \frac{N}{4}\right) = \frac{\sqrt{\pi}\Gamma\left(M+\frac{1}{2}\right)}{2^{\left(M-\frac{1}{2}\right)}(M!)} = \frac{2\sqrt{\pi}\Gamma\left(\frac{N}{2}\right)}{2^{\frac{N}{2}}\left(\frac{N-1}{2}\right)!} \qquad (4.211)$$

Since $N/2$ is an odd integer, the exact value of $\Gamma(N/2)$, and therefore the exact value of $\beta[(N+2)/4, N/4]$ can be determined. □

4.8   Integration Using Heaviside Operators

The *Heaviside operator* D_x is defined as a shorthand notation for the derivative.
That is,

$$D_x \equiv \frac{d}{dx} \tag{4.212}$$

With this definition, it is possible to evaluate integrals (and solve many types of
differential equations) by treating D_x as if it was an algebraic quantity (see, for
example, Kells, L., 1954, pp. 74 et. seq.).

If $P(z)$ is a function that is analytic at z_0, its Taylor series representation is
given by

$$P(z) = \sum_{k=0}^{\infty} \frac{P^{(k)}(z_0)}{k!} (z - z_0)^k \tag{4.213a}$$

By replacing the algebraic quantity z by the operator D_x we obtain the operator
function

$$P(D_x) = \sum_{k=0}^{\infty} \frac{P^{(k)}(z_0)}{k!} (D_x - z_0)^k \tag{4.213b}$$

An integral, being the inverse operation of a derivative, can be expressed in
Heaviside operator form as

$$\int f(x)dx \equiv D_x^{-1} f(x) \tag{4.214a}$$

from which the definite integral is

$$\int_a^b f(x)dx \equiv D_x^{-1} f(x) \Big|_a^b \tag{4.214b}$$

Integral of an integrand containing $e^{\mu x}$

A linear exponential function is one in which the exponent is linear in the variable.
It is of the form $e^{\mu x}$ where μ is a constant. The integral of a function with a linear
exponential is of the form

$$\int e^{\mu x} F(x)dx = D_x^{-1} [e^{\mu x} F(x)] \tag{4.215}$$

Consider

$$D_x[e^{\mu x}F(x)] = e^{\mu x}[D_xF(x) + \mu F(x)] = e^{\mu x}[D_x + \mu]F(x) \qquad (4.216a)$$

$$D_x^2[e^{\mu x}F(x)] = e^{\mu x}[D_x + \mu]^2 F(x) \qquad (4.216b)$$

$$D_x^k[e^{\mu x}F(x)] = e^{\mu x}[D_x + \mu]^k F(x) \qquad (4.216c)$$

Therefore, if the function $P(z)$ has a Taylor series

$$P(z) = \sum_{k=0}^{\infty} \frac{P^{(k)}(z_0)}{k!}(z - z_0)^k \qquad (\mathit{4.213a})$$

we can write

$$P(D_x)[e^{\mu x}F(x)] = e^{\mu x}P(D_x + \mu)F(x)$$
$$= e^{\mu x}\sum_{k=0}^{\infty} \frac{P^{(k)}(z_0)}{k!}(D_x + \mu)^k F(x) \qquad (4.217)$$

Therefore,

$$\int e^{\mu x}F(x)dx = e^{\mu x}[D_x + \mu]^{-1}F(x) = \frac{e^{\mu x}}{\mu}\left[1 + \frac{D_x}{\mu}\right]^{-1}F(x) \qquad (4.218)$$

The geometric series for $-1 < z < 1$ is given by

$$P(z) = \left[1 + \frac{z}{\mu}\right]^{-1} = \sum_{k=0}^{\infty} \frac{(-1)^k}{\mu^k}z^k \qquad (4.219)$$

Therefore, eq. 4.218 can be written

$$\int e^{\mu x}F(x)dx = \frac{e^{\mu x}}{\mu}\sum_{k=0}^{\infty} \frac{(-1)^k}{\mu^k}D_x^k F(x) \qquad (4.220)$$

If $F(x)$ is a finite polynomial of order N,

$$F(x) = \sum_{m=0}^{N} f_m(x - x_0)^m \qquad (4.221)$$

then the derivative D_x^k operating on $F(x)$ will be zero for $k > N$ and an exact result can be obtained. If $F(x)$ is not a polynomial, it is possible that it can be approximated by a polynomial by truncating its Taylor series as in eq. 4.221. Then, the series in eq. 4.220 becomes a finite sum.

Example 4.21: Integral of a product of an exponential and a polynomial by Heaviside methods

Using the Heaviside operator, we consider the integral

$$
I_e(\mu) \equiv \int e^{\mu x} x^2 dx = D_x^{-1} \left[e^{\mu x} x^2 \right] = e^{\mu x} \left[(D_x + \mu)^{-1} x^2 \right]
$$
$$
= \left(1 + \frac{D_x}{\mu} \right)^{-1} x^2 = \frac{e^{\mu x}}{\mu} \left(1 - \frac{D_x}{\mu} + \frac{D_x^2}{\mu^2} \right) x^2 = \frac{e^{\mu x}}{\mu} \left(x^2 - \frac{2x}{\mu} + \frac{2}{\mu^2} \right) \tag{4.222}
$$

This solution does not include the constant of integration. Therefore, the complete solution is the expression given in eq. 4.222 plus a constant. □

Integral of an integrand containing cos(μx) and/or sin(μx)

Integrals of the form

$$
I_c = \int \cos(\mu x) F(x) dx \tag{4.223a}
$$

and

$$
I_s = \int \sin(\mu x) F(x) dx \tag{4.223b}
$$

are evaluate in essentially the same way as those containing an exponential (see eq. 4.215).

If $F(x)$ is real, then eqs. 4.224 can be written as

$$
I_c = \mathrm{Re} \int e^{i\mu x} F(x) dx \tag{4.224a}
$$

and

$$
I_s = \mathrm{Im} \int e^{i\mu x} F(x) dx \tag{4.224b}
$$

and if $F(x)$ is imaginary ($F(x) = iG(x)$ with $G(x)$ real), then

$$
I_c = \mathrm{Im} \int e^{i\mu x} F(x) dx \tag{4.225a}
$$

and

$$
I_s = -\mathrm{Re} \int e^{i\mu x} F(x) dx \tag{4.225b}
$$

If $F(x)$ is complex, then

$$I_c = \int \frac{(e^{i\mu x} + e^{-i\mu x})}{2} F(x)dx \tag{4.226a}$$

and

$$I_s = \int \frac{(e^{i\mu x} - e^{-i\mu x})}{2i} F(x)dx \tag{4.226b}$$

Example 4.22: Integral of a product of a sine and a polynomial by Heaviside methods

We consider the sine integral

$$I_s(\mu) \equiv \int \sin(\mu x)x^2 dx = \mathrm{Im} \int e^{i\mu x}x^2 dx = \mathrm{Im}D_x^{-1}\left[e^{i\mu x}x^2\right]$$

$$= \mathrm{Im}e^{i\mu x}\left[(D_x + i\mu)^{-1}x^2\right] = \mathrm{Im}\left[\frac{e^{i\mu x}}{i\mu}\left(1 - \frac{D_x}{i\mu} + \frac{D_x^2}{(i\mu)^2}\right)x^2\right] \tag{2.227a}$$

from which we have

$$I_s(\mu) = \mathrm{Im}\left[\frac{e^{i\mu x}}{i\mu}\left(x^2 + \frac{2ix}{\mu} - \frac{2}{\mu^2}\right)\right] =$$

$$\frac{1}{\mu}\mathrm{Im}\left[(-i\cos(\mu x) + \sin(\mu x))\left(x^2 - \frac{2}{\mu^2} + \frac{2ix}{\mu}\right)\right]$$

$$= \frac{1}{\mu}\left[-\left(x^2 - \frac{2}{\mu^2}\right)\cos(\mu x) + \frac{2x}{\mu}\sin(\mu x)\right] \tag{4.227b}$$

The complete solution is obtained by adding a constant of integration to this result. □

Problems

. Problems 1, 2, 5, and 6 refer to the following integrals:

(I) $I_1 = \int_{-1}^{3} xe^x dx = (x - 1)e^x|_{-1}^{3} = 40.90683$

(II) $I_{II} = \int_{-\infty}^{3} xe^x dx = (x - 1)e^x|_{-\infty}^{3} = 40.17107$

(III) $I_{III} = \int_{0}^{\infty} \frac{1}{(2 + x)^3} dx = \frac{-1}{2(2 + x)^2}\Big|_{0}^{\infty} = 0.12500$

(IV) $I_{IV} = \int_{-\infty}^{\infty} x^2 e^{-x^2} dx = \frac{\sqrt{\pi}}{2} = 0.88623$

1. Use the first three non-zero terms in the Euler–MacLaurin approximation method to estimate the value of each of the above integrals.
2. For each of the integrals above, transform the range of integration to [0, 2]. Then approximate each integral by

   (a) a trapezoidal rule using two segments. Repeat the procedure using three segments.
   (b) a Simpson's 1/3 rule using two segments. Repeat the procedure using three segments.
   (c) a Simpson's 3/8 rule using two segments.

   For each integral, approximated by each of these Newton-Cotes quadratures, estimate the error from the term that contains the lowest derivative of $f(x)$.
3. Prove that Simpson's 1/3 rule does not yield an exact value of $\int_a^b x^4 dx = \frac{1}{5}\left(b^5 - a^5\right)$
4. Develop the Newton-Cotes quadrature rule for the integral of a function that is approximated by a fifth order polynomial over each segment. Specify the number of intervals that make up one segment and determine the weights of the quadrature rule.
5. For

   (a) I_{II}, transform the range of integration to $[0, \infty]$ and approximate the integral using a 4-point Gauss–Laguerre quadrature.
   (b) I_{III}, transform the range of integration to $[-1, 1]$ and approximate the integral using a 4-point Gauss–Legendre quadrature.

6. For I_{IV},

   (a) approximate the integral using a 4-point Gauss–Hermite quadrature.
   (b) transform the range of integration to $[-1, 1]$ and approximate the integral using a 4-point Gauss–Legendre quadrature.

7. Find an approximate value of $\Gamma\left(\frac{5}{4}\right) = \int_0^\infty x^{1/4} e^{-x} dx$ using

   (a) a two point Gauss–Laguerre quadrature rule.
   (b) a five-point Gauss–Laguerre quadrature rule.

8. (a) Evaluate

$$\int_{-1}^{1} x \ell n|x - 0.2| dx$$

   exactly
   (b) Estimate the value of this integral by writing

$$\int_{-1}^{1} x \ell n|x - 0.2| dx = \int_{-1}^{1} (x - 0.2)\ell n|x - 0.2| dx + 0.2 \int_{-1}^{1} \ell n|x - 0.2| dx$$

   then approximating the first integral by a four-point Gauss–Legendre quadrature rule and evaluating the second integral exactly.

9. Estimate $P \int_0^4 \frac{x^2}{(x-1)} dx = 12 + \ell n(3) = 13.09861$

   (a) by writing the integral as $\int_0^4 \frac{x^2-1}{(x-1)} dx + P \int_0^4 \frac{1}{(x-1)} dx$ and estimating the first integral by a 4-point Gauss–Legendre quadrature.

   (b) by dividing the range $[0, 4]$ into three segments, two of which are separated by $x_0 = 1$ and are the same width. Use a 4-point Gauss–Legendre quadrature to estimate the integrals over each of the three segments.

10. The polynomials $G_k(x)$ have the following properties:

    Weighting function:

    $$\rho(x) = \frac{1}{\sqrt{1-x^2}}$$

    Orthogonality and normalization:

    $$\int_{-1}^1 \frac{G_k(x)G_m(x)}{\sqrt{1-x^2}} dx = \begin{cases} \frac{2\pi}{k^2} \delta_{mk} & k \neq 0 \\ \pi \delta_{mk} & k = 0 \end{cases}$$

    Summation representation:

    $$G_k(x) = \sum_{n=0}^{k/2} (-1)^n \frac{(k-n-1)!}{n!\,(k-2n)!} (2x)^{k-2n} \quad k \text{ even}$$

    $$G_k(x) = \sum_{n=0}^{(k-1)/2} (-1)^n \frac{(k-n-1)!}{n!\,(k-2n)!} (2x)^{k-2n} \quad k \text{ odd}$$

    Five lowest order polynomials:

    $$G_0(x) = 1 \qquad G_1(x) = 2x \qquad G_2(x) = 2x^2 - 1$$
    $$G_3(x) = \tfrac{8}{3}x^3 - 2x \qquad G_4(x) = 16x^4 - 16x^2 + 2$$

    (a) Find the abscissae and weights for a two point and four point rule.

    (b) Find the exact value of $I = \int_{-1}^1 \frac{x^2}{\sqrt{1-x^2}} dx$ with the (a) two-point quadrature rule and (b) four-point quadrature rule.

11. Assume the random number generator in the computer language you use generates integers between 0 and an integer N that you specify. Write a computer program to approximate the following integral by the Monte Carlo methods.

$$\int_1^4 e^{\sqrt{x}} dx = 2e^2 = 14.77811$$

12. If N is a positive integer, determine the exact value of
$$\int_0^\infty \int_0^\infty (x+y)^2 (xy)^{N+\frac{1}{2}} e^{-(x+y)} dx dy$$

13. Find the exact value of $\displaystyle\int_0^\infty x^{\frac{3}{4}} y^{\frac{5}{4}} e^{-(x+y)} dx dy$ integrated over the first quadrant of the x–y plane.

14. Find the exact value of $\displaystyle\int_0^{\frac{\pi}{2}} \cos^{\frac{1}{4}}\phi \sin^{\frac{7}{4}}\phi \, d\phi$

15. (a) Estimate the value of $\Gamma(0.8)$ by the first three terms in the MacLaurin series for $\ell n[\Gamma(1-0.2)]$.
    (b) Estimate the value of $\Gamma(0.8)$ by assuming that $\Gamma(5.8)$ is reasonably well approximated by the Stirling approximation
    (c) Use either of these results to estimate the value of $\int_0^1 u^{-0.2}(1-u)^{4.8} du$

16. Estimate the value of $\displaystyle\int_0^\infty \frac{1}{(1+x)^{2.2}\sqrt{x}} dx$

17. Use the Heaviside differential operator methods to evaluate

    (a) $\int_0^1 x^3 e^{-2x} dx$

    (b) $\int_0^1 x^3 \cos(2x) dx$

18. By approximating $1/(1+x)$ by the first terms of its MacLaurin series, use Heaviside methods to approximate $\displaystyle\int_0^1 \frac{e^{-x}}{(1+x)} dx$

Chapter 5
DETERMINANTS AND MATRICES

5.1 Evaluation of a Determinant

It is assumed that the reader has been introduced to the fundamental properties of determinants. Appendix 4 presents many of these properties that will be used to develop methods of evaluating determinants.

If a determinant is a 2×2 or a 3×3 array, it can be evaluated by multiplication along diagonals as shown in eqs. 5.1 and 5.2.

$$A_2 \equiv \begin{vmatrix} a_{11} & a_{12} \\ a_{21} & a_{22} \end{vmatrix} = a_{11}a_{22} - a_{12}a_{21} \tag{5.1}$$

and

$$A_3 \equiv \begin{vmatrix} a_{11} & a_{12} & a_{13} \\ a_{21} & a_{22} & a_{23} \\ a_{31} & a_{32} & a_{33} \end{vmatrix} = \begin{aligned} &(a_{11}a_{22}a_{33} + a_{12}a_{23}a_{31} + a_{13}a_{21}a_{32}) \\ &-(a_{13}a_{22}a_{31} + a_{12}a_{21}a_{33} + a_{11}a_{23}a_{32}) \end{aligned} \tag{5.2}$$

For arrays larger than 3×3, multiplication along diagonals yields incorrect results, and other techniques must be used. All these other methods are designed to reduce the order of an $N \times N$ determinant to one of lower order.

Laplace expansion

The methods we will develop for evaluating determinants of order larger than 3×3 depend on an operation called *Laplace expansion* about any element. Let

H. Cohen, *Numerical Approximation Methods*, DOI 10.1007/978-1-4419-9837-8_5,
© Springer Science+Business Media, LLC 2011

$$A_N \equiv \begin{vmatrix} a_{11} & \bullet\bullet & a_{1(n-1)} & a_{1n} & a_{1(n+1)} & \bullet\bullet & a_{1N} \\ \bullet & & \bullet & & \bullet & & \bullet \\ \bullet & & \bullet & & \bullet & & \bullet \\ a_{(m-1)1} & \bullet\bullet & a_{(m-1)(n-1)} & a_{(m-1)n} & a_{(m-1)(n+1)} & \bullet\bullet & a_{(m-1)N} \\ a_{m1} & \bullet\bullet & a_{m(n-1)} & a_{mn} & a_{m(n+1)} & \bullet\bullet & a_{mN} \\ a_{(m+1)1} & \bullet\bullet & a_{(m+1)(n-1)} & a_{(m+1)n} & a_{(m+1)(n+1)} & \bullet\bullet & a_{(m+1)N} \\ \bullet & & \bullet & & \bullet & & \bullet \\ \bullet & & \bullet & & \bullet & & \bullet \\ a_{N1} & \bullet\bullet & a_{N(n-1)} & a_{Nn} & a_{N(n+1)} & \bullet\bullet & a_{NN} \end{vmatrix} \tag{5.3}$$

By eliminating the mth row and nth column from this determinant, we obtain the $(N-1) \times (N-1)$ determinant M_{mn}, called the m–n minor of the element a_{mn}:

$$M_{mn} \equiv \begin{vmatrix} a_{11} & \bullet\bullet & a_{1(n-1)} & a_{1(n+1)} & \bullet\bullet & a_{1N} \\ \bullet & & \bullet & \bullet & & \bullet \\ \bullet & & \bullet & \bullet & & \bullet \\ a_{(m-1)1} & \bullet\bullet & a_{(m-1)(n-1)} & a_{(m-1)(n+1)} & \bullet\bullet & a_{(m-1)N} \\ a_{(m+1)1} & \bullet\bullet & a_{(m+1)(n-1)} & a_{(m+1)(n+1)} & \bullet\bullet & a_{(m+1)N} \\ \bullet & & \bullet & \bullet & & \bullet \\ \bullet & & \bullet & \bullet & & \bullet \\ a_{N1} & \bullet\bullet & a_{N(n-1)} & a_{N(n+1)} & \bullet\bullet & a_{NN} \end{vmatrix} \tag{5.4}$$

The Laplace expansion of A_N is a sum involving these minors. The m–n cofactor of a_{mn}, also called the m–n signed minor of a_{mn}, is defined by

$$cof(a_{mn}) \equiv (-1)^{m+n} M_{mn} \tag{5.5}$$

The determinant can then be written as a sum over the row indices for a fixed column index or a sum over column indices for a given row index:

$$A_N = \sum_{m=1}^{N} a_{mn} cof(a_{mn}) = \sum_{n=1}^{N} a_{mn} cof(a_{mn}) \tag{5.6}$$

Then, one is left with a sum of $(N-1) \times (N-1)$ determinants. One repeats this process until all determinants in the sum are of order 3×3. The 3×3 determinants are then evaluated by multiplication along diagonals as in eq. 5.2.

Example 5.1: Laplace expansion of a determinant

The Laplace expansion about the elements in the second row of

$$A = \begin{vmatrix} 0 & 5 & 1 & 3 \\ 3 & 2 & 7 & 4 \\ 2 & 2 & 5 & 7 \\ 1 & 4 & 9 & 3 \end{vmatrix} \tag{5.7}$$

is given by

$$A = 3(-1)^{2+1} \begin{vmatrix} 5 & 1 & 3 \\ 2 & 5 & 7 \\ 4 & 9 & 3 \end{vmatrix} + 2(-1)^{2+2} \begin{vmatrix} 0 & 1 & 3 \\ 2 & 5 & 7 \\ 1 & 9 & 3 \end{vmatrix}$$

$$+ 7(-1)^{2+3} \begin{vmatrix} 0 & 5 & 3 \\ 2 & 2 & 7 \\ 1 & 4 & 3 \end{vmatrix} + 4(-1)^{2+4} \begin{vmatrix} 0 & 5 & 1 \\ 2 & 2 & 5 \\ 1 & 4 & 9 \end{vmatrix}$$

$$= (-3)(-224) + (2)(40) + (-7)(23) + (4)(-59) = 355 \qquad (5.8)\square$$

It is clear that this can be a very cumbersome operation, particularly when the order of A_N is larger than 4.

However, we see that in this example, $a_{11} = 0$. Therefore, expansion about the first row or the first column results in evaluating three minor determinants instead of four.

With this as a guide, it is clear that if all but one of the elements in any one row or any one column of A_N were zero, the Laplace expansion would result in a single $(N-1) \times (N-1)$ determinant. It is possible to accomplish this by manipulating the elements of A_N to replace all but one of the elements in any row or any column. As stated in Appendix 4, (A4.10), a determinant is unchanged if a multiple of any one row is added to another row, or a multiple of any one column is added to another column.

With this property, it is possible to manipulate a determinant so that all but one of the elements in the any specified column are zero. For example, we can manipulate A_N into the form

$$A_N = \begin{vmatrix} \alpha_{11} & \bullet\bullet & \alpha_{1(n-1)} & 0 & \alpha_{1(n+1)} & \bullet\bullet & \alpha_{1N} \\ \alpha_{21} & \bullet\bullet & \alpha_{2(n-1)} & \alpha_{2n} & \alpha_{2(n+1)} & \bullet\bullet & \alpha_{2N} \\ \alpha_{31} & \bullet\bullet & \alpha_{3(n-1)} & 0 & \alpha_{3(n+1)} & \bullet\bullet & \alpha_{3N} \\ \bullet & & \bullet & \bullet & \bullet & & \bullet \\ \bullet & & \bullet & \bullet & \bullet & & \bullet \\ \alpha_{N1} & \bullet\bullet & \alpha_{N(n-1)} & 0 & \alpha_{N(n+1)} & \bullet\bullet & \alpha_{NN} \end{vmatrix} \qquad (5.9)$$

with $\alpha_{2n} \neq 0$.

To achieve this, for each row defined by the index k with $k \neq 2$, we multiply α_{2n} by a constant C_k such that

$$a_{kn} + C_k a_{2n} = 0 \quad k \neq 2 \qquad (5.10)$$

We define the row in which the element will be non-zero [row_2 in the determinant of eq. 5.9] as the *working row*. The operations performed to cast A_N in the form given in eq. 5.9, which replace each a_{km} (for $1 \leq m \leq N$ and $k \neq$ working row) by $a_{km} + C_k \times a_{(working\ row),m}$, are denoted by

$$row_k \rightarrow row_k + C_k \times (working\ row) \tag{5.11}$$

Identical operations can be performed with columns to obtain a determinant in which all elements but one are zero in a specified row.

Expanding A_N of eq. 5.9 around column n, the Laplace expansion of eq. 5.6 is

$$A_N = \sum_{m=1}^{N} \alpha_{mn} cof(\alpha_{mn}) = \alpha_{2n} cof(\alpha_{2n}) \tag{5.12}$$

Thus, A_N is reduced to a single $(N-1) \times (N-1)$ determinant.

Example 5.2: Manipulating elements of a 4×4 determinant to reduce it to a single 3×3 determinant

We again consider

$$A = \begin{vmatrix} 0 & 5 & 1 & 3 \\ 3 & 2 & 7 & 4 \\ 2 & 2 & 5 & 7 \\ 1 & 4 & 9 & 3 \end{vmatrix} \tag{5.7}$$

Since a_{11} is already zero, we manipulate the elements so as to change A into a determinant for which all elements in the first row except a_{13} are zero. To do so, we add a multiple of each element in the third column to the corresponding elements in the other columns such that the elements in the first row of those other columns is zero. Thus, the third column is the working column. We perform the operations $col_2 \rightarrow col_2 - 5 \times col_3$ and $col_4 \rightarrow col_4 - 3 \times col_3$. With these operations, we obtain

$$A = \begin{vmatrix} 0 & 0 & 1 & 0 \\ 3 & -33 & 7 & -17 \\ 2 & -23 & 5 & -8 \\ 1 & -41 & 9 & -24 \end{vmatrix} \tag{5.13}$$

The Laplace expansion of this determinant about the elements of the first row results in

$$A = 1 \times (-1)^{1+3} \begin{vmatrix} 3 & -33 & -17 \\ 2 & -23 & -8 \\ 1 & -41 & -24 \end{vmatrix} \tag{5.14}$$

This can now be evaluated as given in eq. 5.2 to obtain $A = 355$. □

Pivotal condensation

Pivotal condensation (Chio, F., 1853) is a method of reducing an $N \times N$ determinant to a single $(N-1) \times (N-1)$ determinant without manipulating the elements of A_N. Although any element can be used as the *pivotal element*, (see, for example, Fuller, L., and Logan, J.,.1975), for ease of use it is recommended that a_{11} be the pivotal element.

The only requirement necessary for this method is at $a_{11} \neq 0$. If $a_{11} = 0$, then one can exchange the first row (or the first column) with another row (or column) so that the (1,1) element is not zero. Then $A_N \rightarrow - A_N$ [see Appendix 4, eqs. A4.2]. Or, one can replace the first row (or column) with the sum of the first row (or column) and another row (or column). Then A_N does not change sign [see Appendix 4, eqs. A4.10].

Then with $a_{11} \neq 0$, we multiply

$$A = \begin{vmatrix} a_{11} & a_{12} & a_{13} & \bullet\bullet & a_{1N} \\ a_{21} & a_{22} & a_{23} & \bullet\bullet & a_{2N} \\ a_{31} & a_{32} & a_{33} & \bullet\bullet & a_{3N} \\ \bullet & & & & \\ \bullet & & & & \\ a_{N1} & a_{N2} & a_{N3} & \bullet\bullet & a_{NN} \end{vmatrix} \tag{5.15}$$

by $a_{11}{}^{N-1}$ to obtain

$$a_{11}^{N-1}A = a_{11}^{N-1} \begin{vmatrix} a_{11} & a_{12} & a_{13} & \bullet\bullet & a_{1N} \\ a_{21} & a_{22} & a_{23} & \bullet\bullet & a_{2N} \\ a_{31} & a_{32} & a_{33} & \bullet\bullet & a_{3N} \\ \bullet & & & & \\ \bullet & & & & \\ a_{N1} & a_{N2} & a_{N3} & \bullet\bullet & a_{NN} \end{vmatrix} \tag{5.16a}$$

We multiply one factor of a_{11} into each row except for row_1 to obtain

$$a_{11}^{N-1}A = \begin{vmatrix} a_{11} & a_{12} & a_{13} & \bullet\bullet & a_{1N} \\ a_{11}a_{21} & a_{11}a_{22} & a_{11}a_{23} & \bullet\bullet & a_{11}a_{2N} \\ a_{11}a_{31} & a_{11}a_{32} & a_{11}a_{33} & \bullet\bullet & a_{11}a_{3N} \\ \bullet & & & & \\ \bullet & & & & \\ a_{11}a_{N1} & a_{11}a_{N2} & a_{11}a_{N3} & \bullet\bullet & a_{11}a_{NN} \end{vmatrix} \tag{5.16b}$$

We then perform the following operations on this determinant:

$$row_2 \rightarrow row_2 - a_{21} \times row_1 \qquad (5.17a)$$

$$row_3 \rightarrow row_3 - a_{31} \times row_1 \qquad (5.17b)$$

$$\vdots$$

$$row_N \rightarrow row_N - a_{N1} \times row_1 \qquad (5.17c)$$

to obtain

$$a_{11}^{N-1} A =$$

$$
\begin{vmatrix}
a_{11} & a_{12} & a_{13} & \bullet\bullet & a_{1N} \\
0 & (a_{11}a_{22} - a_{12}a_{21}) & (a_{11}a_{23} - a_{13}a_{21}) & \bullet\bullet & (a_{11}a_{2N} - a_{1N}a_{21}) \\
0 & (a_{11}a_{32} - a_{12}a_{31}) & (a_{11}a_{33} - a_{13}a_{31}) & \bullet\bullet & (a_{11}a_{3N} - a_{1N}a_{31}) \\
\bullet & & & & \\
\bullet & & & & \\
0 & (a_{11}a_{N2} - a_{12}a_{N1}) & (a_{11}a_{N3} - a_{13}a_{N1}) & \bullet\bullet & (a_{11}a_{NN} - a_{1N}a_{N1})
\end{vmatrix}
\qquad (5.18)
$$

We note that the quantities in the parentheses are 2×2 determinants. For example,

$$(a_{11}a_{22} - a_{12}a_{21}) = \begin{vmatrix} a_{11} & a_{12} \\ a_{21} & a_{22} \end{vmatrix} \qquad (5.19a)$$

$$(a_{11}a_{33} - a_{13}a_{31}) = \begin{vmatrix} a_{11} & a_{13} \\ a_{31} & a_{33} \end{vmatrix} \qquad (5.19b)$$

and

$$(a_{11}a_{NN} - a_{1N}a_{N1}) = \begin{vmatrix} a_{11} & a_{1N} \\ a_{N1} & a_{NN} \end{vmatrix} \qquad (5.19c)$$

Therefore, we can write eq. 5.18 as

$$
a_{11}^{N-1} A =
\begin{vmatrix}
a_{11} & a_{12} & a_{13} & \bullet\bullet & a_{1N} \\
0 & \begin{vmatrix} a_{11} & a_{12} \\ a_{21} & a_{22} \end{vmatrix} & \begin{vmatrix} a_{11} & a_{13} \\ a_{21} & a_{23} \end{vmatrix} & \bullet\bullet & \begin{vmatrix} a_{11} & a_{1N} \\ a_{21} & a_{2N} \end{vmatrix} \\
0 & \begin{vmatrix} a_{11} & a_{12} \\ a_{31} & a_{32} \end{vmatrix} & \begin{vmatrix} a_{11} & a_{13} \\ a_{31} & a_{33} \end{vmatrix} & \bullet\bullet & \begin{vmatrix} a_{11} & a_{1N} \\ a_{31} & a_{3N} \end{vmatrix} \\
\bullet & & & & \\
\bullet & & & & \\
0 & \begin{vmatrix} a_{11} & a_{12} \\ a_{N1} & a_{N2} \end{vmatrix} & \begin{vmatrix} a_{11} & a_{13} \\ a_{N1} & a_{N3} \end{vmatrix} & \bullet\bullet & \begin{vmatrix} a_{11} & a_{1N} \\ a_{N1} & a_{NN} \end{vmatrix}
\end{vmatrix}
\qquad (5.20)
$$

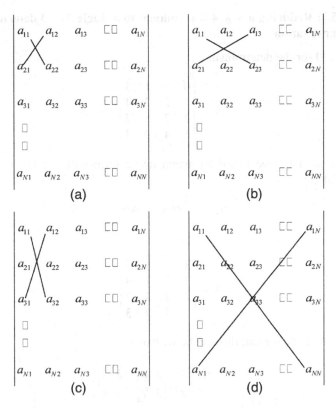

Fig. 5.1 Patterns for 2×2 determinants that are the elements of the single $(N{-}1) \times (N{-}1)$ determinant reduced from an $N \times N$ determinant by the method of pivotal condensation: The (**a**) (1,1) element, (**b**) (1,2) element, (**c**) (2,1) element, (**d**) $(N{-}1, N{-}1)$ element of the reduced determinant

Performing an expansion of this determinant about the first column, we obtain

$$
A = \frac{1}{a_{11}^{N-2}}
\begin{vmatrix}
\begin{vmatrix} a_{11} & a_{12} \\ a_{21} & a_{22} \end{vmatrix} & \begin{vmatrix} a_{11} & a_{13} \\ a_{21} & a_{23} \end{vmatrix} & \bullet\ \bullet & \begin{vmatrix} a_{11} & a_{1N} \\ a_{21} & a_{2N} \end{vmatrix} \\
\begin{vmatrix} a_{11} & a_{12} \\ a_{31} & a_{32} \end{vmatrix} & \begin{vmatrix} a_{11} & a_{13} \\ a_{31} & a_{33} \end{vmatrix} & \bullet\ \bullet & \begin{vmatrix} a_{11} & a_{1N} \\ a_{31} & a_{3N} \end{vmatrix} \\
\begin{matrix} \bullet \\ \bullet \\ \bullet \end{matrix} & & & \\
\begin{vmatrix} a_{11} & a_{12} \\ a_{N1} & a_{N2} \end{vmatrix} & \begin{vmatrix} a_{11} & a_{13} \\ a_{N1} & a_{N3} \end{vmatrix} & \bullet\ \bullet & \begin{vmatrix} a_{11} & a_{1N} \\ a_{N1} & a_{NN} \end{vmatrix}
\end{vmatrix}
\tag{5.21}
$$

Thus, a determinant of order $N \times N$ can be reduced to a single $(N-1) \times (N{-}1)$ determinant by evaluating 2×2 determinants. All that is needed is that the reader becomes familiar with the pattern of evaluating those 2×2 determinants. Some of the patterns are shown in Fig. 5.1.

Example 5.3: Reducing a 4 × 4 determinant to a single 3 × 3 determinant by pivotal condensation

Since $a_{11} = 0$ for the determinant

$$A = \begin{vmatrix} 0 & 5 & 1 & 3 \\ 3 & 2 & 7 & 4 \\ 2 & 2 & 5 & 7 \\ 1 & 4 & 9 & 3 \end{vmatrix} \tag{5.7}$$

we manipulate A to cast this determinant into a form with the (1,1) element is not zero. Taking

$$row_1 \rightarrow row_1 + row_2 \tag{5.22}$$

Then

$$A = \begin{vmatrix} 3 & 7 & 8 & 7 \\ 3 & 2 & 7 & 4 \\ 2 & 2 & 5 & 7 \\ 1 & 4 & 9 & 3 \end{vmatrix} \tag{5.23}$$

Using pivotal condensation, this can be written as

$$A = \begin{vmatrix} 3 & 7 & 8 & 7 \\ 3 & 2 & 7 & 4 \\ 2 & 2 & 5 & 7 \\ 1 & 4 & 9 & 3 \end{vmatrix} = \frac{1}{3^{4-2}} \begin{vmatrix} \begin{vmatrix} 3 & 7 \\ 3 & 2 \end{vmatrix} & \begin{vmatrix} 3 & 8 \\ 3 & 7 \end{vmatrix} & \begin{vmatrix} 3 & 7 \\ 3 & 4 \end{vmatrix} \\ \begin{vmatrix} 3 & 7 \\ 2 & 2 \end{vmatrix} & \begin{vmatrix} 3 & 8 \\ 2 & 5 \end{vmatrix} & \begin{vmatrix} 3 & 7 \\ 2 & 7 \end{vmatrix} \\ \begin{vmatrix} 3 & 7 \\ 1 & 4 \end{vmatrix} & \begin{vmatrix} 3 & 8 \\ 1 & 9 \end{vmatrix} & \begin{vmatrix} 3 & 7 \\ 1 & 3 \end{vmatrix} \end{vmatrix} \tag{5.24}\square$$

$$= \frac{1}{9} \begin{vmatrix} -15 & -3 & -9 \\ -8 & -1 & 7 \\ 5 & 19 & 2 \end{vmatrix} = 355$$

Triangularization

All the elements below the main diagonal of a determinant in *upper triangular form* are zero. Such a determinant is of the form

$$A = \begin{vmatrix} a_{11} & a_{12} & a_{13} & \bullet\bullet & a_{1N} \\ 0 & a_{22} & a_{23} & \bullet\bullet & a_{2N} \\ 0 & 0 & a_{33} & \bullet\bullet & a_{3N} \\ \bullet & & & & \bullet \\ \bullet & & & & \\ 0 & 0 & 0 & 0 & a_{NN} \end{vmatrix} \tag{5.25a}$$

If all the elements above the main diagonal are zero, the determinant is in *lower triangular form*

$$A = \begin{vmatrix} a_{11} & 0 & 0 & \bullet \bullet & 0 \\ a_{21} & a_{22} & 0 & \bullet \bullet & 0 \\ a_{31} & a_{32} & a_{33} & \bullet \bullet & 0 \\ \bullet & & & \bullet & \bullet \\ \bullet & & & & \bullet \\ a_{N1} & a_{N2} & a_{N3} & \bullet \bullet & a_{NN} \end{vmatrix} \qquad (5.25b)$$

The value of a determinant in triangular form is the product of the diagonal elements:

$$A = \prod_{n=1}^{N} a_{nn} \qquad (5.26)$$

This is found straightforwardly by expanding the determinant around the elements in successive columns.

For example, if the determinant is in upper triangular form, we first expand about elements of the column headed by a_{11} to obtain

$$A = a_{11} \begin{vmatrix} a_{22} & a_{23} & a_{24} & \bullet \bullet & a_{2N} \\ 0 & a_{33} & a_{34} & \bullet \bullet & a_{3N} \\ 0 & 0 & a_{44} & \bullet \bullet & a_{4N} \\ \bullet & & & & \bullet \\ \bullet & & & & \\ 0 & 0 & 0 & 0 & a_{NN} \end{vmatrix} \qquad (5.27a)$$

The second expansion, about the column headed by a_{22} yields

$$A = a_{11}a_{22} \begin{vmatrix} a_{33} & a_{34} & a_{35} & \bullet \bullet & a_{3N} \\ 0 & a_{44} & a_{45} & \bullet \bullet & a_{4N} \\ 0 & 0 & a_{55} & \bullet \bullet & a_{5N} \\ \bullet & & & & \bullet \\ \bullet & & & & \\ 0 & 0 & 0 & 0 & a_{NN} \end{vmatrix} \qquad (5.27b)$$

and so on eventually obtaining eq. 5.26.

If the determinant is in lower triangular form, the first expansion is about the elements in the column footed by a_{NN}. The second expansion is about the elements in column footed by $a_{(N-1)(N-1)}$, and so on.

Therefore, if a determinant can be triangularized, it can be evaluated without successive reductions to one or more 3×3 determinants.

Triangularization is accomplished using the property that a determinant is unchanged, if a multiple of any row (or column) is added to any other row (or column) [Appendix 4, eq. A4.10].

To triangularize the determinant, we use each successive row as a working row, and replace all the elements in a given column by zero by adding to that row, a multiple of the working row. For example, to upper triangularize a determinant, we first replace all elements in the first column (except a_{11}), by zero by using row_1 as the working row and performing the manipulations

$$row_k \rightarrow row_k - \frac{a_{k1}}{a_{11}} row_1 \quad k \geq 2 \tag{5.28a}$$

Then, using row_2 as the working row, we perform the replacements

$$row_k \rightarrow row_k - \frac{a_{k2}}{a_{22}} row_2 \quad k \geq 3 \tag{5.28b}$$

and so on.

We see that at each step, the resulting diagonal element in the working row must be non-zero. If $a_{kk} = 0$, one must interchange the working row with some other row with row index $> k$. If the triangularized form of a determinant has one or more zeros on the diagonal, the determinant is zero.

Example 5.4: Triangularizing a determinant

We consider the determinant

$$A = \begin{vmatrix} 2 & 3 & 1 & 1 & 0 \\ 4 & 6 & 0 & 1 & 1 \\ 1 & 0 & 4 & 6 & 2 \\ 0 & 1 & 1 & 5 & 2 \\ 6 & 2 & 1 & 1 & 1 \end{vmatrix} \tag{5.29}$$

To "zero out" the elements of the first column, we use row_1 as the working row and perform the following manipulations:

$$row_2 \rightarrow row_2 - 2row_1 \tag{5.30a}$$

$$row_3 \rightarrow row_3 - \tfrac{1}{2}row_1 \tag{5.30b}$$

and

$$row_5 \rightarrow row_5 - 3row_1 \tag{5.30c}$$

to obtain

$$A = \begin{vmatrix} 2 & 3 & 1 & 1 & 0 \\ 0 & 0 & -2 & -1 & 1 \\ 0 & -\tfrac{3}{2} & \tfrac{7}{2} & \tfrac{11}{2} & 2 \\ 0 & 1 & 1 & 5 & 2 \\ 0 & -7 & -2 & -2 & 1 \end{vmatrix} \tag{5.31}$$

We see that the (2,2) element of this resultant determinant is zero. To obtain a second row without a (2,2) element of zero, we make the interchange

$$row_2 \leftrightarrow row_4 \tag{5.32}$$

so that

$$A = - \begin{vmatrix} 2 & 3 & 1 & 1 & 0 \\ 0 & 1 & 1 & 5 & 2 \\ 0 & -\frac{3}{2} & \frac{7}{2} & \frac{11}{2} & 2 \\ 0 & 0 & -2 & -1 & 1 \\ 0 & -7 & -2 & -2 & 1 \end{vmatrix} \tag{5.33}$$

Then, with

$$row_3 \to row_3 + \tfrac{3}{2}row_2 \tag{5.34a}$$

and

$$row_5 \to row_5 + 7row_2 \tag{5.34b}$$

eq. 5.33 becomes

$$A = - \begin{vmatrix} 2 & 3 & 1 & 1 & 0 \\ 0 & 1 & 1 & 5 & 2 \\ 0 & 0 & 5 & 13 & 5 \\ 0 & 0 & -2 & -1 & 1 \\ 0 & 0 & -5 & -33 & 15 \end{vmatrix} \tag{5.35}$$

With

$$row_4 \to row_4 + \tfrac{2}{5}row_3 \tag{5.36a}$$

and

$$row_5 \to row_5 + row_3 \tag{5.36b}$$

we obtain

$$A = - \begin{vmatrix} 2 & 3 & 1 & 1 & 0 \\ 0 & 1 & 1 & 5 & 2 \\ 0 & 0 & 5 & 13 & 5 \\ 0 & 0 & 0 & \frac{21}{5} & 3 \\ 0 & 0 & 0 & -20 & 20 \end{vmatrix} \tag{5.37}$$

Then

$$row_5 \rightarrow row_5 + \frac{100}{21} row_4 \tag{5.38}$$

casts A into the upper triangular determinant

$$A = - \begin{vmatrix} 2 & 3 & 1 & 1 & 0 \\ 0 & 1 & 1 & 5 & 2 \\ 0 & 0 & 5 & 13 & 5 \\ 0 & 0 & 0 & \frac{21}{5} & 3 \\ 0 & 0 & 0 & 0 & -\frac{40}{7} \end{vmatrix} \tag{5.39}$$

from which

$$A = -(2)(1)(5)\left(\frac{21}{5}\right)\left(-\frac{40}{7}\right) = \frac{1680}{7} \tag{5.40} \square$$

Cramer's Rule

One application using determinants arises in the solution of a set of N simultaneous linear equations of the form

$$a_{11}x_1 + a_{12}x_2 + \dots + a_{1N}x_N = c_1 \tag{5.41a}$$

$$a_{21}x_1 + a_{22}x_2 + \dots + a_{2N}x_N = c_2 \tag{5.41b}$$

$$\bullet$$
$$\bullet$$

$$a_{N1}x_1 + a_{N2}x_2 + \dots + a_{NN}x_N = c_N \tag{5.41c}$$

Let A be the determinant of the coefficients, and consider

$$x_1 A = x_1 \begin{vmatrix} a_{11} & a_{12} & \bullet\bullet & a_{1N} \\ a_{21} & a_{22} & \bullet\bullet & a_{2N} \\ & \bullet & & \\ & \bullet & & \\ a_{N1} & a_{N2} & \bullet\bullet & a_{NN} \end{vmatrix} = \begin{vmatrix} a_{11}x_1 & a_{12} & \bullet\bullet & a_{1N} \\ a_{21}x_1 & a_{22} & \bullet\bullet & a_{2N} \\ & \bullet & & \\ & \bullet & & \\ a_{N1}x_1 & a_{N2} & \bullet\bullet & a_{NN} \end{vmatrix} \tag{5.42}$$

Since a determinant is zero if all the elements in any two columns are identical, a determinant with column 1 replaced by any other column is zero. That is, for $2 \leq k \leq N$,

$$\begin{vmatrix} a_{1k} & a_{12} & \bullet\bullet & a_{1N} \\ a_{2k} & a_{22} & \bullet\bullet & a_{2N} \\ \bullet \\ \bullet \\ a_{Nk} & a_{N2} & \bullet\bullet & a_{NN} \end{vmatrix} = 0 \qquad (5.43)$$

Therefore,

$$\sum_{k=2}^{N} x_k \begin{vmatrix} a_{1k} & a_{12} & \bullet\bullet & a_{1N} \\ a_{2k} & a_{22} & \bullet\bullet & a_{2N} \\ \bullet \\ \bullet \\ a_{Nk} & a_{N2} & \bullet\bullet & a_{NN} \end{vmatrix} = \sum_{k=2}^{N} \begin{vmatrix} a_{1k}x_k & a_{12} & \bullet\bullet & a_{1N} \\ a_{2k}x_k & a_{22} & \bullet\bullet & a_{2N} \\ \bullet \\ \bullet \\ a_{Nk}x_k & a_{N2} & \bullet\bullet & a_{NN} \end{vmatrix} = 0 \qquad (5.44a)$$

and thus

$$x_1 A = \begin{vmatrix} a_{11}x_1 & a_{12} & \bullet\bullet & a_{1N} \\ a_{21}x_1 & a_{22} & \bullet\bullet & a_{2N} \\ \bullet \\ \bullet \\ a_{N1}x_1 & a_{N2} & \bullet\bullet & a_{NN} \end{vmatrix} + \sum_{k=2}^{N} \begin{vmatrix} a_{1k}x_k & a_{12} & \bullet\bullet & a_{1N} \\ a_{2k}x_k & a_{22} & \bullet\bullet & a_{2N} \\ \bullet \\ \bullet \\ a_{Nk}x_k & a_{N2} & \bullet\bullet & a_{NN} \end{vmatrix}$$

$$= \begin{vmatrix} \sum_{k=1}^{N} a_{1k}x_k & a_{12} & \bullet\bullet & a_{1N} \\ \sum_{k=1}^{N} a_{2k}x_k & a_{22} & \bullet\bullet & a_{2N} \\ \bullet \\ \bullet \\ \sum_{k=1}^{N} a_{Nk}x_k & a_{N2} & \bullet\bullet & a_{NN} \end{vmatrix} = \begin{vmatrix} c_1 & a_{12} & \bullet\bullet & a_{1N} \\ c_2 & a_{22} & \bullet\bullet & a_{2N} \\ \bullet \\ \bullet \\ c_N & a_{N2} & \bullet\bullet & a_{NN} \end{vmatrix} \equiv A_1 \qquad (5.44b)$$

That is, A_1 is the determinant of the coefficients with col_1 replaced by a column of the inhomogeneous quantities on the right sides of eqs. 5.41. Therefore, in general, to solve for x_k by *Cramer's rule*, we obtain the determinant A_k by replacing col_k by a column of these inhomogeneous quantities. Then

$$x_k = \frac{A_k}{A} \qquad (5.45)$$

If the coefficient determinant is zero, there is no solution to the set of linear inhomogeneous equations.

Example 5.5: Cramer's rule for linear inhomogeneous equations

Let us solve for $\{x_1, x_2, x_3\}$ by Cramer's rule, for

$$5x_1 + 2x_2 + x_3 = 2 \qquad (5.46a)$$

$$2x_1 - 3x_2 + 3x_3 = 1 \qquad\qquad (5.46b)$$

and

$$7x_1 + 4x_2 - 2x_3 = -5 \qquad\qquad (5.46c)$$

With

$$A = \begin{vmatrix} 5 & 2 & 1 \\ 2 & -3 & 3 \\ 7 & 4 & -2 \end{vmatrix} = 49 \qquad\qquad (5.47)$$

$$A_1 = \begin{vmatrix} 2 & 2 & 1 \\ 1 & -3 & 3 \\ -5 & 4 & -2 \end{vmatrix} = -49 \qquad\qquad (5.48a)$$

$$A_2 = \begin{vmatrix} 5 & 2 & 1 \\ 2 & 1 & 3 \\ 7 & -5 & -2 \end{vmatrix} = 98 \qquad\qquad (5.48b)$$

and

$$A_3 = \begin{vmatrix} 5 & 2 & 2 \\ 2 & -3 & 1 \\ 7 & 4 & -5 \end{vmatrix} = 147 \qquad\qquad (5.48c)$$

We see from eq. 5.47 that since $A \neq 0$, a solution exists, and from eqs. 5.48 that solution is

$$\{x_1, x_2, x_3\} = \{-1, 2, 3\} \qquad\qquad (5.49)\square$$

5.2   Matrices

It is assumed that the reader knows how to add, subtract, and multiply matrices, is aware of the property that the determinant of a product of matrices is the product of the individual determinants, and knows that the *trace* or *spur* of a matrix is the sum of its diagonal elements.

Matrix inversion

The *unit* or *identity* matrix has a 1 for each element along the main diagonal and 0 for each element not on the main diagonal. We denote this unit matrix as I.

The inverse of a matrix, denoted by A^{-1} satisfies

$$AA^{-1} = A^{-1}A = 1 \tag{5.50}$$

A standard approach for inverting a matrix, given extensively in the literature, is to replace each element of A by its cofactor [see eqs. 5.4 and 5.5] to form the *cofactor* matrix C. One then forms C^T, the transpose of the cofactor matrix (obtained by interchanging the rows and columns of the matrix). Then

$$A^{-1} = \frac{C^T}{|A|} \tag{5.51}$$

(See, for example, Cohen, H., 1992, pp. 439, 440.) If the determinant of A is zero, then A is said to be a *singular* matrix which has no inverse.

Example 5.6: Matrix inversion

It is straightforward to ascertain that for

$$A = \begin{pmatrix} 5 & 2 & 1 \\ 2 & -3 & 3 \\ 7 & 4 & -2 \end{pmatrix} \tag{5.52a}$$

the determinant of A is

$$|A| = \begin{vmatrix} 5 & 2 & 1 \\ 2 & -3 & 3 \\ 7 & 4 & -2 \end{vmatrix} = 49 \tag{5.52b}$$

Thus, A is not singular and A^{-1} exists.

The cofactor matrix of A is

$$C = \begin{pmatrix} \begin{vmatrix} -3 & 3 \\ 4 & -2 \end{vmatrix} & -\begin{vmatrix} 2 & 3 \\ 7 & -2 \end{vmatrix} & \begin{vmatrix} 2 & -3 \\ 7 & 4 \end{vmatrix} \\ -\begin{vmatrix} 2 & 1 \\ 4 & -2 \end{vmatrix} & \begin{vmatrix} 5 & 1 \\ 7 & -2 \end{vmatrix} & -\begin{vmatrix} 5 & 2 \\ 7 & 4 \end{vmatrix} \\ \begin{vmatrix} 2 & 1 \\ -3 & 3 \end{vmatrix} & -\begin{vmatrix} 5 & 1 \\ 2 & 3 \end{vmatrix} & \begin{vmatrix} 5 & 2 \\ 2 & -3 \end{vmatrix} \end{pmatrix} = \begin{pmatrix} -6 & 25 & 29 \\ 8 & -17 & -6 \\ 9 & -13 & -19 \end{pmatrix}$$
$$\tag{5.53a}$$

from which

$$C^T = \begin{pmatrix} -6 & 8 & 9 \\ 25 & -17 & -13 \\ 29 & -6 & -19 \end{pmatrix} \tag{5.53b}$$

Therefore,

$$A^{-1} = \frac{1}{49} \begin{pmatrix} -6 & 8 & 9 \\ 25 & -17 & -13 \\ 29 & -6 & -19 \end{pmatrix} \tag{5.54}$$

It is straightforward to verify that

$$AA^{-1} = \frac{1}{49} \begin{pmatrix} 5 & 2 & 1 \\ 2 & -3 & 3 \\ 7 & 4 & -2 \end{pmatrix} \begin{pmatrix} -6 & 8 & 9 \\ 25 & -17 & -13 \\ 29 & -6 & -19 \end{pmatrix} = \begin{pmatrix} 1 & 0 & 0 \\ 0 & 1 & 0 \\ 0 & 0 & 1 \end{pmatrix} \tag{5.55a}$$

and

$$A^{-1}A = \frac{1}{49} \begin{pmatrix} -6 & 8 & 9 \\ 25 & -17 & -13 \\ 29 & -6 & -19 \end{pmatrix} \begin{pmatrix} 5 & 2 & 1 \\ 2 & -3 & 3 \\ 7 & 4 & -2 \end{pmatrix} = \begin{pmatrix} 1 & 0 & 0 \\ 0 & 1 & 0 \\ 0 & 0 & 1 \end{pmatrix} \tag{5.55b} \square$$

Gauss–Jordan elimination method for A^{-1}

It is easy to see that this approach to finding A^{-1} can be quite cumbersome for large matrices. The *Gauss–Jordan elimination* method, an arithmetically simpler method for determining A^{-1}, involves performing the following manipulations of all elements in a row (or column) of a matrix:

1. One can add a constant multiple of the elements in any row (or column) to the corresponding elements of any other row (or column).
2. One can multiply each element in any row (or column) by a constant.
3. One can interchange any two rows (or columns).

These manipulations of the elements of a matrix A can be achieved by premultiplying (for manipulation of elements in rows) or postmultiplying (for manipulation of elements in columns) by a matrix M. To describe these manipulations in context, we will use 5×5 matrices as examples.

Let

$$
A = \begin{pmatrix}
a_{11} & a_{12} & a_{13} & a_{14} & a_{15} \\
a_{21} & a_{22} & a_{23} & a_{24} & a_{25} \\
a_{31} & a_{32} & a_{33} & a_{34} & a_{35} \\
a_{41} & a_{42} & a_{43} & a_{44} & a_{45} \\
a_{51} & a_{52} & a_{53} & a_{54} & a_{55}
\end{pmatrix}
\tag{5.56}
$$

(1) To create a matrix M that adds a multiple of each element in row_q to the corresponding element of row_p, we start with the unit matrix and replace the 0 in the (p, q) off-diagonal element with λ. Thus, when the *adding matrix*

$$
M_a = \begin{pmatrix}
1 & 0 & 0 & 0 & 0 \\
0 & 1 & 0 & 0 & 0 \\
0 & \lambda & 1 & 0 & 0 \\
0 & 0 & 0 & 1 & 0 \\
0 & 0 & 0 & 0 & 1
\end{pmatrix}
\tag{5.57}
$$

premultiplies A of eq. 5.56, it produces

$$
M_a A = \begin{pmatrix}
1 & 0 & 0 & 0 & 0 \\
0 & 1 & 0 & 0 & 0 \\
0 & \lambda & 1 & 0 & 0 \\
0 & 0 & 0 & 1 & 0 \\
0 & 0 & 0 & 0 & 1
\end{pmatrix}
\begin{pmatrix}
a_{11} & a_{12} & a_{13} & a_{14} & a_{15} \\
a_{21} & a_{22} & a_{23} & a_{24} & a_{25} \\
a_{31} & a_{32} & a_{33} & a_{34} & a_{35} \\
a_{41} & a_{42} & a_{43} & a_{44} & a_{45} \\
a_{51} & a_{52} & a_{53} & a_{54} & a_{55}
\end{pmatrix}
$$

$$
= \begin{pmatrix}
a_{11} & a_{12} & a_{13} & a_{14} & a_{15} \\
a_{21} & a_{22} & a_{23} & a_{24} & a_{25} \\
a_{31} + \lambda a_{21} & a_{32} + \lambda a_{22} & a_{33} + \lambda a_{23} & a_{34} + \lambda a_{24} & a_{35} + \lambda a_{25} \\
a_{41} & a_{42} & a_{43} & a_{44} & a_{45} \\
a_{51} & a_{52} & a_{53} & a_{54} & a_{55}
\end{pmatrix}
\tag{5.58}
$$

That is, premultiplying A by M_a of eq. 5.57 generates the transformation

$$
row_3 \to row_3 + \lambda row_2
\tag{5.59}
$$

Transformations of this type allow one to diagonalize the matrix A by transforming its off-diagonal elements to zero. $\square$

(2) To multiply every element in row_p by the constant λ, we start with the unit matrix and replace the 1 in the (p, p) element by λ. Thus, when the *multiplication matrix*

$$M_m = \begin{pmatrix} 1 & 0 & 0 & 0 & 0 \\ 0 & \lambda & 0 & 0 & 0 \\ 0 & 0 & 1 & 0 & 0 \\ 0 & 0 & 0 & 1 & 0 \\ 0 & 0 & 0 & 0 & 1 \end{pmatrix} \qquad (5.60)$$

premultiplies A of eq. 5.56, it produces

$$M_m A = \begin{pmatrix} 1 & 0 & 0 & 0 & 0 \\ 0 & \lambda & 0 & 0 & 0 \\ 0 & 0 & 1 & 0 & 0 \\ 0 & 0 & 0 & 1 & 0 \\ 0 & 0 & 0 & 0 & 1 \end{pmatrix} \begin{pmatrix} a_{11} & a_{12} & a_{13} & a_{14} & a_{15} \\ a_{21} & a_{22} & a_{23} & a_{24} & a_{25} \\ a_{31} & a_{32} & a_{33} & a_{34} & a_{35} \\ a_{41} & a_{42} & a_{43} & a_{44} & a_{45} \\ a_{51} & a_{52} & a_{53} & a_{54} & a_{55} \end{pmatrix}$$

$$= \begin{pmatrix} a_{11} & a_{12} & a_{13} & a_{14} & a_{15} \\ \lambda a_{21} & \lambda a_{22} & \lambda a_{23} & \lambda a_{24} & \lambda a_{25} \\ a_{31} & a_{32} & a_{33} & a_{34} & a_{35} \\ a_{41} & a_{42} & a_{43} & a_{44} & a_{45} \\ a_{51} & a_{52} & a_{53} & a_{54} & a_{55} \end{pmatrix} \qquad (5.61)$$

That is, premultiplying A by M_m of eq. 5.60 generates the transformation

$$row_2 \rightarrow \lambda row_2 \qquad (5.62)$$

Transformations of this type allow one to transform an arithmetically cumbersome number to a number that is easier to work with, (e.g., to convert diagonal elements to 1).

(3) To interchange all the element in row_p and row_q, we start with the unit matrix. We replace the 1 in the (p, p) and (q, q) elements by 0, and the 0 in the (p, q) and (q, p) elements by 1. Therefore, when the *interchange matrix*

$$M_i = \begin{pmatrix} 1 & 0 & 0 & 0 & 0 \\ 0 & 0 & 0 & 1 & 0 \\ 0 & 0 & 1 & 0 & 0 \\ 0 & 1 & 0 & 0 & 0 \\ 0 & 0 & 0 & 0 & 1 \end{pmatrix} \qquad (5.63)$$

premultiplies A of eq. 5.56, it produces

$$M_i A = \begin{pmatrix} 1 & 0 & 0 & 0 & 0 \\ 0 & 0 & 0 & 1 & 0 \\ 0 & 0 & 1 & 0 & 0 \\ 0 & 1 & 0 & 0 & 0 \\ 0 & 0 & 0 & 0 & 1 \end{pmatrix} \begin{pmatrix} a_{11} & a_{12} & a_{13} & a_{14} & a_{15} \\ a_{21} & a_{22} & a_{23} & a_{24} & a_{25} \\ a_{31} & a_{32} & a_{33} & a_{34} & a_{35} \\ a_{41} & a_{42} & a_{43} & a_{44} & a_{45} \\ a_{51} & a_{52} & a_{53} & a_{54} & a_{55} \end{pmatrix}$$

$$= \begin{pmatrix} a_{11} & a_{12} & a_{13} & a_{14} & a_{15} \\ a_{41} & a_{42} & a_{43} & a_{44} & a_{45} \\ a_{31} & a_{32} & a_{33} & a_{34} & a_{35} \\ a_{21} & a_{22} & a_{23} & a_{24} & a_{25} \\ a_{51} & a_{52} & a_{53} & a_{54} & a_{55} \end{pmatrix} \tag{5.64}$$

That is, premultiplying A by M_i of eq. 5.63 generates the transformation

$$row_2 \leftrightarrow row_4 \tag{5.65}$$

Such a transformation is needed when one of the diagonal elements in some row is zero and one cannot perform a transformation as in (I) with zero as that element.

Let a product of several such matrices M premultiplying A convert A into the unit matrix:

$$M_1 M_2 M_3 ... M_N A = (M_1 M_2 M_3 ... M_N 1)A = 1 \tag{5.66}$$

Therefore,

$$M_1 M_2 M_3 ... M_N 1 = A^{-1} \tag{5.67}$$

That is, the same set of manipulations that convert A to the unit matrix will convert the unit matrix to A^{-1}.

If one postmultiplies A by a product of several of these matrices, one manipulates the columns of A. By the same analysis as above, performing the same manipulations on the columns of the unit matrix converts it to A^{-1}.

If we premultiply A by one set of M matrices and postmultiply A by another set,

$$M_1 M_2 .. M_N A M_{N+1} M_{N+2} .. M_{N+L} = 1 \tag{5.68}$$

it is not possible to write this as a product of matrices M multiplying 1 that can be identified as A^{-1}. Therefore, one cannot manipulate both rows and columns in using the Gauss–Jordan method. Either one manipulates rows only or one manipulates columns only.

Example 5.7: Matrix inversion by the Gauss–Jordan method

We again consider the nonsingular matrix

$$A = \begin{pmatrix} 5 & 2 & 1 \\ 2 & -3 & 3 \\ 7 & 4 & -2 \end{pmatrix} \qquad (5.52a)$$

When applying the Gauss–Jordan method it is convenient to write the matrix A to be inverted and the unit matrix side by side, then manipulate the rows of these matrices as if the two were combined into one six column matrix. If one wishes to manipulate the columns of A and I, an equivalent approach is to write I directly beneath A and manipulate the columns of the two matrices as if they were a single six row matrix.
    We write

$$A = \begin{pmatrix} 5 & 2 & 1 \\ 2 & -3 & 3 \\ 7 & 4 & -2 \end{pmatrix} \qquad 1 = \begin{pmatrix} 1 & 0 & 0 \\ 0 & 1 & 0 \\ 0 & 0 & 1 \end{pmatrix} \qquad (5.69)$$

and perform the following manipulations on the rows of these matrices:
    Using row_1 as the working row, we take

$$row_2 \rightarrow row_2 - \frac{2}{5} row_1 \qquad (5.70a)$$

and

$$row_3 \rightarrow row_3 - \frac{7}{5} row_1 \qquad (5.70b)$$

to obtain

$$A \rightarrow \begin{pmatrix} 5 & 2 & 1 \\ 0 & -\frac{19}{5} & \frac{13}{5} \\ 0 & \frac{6}{5} & -\frac{17}{5} \end{pmatrix} \qquad 1 \rightarrow \begin{pmatrix} 1 & 0 & 0 \\ -\frac{2}{5} & 1 & 0 \\ -\frac{7}{5} & 0 & 1 \end{pmatrix} \qquad (5.71)$$

Then, with row_2 as the working row, we replace

$$row_1 \rightarrow row_1 + \frac{10}{19} row_2 \qquad (5.72a)$$

and

$$row_3 \rightarrow row_3 + \frac{6}{19} row_2 \qquad (5.72b)$$

We find

$$A \rightarrow \begin{pmatrix} 5 & 0 & \frac{45}{19} \\ 0 & -\frac{19}{5} & \frac{13}{5} \\ 0 & 0 & -\frac{49}{19} \end{pmatrix} \qquad 1 \rightarrow \begin{pmatrix} \frac{15}{19} & \frac{10}{19} & 0 \\ -\frac{2}{5} & 1 & 0 \\ -\frac{29}{19} & \frac{6}{19} & 1 \end{pmatrix} \qquad (5.73)$$

Finally, with row_3 as the working row, we perform

$$row_1 \rightarrow row_1 + \frac{45}{49} row_3 \qquad (5.74a)$$

and

$$row_2 \rightarrow row_2 + \frac{247}{245} row_3 \qquad (5.74b)$$

We find

$$A \rightarrow \begin{pmatrix} 5 & 0 & 0 \\ 0 & -\frac{19}{5} & 0 \\ 0 & 0 & -\frac{49}{19} \end{pmatrix} \qquad 1 \rightarrow \begin{pmatrix} -\frac{30}{49} & \frac{40}{49} & \frac{45}{49} \\ -\frac{95}{49} & \frac{323}{245} & \frac{247}{245} \\ -\frac{29}{19} & \frac{6}{19} & 1 \end{pmatrix} \qquad (5.75)$$

We then perform

$$row_1 \rightarrow \frac{1}{5} row_1 \qquad (5.76a)$$

$$row_2 \rightarrow -\frac{5}{19} row_2 \qquad (5.76b)$$

and

$$row_3 \rightarrow -\frac{19}{49} row_3 \qquad (5.76c)$$

Then

$$A \rightarrow \begin{pmatrix} 1 & 0 & 0 \\ 0 & 1 & 0 \\ 0 & 0 & 1 \end{pmatrix} = 1 \qquad (5.77a)$$

and

$$1 \rightarrow \frac{1}{49} \begin{pmatrix} -6 & 8 & 9 \\ 25 & -17 & -13 \\ 29 & -6 & -19 \end{pmatrix} = A^{-1} \qquad (5.77b)$$

Referring to eq. 5.54, we see that the matrix of eq. 5.77b is A^{-1}. $\square$

Gauss–Jordan method to solve a set of simultaneous equations

A set of simultaneous linear equations of the form given in eqs. 5.41 can be solved by matrix inversion. By defining $N \times 1$ matrices (which are also referred to as *column vectors*), containing the unknown values x_n

$$X = \begin{pmatrix} x_1 \\ x_2 \\ \bullet \\ \bullet \\ x_N \end{pmatrix} \tag{5.78a}$$

and the known constants c_n

$$C = \begin{pmatrix} c_1 \\ c_2 \\ \bullet \\ \bullet \\ c_N \end{pmatrix} \tag{5.78b}$$

and with a $N \times N$ matrix containing the coefficients of the unknowns x_n

$$A = \begin{pmatrix} a_{11} & a_{12} & \bullet\bullet & a_{1N} \\ a_{21} & a_{22} & \bullet\bullet & a_{2N} \\ \bullet \\ \bullet \\ a_{N1} & a_{N2} & \bullet\bullet & a_{NN} \end{pmatrix} \tag{5.78c}$$

eqs. 5.41 can be written as

$$AX = C \tag{5.79a}$$

from which

$$X = A^{-1}C \tag{5.79b}$$

With

$$M_1 M_2 M_3 ... M_N 1 = A^{-1} \tag{5.67}$$

eq. 5.79b becomes

$$X = M_1 M_2 M_3 ... M_N 1 C = M_1 M_2 M_3 ... M_N C \tag{5.80}$$

These M matrices multiplying C perform the same manipulations on the rows (the elements) of C that the Gauss–Jordan operations perform on the rows of A to invert it. Therefore, by placing A and C side by side to create a quasi-$N \times (N + 1)$ matrix, then manipulating the rows of this $N \times (N + 1)$ matrix to convert A to I, the $(N + 1)$th column is converted to the column vector of solutions.

Example 5.8: Solution to a set of linear equations by Gauss–Jordan manipulations

We consider the set of equations

$$5x + 2y + z = 9 \tag{5.81a}$$

$$2x - 3y + 3z = 10 \tag{5.81b}$$

and

$$7x + 4y - 2z = 8 \tag{5.81c}$$

Writing these equations in matrix form, we have

$$\begin{pmatrix} 5 & 2 & 1 \\ 2 & -3 & 3 \\ 7 & 4 & -2 \end{pmatrix} \begin{pmatrix} x \\ y \\ z \end{pmatrix} = \begin{pmatrix} 9 \\ 10 \\ 8 \end{pmatrix} \tag{5.82}$$

We see from example 5.7 that the operations

$$row_2 \rightarrow row_2 - \frac{2}{5} row_1 \tag{5.70a}$$

$$row_3 \rightarrow row_3 - \frac{7}{5} row_1 \tag{5.70b}$$

$$row_1 \rightarrow row_1 + \frac{10}{19} row_2 \tag{5.72a}$$

$$row_3 \rightarrow row_3 + \frac{6}{19} row_2 \tag{5.72b}$$

$$row_1 \rightarrow row_1 + \frac{45}{49} row_3 \tag{5.74a}$$

$$row_2 \rightarrow row_2 + \frac{247}{245} row_3 \tag{5.74b}$$

$$row_1 \rightarrow \frac{1}{5} row_1 \tag{5.76a}$$

$$row_2 \rightarrow -\frac{5}{19} row_2 \tag{5.76b}$$

and

$$row_3 \rightarrow -\frac{19}{49} row_3 \qquad\qquad (5.76c)$$

transform the coefficient matrix into the unit matrix. Performing these operations on the rows (elements) of the column

$$C = \begin{pmatrix} 9 \\ 10 \\ 8 \end{pmatrix} \qquad\qquad (5.83)$$

we obtain

$$\begin{pmatrix} x \\ y \\ z \end{pmatrix} = \begin{pmatrix} 2 \\ -1 \\ 1 \end{pmatrix} \qquad\qquad (5.84)\square$$

Choleski-Turing method

The same manipulations that are performed in the Gauss–Jordan method of inverting a matrix A can also cast A into upper or lower triangular form.

Let a product of M matrices premultiplying the matrix equation

$$AX = C \qquad\qquad (5.79a)$$

transform A into an upper triangular or lower triangular matrix. Thus, either

$$M_1 M_2 ... M_n AX = \begin{pmatrix} a_{11} & a_{12} & a_{13} & \bullet\;\bullet & a_{1N} \\ 0 & a_{22} & a_{23} & \bullet\;\bullet & a_{2N} \\ \bullet & & & & \bullet \\ \bullet & & & & \bullet \\ 0 & 0 & 0 & a_{(N-1)(N-1)} & a_{(N-1)N} \\ 0 & 0 & 0 & \bullet\;\bullet & a_{NN} \end{pmatrix} \begin{pmatrix} x_1 \\ x_2 \\ \bullet \\ \bullet \\ x_{N-1} \\ x_N \end{pmatrix}$$

$$= M_1 M_2 ... M_n C \equiv \Gamma = \begin{pmatrix} \gamma_1 \\ \gamma_2 \\ \bullet \\ \bullet \\ \gamma_{N-1} \\ \gamma_N \end{pmatrix} \qquad\qquad (5.85a)$$

or

$$M'_1 M'_2 ... M'_n A X = \begin{pmatrix} a_{11} & 0 & \bullet \bullet & 0 & 0 \\ a_{21} & a_{22} & \bullet \bullet & 0 & 0 \\ \bullet & & & & \bullet \\ \bullet & & & & \bullet \\ a_{(N-1)1} & a_{(N-1)2} & \bullet \bullet & a_{(N-1)(N-1)} & 0 \\ a_{N1} & a_{N2} & \bullet \bullet & a_{N(N-1)} & a_{NN} \end{pmatrix} \begin{pmatrix} x_1 \\ x_2 \\ \bullet \\ \bullet \\ x_{N-1} \\ x_N \end{pmatrix}$$

$$= M'_1 M'_2 ... M'_n C \equiv \Gamma' = \begin{pmatrix} \gamma'_1 \\ \gamma'_2 \\ \bullet \\ \bullet \\ \gamma'_{N-1} \\ \gamma'_N \end{pmatrix} \tag{5.85b}$$

If we upper triangularize the coefficient matrix as in eq. 5.85a we see that by multiplying the X column by the last row of the triangular coefficient matrix, we obtain

$$a_{NN} x_N = \gamma_N \tag{5.86a}$$

from which we find the value of x_N. Then we multiply the $(N-1)$th row into the X column to obtain

$$a_{(N-1)(N-1)} x_{N-1} + a_{(N-1)N} x_N = \gamma_{N-1} \tag{5.86b}$$

With the value of x_N that was determined in eq. 5.86a this yields the value of x_{N-1}. Continuing this process for each row, we obtain equations containing only one unknown x_k.

If the coefficient matrix is lower triangularized as in eq. 5.85b, we begin by multiplying the first row into X to obtain

$$a_{11} x_1 = \gamma_1 \tag{5.87a}$$

to solve for x_1. Then, multiplying the second row into X yields

$$a_{21} x_1 + a_{22} x_2 = \gamma_2 \tag{5.87b}$$

which, with eq. 5.87a yields the value of x_2, and so on. This is the *Choleski-Turing* method of solving a set of simultaneous equations using the same manipulations specified for inverting a matrix by the Gauss–Jordan method.

Example 5.9: Solution to a set of linear equations by the Choleski-Turing method

We again consider the set of equations of example 5.8, given in matrix form as

$$\begin{pmatrix} 5 & 2 & 1 \\ 2 & -3 & 3 \\ 7 & 4 & -2 \end{pmatrix} \begin{pmatrix} x \\ y \\ z \end{pmatrix} = \begin{pmatrix} 9 \\ 10 \\ 8 \end{pmatrix} \qquad (5.82)$$

Referring to example 5.7, we see that using row_1 as the working row, the manipulations

$$row_2 \rightarrow row_2 - \frac{2}{5} row_1 \qquad (5.70a)$$

and

$$row_3 \rightarrow row_3 - \frac{7}{5} row_1 \qquad (5.70b)$$

yield

$$A \rightarrow \begin{pmatrix} 5 & 2 & 1 \\ 0 & -\frac{19}{5} & \frac{13}{5} \\ 0 & \frac{6}{5} & -\frac{17}{5} \end{pmatrix} \qquad C \rightarrow \begin{pmatrix} 9 \\ \frac{32}{5} \\ -\frac{23}{5} \end{pmatrix} \qquad (5.88)$$

Then, with

$$row_3 \rightarrow row_3 + \frac{6}{19} row_2 \qquad (5.72b)$$

we obtain

$$A \rightarrow \begin{pmatrix} 5 & 2 & 1 \\ 0 & -\frac{19}{5} & \frac{13}{5} \\ 0 & 0 & -\frac{49}{19} \end{pmatrix} \qquad C \rightarrow \begin{pmatrix} 9 \\ \frac{32}{5} \\ -\frac{49}{19} \end{pmatrix} \qquad (5.89)$$

Multiplying the third row of this modified A into the X column, we have

$$-\frac{49}{19} z = -\frac{49}{19} \Rightarrow z = 1 \qquad (5.90a)$$

Then multiplying the second row into the X column, we obtain

$$-\frac{19}{5} y + \frac{13}{5} z = -\frac{19}{5} y + \frac{13}{5} = \frac{32}{5} \Rightarrow y = -1 \qquad (5.90b)$$

and multiplication of the first row into X yields

$$5x + 2y + z = 5x - 2 + 1 = 9 \Rightarrow x = 2 \qquad (5.90c)\square$$

Eigenvalues and eigenvectors of a real symmetric matrix by the power method

When an $N \times N$ matrix A multiplies a $N \times 1$ column matrix Y, the result is an $N \times 1$ column matrix Z:

$$AY = Z \qquad (5.91)$$

If

$$\begin{pmatrix} z_1 \\ z_2 \\ \bullet \\ \bullet \\ z_N \end{pmatrix} = \lambda \begin{pmatrix} y_1 \\ y_2 \\ \bullet \\ \bullet \\ y_N \end{pmatrix} \qquad (5.92)$$

eq. 5.91 becomes an *eigenvalue equation*, where λ is an *eigenvalue* of the matrix A and Y is the *eigenvector* corresponding to λ. An eigenvalue and its corresponding eigenvector are sometimes referred to as an *eigenpair*.

The eigenvalues of a matrix are (in general, complex) numbers. But in applications to physical problems, where the eigenvalues represent measurable quantities, these eigenvalues are real, and their eigenvectors are naturally orthogonal. That means that for $\lambda_m \neq \lambda_n$,

$$Y_m^\dagger Y_n = \begin{pmatrix} y_{1,m}^* & y_{2,m}^* & \bullet & \bullet & y_{N,m}^* \end{pmatrix} \begin{pmatrix} y_{1,n} \\ y_{2,n} \\ \bullet \\ \bullet \\ y_{N,n} \end{pmatrix} = 0 \qquad (5.93)$$

If two or more eigenvalues have the same value, they are *degenerate* eigenvalues. The corresponding eigenvectors of degenerate eigenvalues are not naturally orthogonal, but can be made so by an orthogonalization scheme (see, for example the Gram–Schmidt method in Cohen, H., 1992, p. 458).

If a matrix is unchanged by transposing it and complex conjugating all its elements, it is a *hermitean* matrix. If A is hermitean, all its eigenvalues are real and its eigenvectors are orthogonal (for nondegenerate eigenvalues) or can be made orthogonal (for eigenvalues that are degenerate). A special case of a hermitean matrix is one for all its elements are real, and the matrix and its transpose are identical. Such a matrix is a *real symmetric* matrix.

Let A be a hermitean matrix for which all eigenvalues are nondegenerate. Let the eigenvalues λ_1, λ_2, ..., λ_N be ordered such that

$$|\lambda_1| > |\lambda_2| > ... > |\lambda_N| \tag{5.94}$$

Because the eigenvectors are mutually orthogonal, any vector X_0 containing N elements can be written as a linear combination of the eigenvectors:

$$X_0 = \sum_{m=1}^{N} \alpha_m Y_m \tag{5.95}$$

To determine the eigenpairs by the power method, one begins by choosing an arbitrary vector X_0 such as the *unit vector*

$$X_0 = \begin{pmatrix} 1 \\ 1 \\ \bullet \\ \bullet \\ 1 \end{pmatrix} \tag{5.96}$$

It may happen that the chosen X_0 may be orthogonal to one or more of the eigenvectors. Then one will not obtain all the eigenpairs of the matrix. When this occurs, one must choose another X_0 and repeat the process to find those eigenpairs that were not obtained with the first choice.

It may also occur that $\lambda = 0$ is an eigenvalue. This will be the eigenvalue with the smallest magnitude. For this eigenvalue, the determinant of A will be zero, so one can test a priori to determine if one of the eigenvalues of A is zero.

Referring to eq. 5.95 let us assume that we choose an X_0 that is not orthogonal to Y_1 so that, with A^k the product of k factors of A, we can write

$$A^k X_0 = \left(\alpha_1 \lambda_1^k Y_1 + \alpha_2 \lambda_2^k Y_2 + \bullet \ \bullet + \alpha_N \lambda_N^k Y_N \right)$$
$$= \lambda_1^k \left(\alpha_1 Y_1 + \alpha_2 \frac{\lambda_2^k}{\lambda_1^k} Y_2 + \bullet \ \bullet + \alpha_N \frac{\lambda_N^k}{\lambda_1^k} Y_N \right) \tag{5.97}$$

Since λ_1 is the largest eigenvalue, we see that for k large, $|\lambda_m/\lambda_1|^k << 1$ for $m \geq 2$. Thus, for some large k (chosen to achieve a required level of precision), eq. 5.97 can be approximated by

$$A^k X_0 \simeq \lambda_1^k \alpha_1 Y_1 \tag{5.98a}$$

With one additional multiplication by A we have

$$A^{k+1} X_0 \simeq \lambda_1^{k+1} \alpha_1 Y_1 \tag{5.98b}$$

Combining eqs. 5.98a and 5.98b, we obtain

$$A\left(A^kX_0\right) \simeq \lambda_1\left(A^kX_0\right) \qquad (5.99)$$

This implies that Y_1 is given approximately as some multiple of A^kX_0. We express this as

$$Y_1 \simeq \beta A^kX_0 \qquad (5.100)$$

where β is a constant of proportionality that cannot be determined from the eigenvalue equation. Once this eigenvector A^kX_0 is determined to a required level of precision, λ_1 is obtained easily from eq. 5.99.

To obtain an approximation to λ_2, the second largest eigenvalue, we must determine β by *normalizing* $\beta(A^kX_0)$ to 1. We do this by imposing the condition

$$Y_1^\dagger Y_1 = |\beta|^2 \left(A^kX_0\right)^\dagger \left(A^kX_0\right) = 1 \qquad (5.101)$$

With Y_1 normalized to 1, we construct the matrix

$$B \equiv A - \lambda_1 Y_1 Y_1^\dagger \qquad (5.102)$$

We see that

$$BY_1 = AY_1 - \lambda_1 Y_1 Y_1^\dagger Y_1 = \lambda_1 Y_1 - \lambda_1 Y_1 = 0 \qquad (5.103a)$$

and for the eigenvectors Y_m, $m > 1$,

$$BY_m = AY_m - \lambda_1 Y_1 Y_1^\dagger Y_m = \lambda_m Y_m \qquad (5.103b)$$

That is, all the eigenvalues of A are also the eigenvalues of B except for λ_1. The largest eigenvalue of A has been replaced by 0 for B. As such, the largest eigenvalue for B is λ_2, and the above process is carried out with B to find this eigenvalue.

Example 5.10: Power method for approximating the eigenpairs of a real symmetric matrix

We consider the real symmetric matrix

$$A = \begin{pmatrix} 6 & 0 & 4 \\ 0 & 1 & 0 \\ 4 & 0 & 6 \end{pmatrix} \qquad (5.104)$$

Solving the *characteristic* equation

$$|A - \lambda 1| = \begin{vmatrix} 6 - \lambda & 0 & 4 \\ 0 & 1 - \lambda & 0 \\ 4 & 0 & 6 - \lambda \end{vmatrix} = (6 - \lambda)^2(1 - \lambda) - 16(1 - \lambda) = 0 \quad (5.105)$$

we obtain the eigenvalues $\{\lambda_1, \lambda_2, \lambda_3\} = \{10, 2, 1\}$.

Starting with the unit vector

$$X_0 \equiv \begin{pmatrix} 1 \\ 1 \\ 1 \end{pmatrix} \tag{5.106}$$

we obtain

$$AX_0 = \begin{pmatrix} 10 \\ 1 \\ 10 \end{pmatrix} \tag{5.107a}$$

$$A^2 X_0 = \begin{pmatrix} 100 \\ 1 \\ 100 \end{pmatrix} \tag{5.107b}$$

and so on. With k iterations,

$$A^k X_0 = \begin{pmatrix} 10^k \\ 1 \\ 10^k \end{pmatrix} = 10^k \begin{pmatrix} 1 \\ 10^{-k} \\ 1 \end{pmatrix} \tag{5.108a}$$

which, when normalized to 1 is

$$Y_1 \simeq \frac{1}{\sqrt{2 + 10^{-2k}}} \begin{pmatrix} 1 \\ 10^{-k} \\ 1 \end{pmatrix} \tag{5.108b}$$

After a large enough number of iterations so that we can take $10^{-k} \cong 0$, we have

$$Y_1 \simeq \frac{1}{\sqrt{2}} \begin{pmatrix} 1 \\ 0 \\ 1 \end{pmatrix} \tag{5.109a}$$

from which we obtain

$$AY_1 = 10Y_1 \tag{5.109b}$$

Thus,

$$\lambda_1 \simeq 10 \tag{5.110}$$

We see that for this example, Y_1 and λ_1 are exact.

To obtain the second largest eigenvalue, we define

$$B = A - \lambda_1 Y_1 Y_1^\dagger = \begin{pmatrix} 6 & 0 & 4 \\ 0 & 1 & 0 \\ 4 & 0 & 6 \end{pmatrix} - 10\frac{1}{2}\begin{pmatrix} 1 \\ 0 \\ 1 \end{pmatrix}(1 \quad 0 \quad 1)$$

$$= \begin{pmatrix} 6 & 0 & 4 \\ 0 & 1 & 0 \\ 4 & 0 & 6 \end{pmatrix} - 5\begin{pmatrix} 1 & 0 & 1 \\ 0 & 0 & 0 \\ 1 & 0 & 1 \end{pmatrix} = \begin{pmatrix} 1 & 0 & -1 \\ 0 & 1 & 0 \\ -1 & 0 & 1 \end{pmatrix} \qquad (5.111)$$

Then, with

$$X_0 \equiv \begin{pmatrix} 1 \\ 1 \\ 1 \end{pmatrix} \qquad (5.106)$$

it is straightforward to show that for $k \geq 2$,

$$B^k = \begin{pmatrix} 2^{k-1} & 0 & -2^{k-1} \\ 0 & 1 & 0 \\ -2^{k-1} & 0 & 2^{k-1} \end{pmatrix} \qquad (5.112)$$

Therefore, for all $k \geq 2$,

$$B^k X_0 = \begin{pmatrix} 0 \\ 1 \\ 0 \end{pmatrix} \equiv Y_2 \qquad (5.113)$$

From

$$AY_2 = Y_2 \qquad (5.114)$$

we see that

$$Y_2 \equiv \begin{pmatrix} 0 \\ 1 \\ 0 \end{pmatrix} \qquad (5.115a)$$

is an eigenvector corresponding to

$$\lambda_2 = 1 \qquad (5.115b)$$

To determine the third eigenpair, we form

$$C = B - \lambda_2 Y_2 Y_2^\dagger = \begin{pmatrix} 1 & 0 & -1 \\ 0 & 1 & 0 \\ -1 & 0 & 1 \end{pmatrix} - \begin{pmatrix} 0 \\ 1 \\ 0 \end{pmatrix} (0 \quad 1 \quad 0)$$

$$= \begin{pmatrix} 1 & 0 & -1 \\ 0 & 0 & 0 \\ -1 & 0 & 1 \end{pmatrix} \tag{5.116}$$

With

$$C^k = \begin{pmatrix} 2^{k-1} & 0 & -2^{k-1} \\ 0 & 0 & 0 \\ -2^{k-1} & 0 & 2^{k-1} \end{pmatrix} \tag{5.117}$$

we see that

$$C^k \begin{pmatrix} 1 \\ 1 \\ 1 \end{pmatrix} = \begin{pmatrix} 0 \\ 0 \\ 0 \end{pmatrix} \tag{5.118}$$

This can be interpreted as X_0 being an eigenvector corresponding to the eigenvalue $\lambda = 0$. But since it is straightforward that the determinant of A is non-zero, $\lambda = 0$ is not an eigenvalue of A. Thus, we have only found two eigenpairs.

To find the third eigenpair, we take another choice for X_0. We choose

$$X_0 = \begin{pmatrix} 1 \\ 0 \\ 0 \end{pmatrix} \tag{5.119}$$

We find that

$$A^5 \begin{pmatrix} 1 \\ 0 \\ 0 \end{pmatrix} = \frac{1}{5 \times 10^4} \begin{pmatrix} 1.00032 \\ 0 \\ 0.99968 \end{pmatrix} \simeq \frac{1}{5 \times 10^4} \begin{pmatrix} 1 \\ 0 \\ 1 \end{pmatrix} \tag{5.120}$$

Comparing this to the results presented in eqs. 5.109, we see that when normalized, this is the eigenvector Y_1 corresponding to $\lambda_1 = 10$. After normalizing this to 1, we again form

$$B = A - \lambda_1 Y_1 Y_1^\dagger = \begin{pmatrix} 1 & 0 & -1 \\ 0 & 1 & 0 \\ -1 & 0 & 1 \end{pmatrix} \tag{5.121}$$

and note that

$$B^k X_0 = \begin{pmatrix} 2^{k-1} & 0 & -2^{k-1} \\ 0 & 1 & 0 \\ -2^{k-1} & 0 & 2^{k-1} \end{pmatrix} \begin{pmatrix} 1 \\ 0 \\ 0 \end{pmatrix} = 2^{k-1} \begin{pmatrix} 1 \\ 0 \\ -1 \end{pmatrix} \tag{5.122}$$

After normalizing, we see that this eigenvector is

$$Y_3 = \frac{1}{\sqrt{2}} \begin{pmatrix} 1 \\ 0 \\ -1 \end{pmatrix} \tag{5.123}$$

and since

$$AY_3 = 2Y_3 \tag{5.124}$$

Y_3 is the eigenvector for

$$\lambda_3 = 2 \tag{5.125}$$

It is straightforward to show that by forming the matrix C, we would obtain the eigenvalue $\lambda_2 = 1$ and its corresponding eigenvector.

We note that the unit vector

$$X_0 \equiv \begin{pmatrix} 1 \\ 1 \\ 1 \end{pmatrix} \tag{5.106}$$

is orthogonal to

$$Y_3 = \frac{1}{\sqrt{2}} \begin{pmatrix} 1 \\ 0 \\ -1 \end{pmatrix} \tag{5.123}$$

That is the reason we are unable to determine the eigenpair λ_3, Y_3 of eqs. 5.123 and 5.125 by taking X_0 to be the unit vector. ☐

Eigenvalues and eigenvectors of a real symmetric matrix

Matrices A and A' are related by a *similarity transformation* if

$$A' = S^{-1}AS \tag{5.126a}$$

or equivalently,

$$A = SA'S^{-1} \tag{5.126b}$$

Eigenvalues and eigenvectors of A are defined by

$$(A - \lambda_r 1)X_r = 0 \tag{5.127}$$

From eq. 5.126b, we can write this as

$$S(A' - \lambda_r 1)S^{-1}X_r = S(A' - \lambda_r 1)X'_r = 0 \tag{5.128a}$$

from which

$$(A' - \lambda_r 1)X'_r = 0 \tag{5.128b}$$

Thus, two matrices, related by a similarity transformation S have the same eigenvalues, and their eigenvectors are related by

$$X' = S^{-1}X \tag{5.129}$$

The elements of a real *orthogonal matrix S* that describes a "rotation through an angle ϕ in the p–q plane $(p \neq q)$" are obtained by starting with the unit matrix, then replacing the elements 1 in the pp and qq locations by

$$s_{pp} = s_{qq} = \cos\phi \tag{5.130a}$$

and replacing the elements 0 in the pq and qp locations by

$$s_{pq} = -s_{qp} = \sin\phi \tag{5.130b}$$

For example, a 5×5 real orthogonal matrix describing a two dimensional "rotation in the 2–4 plane" through an angle ϕ, is of the form

$$S = \begin{pmatrix} 1 & 0 & 0 & 0 & 0 \\ 0 & \cos\phi & 0 & \sin\phi & 0 \\ 0 & 0 & 1 & 0 & 0 \\ 0 & -\sin\phi & 0 & \cos\phi & 0 \\ 0 & 0 & 0 & 0 & 1 \end{pmatrix} \tag{5.131}$$

Such a matrix satisfies

$$SS^T = S^T S = 1 \tag{5.132}$$

Let A be a real *symmetric* matrix, so that its elements satisfy

$$a_{mn}* = a_{mn} = a_{nm} \tag{5.133}$$

A can be diagonalized by the similarity transformation given in eq. 5.126a, if the columns of S are the elements of the eigenvectors of A. For example, let the eigenvectors of a 3×3 matrix A be

$$X_{\lambda_1} = \begin{pmatrix} x_{11} \\ x_{12} \\ x_{13} \end{pmatrix} \tag{5.134a}$$

$$X_{\lambda_2} = \begin{pmatrix} x_{21} \\ x_{22} \\ x_{23} \end{pmatrix} \tag{5.134b}$$

and

$$X_{\lambda_3} = \begin{pmatrix} x_{31} \\ x_{32} \\ x_{33} \end{pmatrix} \tag{5.134c}$$

with each vector normalized to 1 and the vectors mutually orthogonal to one another. Such vectors are said to be *orthonormalized*. They satisfy

$$X_{\lambda_m}^T X_{\lambda_n} = \delta_{mn} = \begin{cases} 1 & m = n \\ 0 & m \neq n \end{cases} \tag{5.135a}$$

from which

$$x_{m1}^2 + x_{m2}^2 + x_{m3}^2 = 1 \tag{5.135b}$$

Using the 3×3 matrix to illustrate, the orthogonal transformation $S^T A S$ results in a diagonal matrix if S is of the form

$$S = \begin{pmatrix} x_{11} & x_{21} & x_{31} \\ x_{12} & x_{22} & x_{32} \\ x_{13} & x_{23} & x_{33} \end{pmatrix} \equiv (X_1 \quad X_2 \quad X_3) \tag{5.136}$$

where X_n ($n = 1, 2, 3$) is a column eigenvector of A forming the nth column of the matrix S.

We consider the transformation $S^T A S$. Since the columns of S are the eigenvectors of A, we have

$$AS = (AX_1 \quad AX_2 \quad AX_3) = (\lambda_1 X_1 \quad \lambda_2 X_2 \quad \lambda_3 X_3) \tag{5.137}$$

Premultiplying this by

$$S^T = \begin{pmatrix} X_1^T \\ X_2^T \\ X_3^T \end{pmatrix} \tag{5.138}$$

where X_n^T is a row eigenvector of A forming the nth row of the matrix S^T. Using the orthonormality of the eigenvectors, this yields

$$S^T A S = \begin{pmatrix} X_1^T \\ X_2^T \\ X_3^T \end{pmatrix} (\lambda_1 X_1 \quad \lambda_2 X_2 \quad \lambda_3 X_3)$$

$$= \begin{pmatrix} \lambda_1 X_1^T X_1 & \lambda_2 X_1^T X_2 & \lambda_3 X_1^T X_3 \\ \lambda_1 X_2^T X_1 & \lambda_2 X_2^T X_2 & \lambda_3 X_2^T X_3 \\ \lambda_1 X_3^T X_1 & \lambda_2 X_3^T X_2 & \lambda_3 X_3^T X_3 \end{pmatrix} = \begin{pmatrix} \lambda_1 & 0 & 0 \\ 0 & \lambda_2 & 0 \\ 0 & 0 & \lambda_3 \end{pmatrix} \tag{5.139}$$

Jacobi method

The Jacobi method for diagonalizing a real symmetric matrix consists of applying several similarity transformations to minimize the off-diagonal elements. This is done by applying an orthogonal transformation that will transform two off-diagonal elements to zero. When the method is applied by hand, it is most efficient to choose the off-diagonal element(s) with the largest magnitude to transform to zero (the number with the smallest possible magnitude). Performing the computations with a computer, one can follow a pattern such as transforming a_{12} and a_{21} to zero, then transforming a_{13} and a_{31} to zero, and so on. Each transformation may destroy the zeros created by the previous transformation, but as will be noted below, that is unimportant. This process is continued until all off-diagonal elements are small enough that they can be approximated by zero. Then the matrix is approximately diagonal, and the diagonal elements are approximations of the eigenvalues.

Let S be the orthogonal matrix that transforms a_{pq} and a_{qp} to zero. It is of the form

$$S = \begin{pmatrix} 1 & \bullet\bullet & 0 & \bullet\bullet & 0 & \bullet\bullet & 0 \\ \bullet & & \bullet & & \bullet & & \bullet \\ \bullet & & \bullet & & \bullet & & \bullet \\ 0 & \bullet\bullet & \cos\phi & \bullet\bullet & \sin\phi & \bullet\bullet & 0 \\ \bullet & & \bullet & & \bullet & & \bullet \\ \bullet & & \bullet & & \bullet & & \bullet \\ 0 & \bullet\bullet & -\sin\phi & \bullet\bullet & \cos\phi & \bullet\bullet & 0 \\ \bullet & & \bullet & & \bullet & & \bullet \\ \bullet & & \bullet & & \bullet & & \bullet \\ 0 & \bullet\bullet & 0 & \bullet\bullet & 0 & \bullet\bullet & 1 \end{pmatrix} \begin{matrix} \\ \\ \\ \leftarrow row_p \\ \\ \\ \leftarrow row_q \\ \\ \\ \end{matrix} \qquad (5.140)$$

$$\uparrow col_p \qquad \uparrow col_q$$

The similarity transformation

$$A' \equiv S^T A S \qquad \textbf{(5.126a)}$$

leaves all elements of A unchanged except a_{pp}, a_{qq}, and $a_{qp} = a_{qp}$. These become

$$a'_{pp} = a_{pp}\cos^2\phi + a_{qq}\sin^2\phi + 2a_{pq}\sin\phi\cos\phi \qquad (5.141a)$$

$$a'_{qq} = a_{pp}\sin^2\phi + a_{qq}\cos^2\phi - 2a_{pq}\sin\phi\cos\phi \qquad (5.141b)$$

and

$$a'_{pq} = a'_{qp} = \left(a_{pp} - a_{qq}\right)\sin\phi\cos\phi - a_{pq}\left(\cos^2\phi - \sin^2\phi\right) \qquad (5.141c)$$

Setting

$$a'_{pq} = a'_{qp} = 0 \qquad (5.142)$$

the angle ϕ is given by

$$\tan(2\phi) = \frac{2a_{pq}}{\left(a_{pp} - a_{qq}\right)} \qquad (5.143)$$

with

$$-\infty \leq \tan(2\phi) \leq \infty \qquad (5.144a)$$

or equivalently

$$-\frac{\pi}{4} \leq \phi \leq \frac{\pi}{4} \qquad (5.144b)$$

It is straightforward to show from (5.141) that

$$\left(a'_{pp}\right)^2 + \left(a'_{qq}\right)^2 = \left(a_{pp}\right)^2 + \left(a_{qq}\right)^2 + 2\left(a_{pq}\right)^2 \tag{5.145}$$

Therefore,

$$\left(a'_{pp}\right)^2 + \left(a'_{qq}\right)^2 > \left(a_{pp}\right)^2 + \left(a_{qq}\right)^2 \tag{5.146a}$$

and

$$\left(a'_{pp}\right)^2 + \left(a'_{qq}\right)^2 > \left(a_{pq}\right)^2 \tag{5.146b}$$

Thus, with each iteration, the magnitudes of the diagonal terms get larger (eq. 5.146a) and the off-diagonal terms get small relative to the diagonal terms (eq. 5.144b).

Therefore, after M such orthogonal transformations, we have

$$A' = \left(S_M^T S_{M-1}^T ... S_2^T S_1^T\right) A \left(S_1 S_2 ... S_{M-1} S_M\right) \simeq \begin{pmatrix} a'_{11} & 0 & \bullet \bullet & 0 \\ 0 & a'_{22} & \bullet \bullet & 0 \\ \bullet & \bullet & \bullet & \bullet \\ \bullet & \bullet & \bullet & \bullet \\ 0 & 0 & \bullet \bullet & a'_{NN} \end{pmatrix} \tag{5.147}$$

where the zeros represent off-diagonal elements that are ignorably small relative to the diagonal elements. In this diagonal form, the eigenvalues are given by

$$\lambda_m = a'_{mm} \tag{5.148}$$

The corresponding orthonormal eigenvectors are the columns of

$$S \equiv S_1 S_2 ... S_{M-1} S_M = \begin{pmatrix} x_{11} & x_{21} & \bullet \bullet & x_{N1} \\ x_{12} & x_{22} & \bullet \bullet & x_{N2} \\ \bullet & \bullet & & \bullet \\ \bullet & \bullet & & \bullet \\ x_{1N} & x_{2N} & \bullet \bullet & x_{NN} \end{pmatrix} \tag{5.149}$$

Example 5.11: Jacobi method for approximating the eigenpairs of a real symmetric matrix

We consider the matrix

$$A = \begin{pmatrix} 4 & 3 & -2 \\ 3 & 2 & 1 \\ -2 & 1 & 1 \end{pmatrix} \tag{5.150}$$

Using the Newton–Raphson method, we solve the characteristic equation

$$\begin{vmatrix} 4 - \lambda & 3 & -2 \\ 3 & 2 - \lambda & 1 \\ -2 & 1 & 1 - \lambda \end{vmatrix} = -\lambda^3 + 7\lambda^2 - 25 = 0 \tag{5.151}$$

to obtain eigenvalues $\{\lambda_1, \lambda_2, \lambda_3\} \cong (6.38720, 2.30839, -1.69559)$.

Writing the eigenvalue equation in the form

$$\begin{pmatrix} 4 - \lambda & 3 & -2 \\ 3 & 2 - \lambda & 1 \\ -2 & 1 & 1 - \lambda \end{pmatrix} \begin{pmatrix} x_1 \\ x_2 \\ x_3 \end{pmatrix} = 0 \tag{5.152}$$

we solve these equations for x_2 and x_3 in terms of x_1 to obtain the general form of the orthonormalized eigenvector

$$X_\lambda = \frac{1}{\sqrt{1 + \rho_2^2 + \rho_3^2}} \begin{pmatrix} 1 \\ \rho_2 \\ \rho_3 \end{pmatrix} \tag{5.153}$$

where

$$\rho_2 \equiv \frac{x_2}{x_1} = \frac{(1 - \lambda)(7 - 2\lambda)}{1 - (1 - \lambda)(2 - \lambda)} + 2 \tag{5.154a}$$

and

$$\rho_3 = \frac{x_3}{x_1} = -\frac{(7 - 2\lambda)}{1 - (1 - \lambda)(2 - \lambda)} \tag{5.154b}$$

From this, we have

$$X_{6.38720} = \begin{pmatrix} 0.82859 \\ 0.51841 \\ -0.21138 \end{pmatrix} \tag{5.155a}$$

$$X_{-1.69559} = \begin{pmatrix} 0.52622 \\ -0.59228 \\ 0.61016 \end{pmatrix} \tag{5.155b}$$

and

$$X_{2.30839} = \begin{pmatrix} 0.19112 \\ -0.61681 \\ -0.76356 \end{pmatrix} \tag{5.155c}$$

We follow the pattern of transforming the off-diagonal elements to zero in the order (1,2), (1,3), and then (2,3). We repeat the method outlined above until the magnitudes of all off-diagonal elements are less than a small number (which we have chosen as 10^{-7} for this example).

For the first iteration, we transform a_{12} and a_{21} to zero by the similarity transformation

$$S_{12}^1 = \begin{pmatrix} \cos(35.78°) & \sin(35.78°) & 0 \\ -\sin(35.78°) & \cos(35.78°) & 0 \\ 0 & 0 & 1 \end{pmatrix} \qquad (5.156a)$$

to obtain

$$A' = \begin{pmatrix} 6.16227 & 0 & -1.03777 \\ 0 & -0.16227 & 1.98066 \\ -1.03777 & 1.98066 & 1 \end{pmatrix} \qquad (5.156b)$$

We then use the similarity transformation

$$S_{13}^1 = \begin{pmatrix} \cos(-10.95°) & 0 & \sin(-10.95°) \\ 0 & 1 & 0 \\ -\sin(-10.95°) & 0 & \cos(-10.95°) \end{pmatrix} \qquad (5.157a)$$

to obtain

$$A'' = \begin{pmatrix} 6.36309 & -0.37628 & 0 \\ -0.37628 & -0.16227 & 1.94459 \\ 0 & 1.94459 & 0.79918 \end{pmatrix} \qquad (5.157b)$$

Then, with

$$S_{23}^1 = \begin{pmatrix} 1 & 0 & 0 \\ 0 & \cos(-38.06°) & \sin(-38.06°) \\ 0 & -\sin(-38.06°) & \cos(-38.06°) \end{pmatrix} \qquad (5.158a)$$

we find

$$A''' = \begin{pmatrix} 6.36309 & -0.29628 & -0.23195 \\ -0.29628 & -1.68467 & 0 \\ -0.23195 & 0 & 2.32158 \end{pmatrix} \qquad (5.158b)$$

We find that we must apply two additional sets of such similarity transformations (a total of nine similarity transformations in all) to obtain the diagonalized matrix

$$A^{(9)} = \begin{pmatrix} 6.38719 & 0 & 0 \\ 0 & -1.69558 & 0 \\ 0 & 0 & 2.30838 \end{pmatrix} \qquad (5.159)$$

with

$$\begin{aligned} S &= S_{23}^3 S_{13}^3 S_{12}^3 S_{23}^2 S_{13}^2 S_{12}^2 S_{23}^1 S_{13}^1 S_{12}^1 \\ &= \begin{pmatrix} 0.82859 & -0.52622 & -0.19112 \\ 0.51841 & 0.59228 & 0.61681 \\ -0.21138 & -0.61016 & 0.76356 \end{pmatrix} \qquad (5.160) \end{aligned}$$

From the diagonalized form of A given in eq. 5.159, the eigenvalues of A are $\{6.38719, -1.69558, 2.30838\}$, and from the columns of S of eq. 5.160, the corresponding eigenvectors are

$$X_{6.38720} = \begin{pmatrix} 0.82859 \\ 0.51841 \\ -0.21138 \end{pmatrix} \qquad (5.155a)$$

$$X_{-1.69559} = \begin{pmatrix} -0.52622 \\ 0.59228 \\ -0.61016 \end{pmatrix} \qquad (5.155b)$$

and

$$X_{2.30839} = \begin{pmatrix} -0.19112 \\ 0.61681 \\ 0.76356 \end{pmatrix} \qquad (5.155c)$$

which are the eigenvectors found above (except for a meaningless multiplier of -1 for the vectors $X_{-1.69559}$ and $X_{2.30839}$). □

Because the similarity transformations used in the Jacobi method destroys off-diagonal zeros determined in a previous iteration of the method, convergence to a diagonal form can be very slow.

Eigenvalues of a symmetric tridiagonal matrix

A symmetric *tridiagonal* matrix is one for which the elements along the main diagonal and those along the diagonals just above and just below the main diagonal do not have to be zero. All other elements are zero. Such a matrix is of the form

$$T \equiv \begin{pmatrix} \alpha_1 & \beta_1 & 0 & 0 & \bullet\bullet & 0 \\ \beta_1 & \alpha_2 & \beta_2 & 0 & \bullet\bullet & 0 \\ 0 & \beta_2 & \alpha_3 & \beta_3 & \bullet\bullet & 0 \\ \bullet & & & & & \\ \bullet & & & & & \\ 0 & 0 & \bullet\bullet & \beta_{N-2} & \alpha_{N-1} & \beta_{N-1} \\ 0 & 0 & 0 & \bullet\bullet & \beta_{N-1} & \alpha_N \end{pmatrix} \tag{5.161}$$

The eigenvalues are then the roots of the Nth order polynomial

$$P_N(\lambda) \equiv \begin{vmatrix} \alpha_1 - \lambda & \beta_1 & 0 & 0 & \bullet\bullet & 0 \\ \beta_1 & \alpha_2 - \lambda & \beta_2 & 0 & \bullet\bullet & 0 \\ 0 & \beta_2 & \alpha_3 - \lambda & \beta_3 & \bullet\bullet & 0 \\ \bullet & & & & & \\ \bullet & & & & & \\ 0 & 0 & \bullet\bullet & \beta_{N-2} & \alpha_{N-1} - \lambda & \beta_{N-1} \\ 0 & 0 & 0 & \bullet\bullet & \beta_{N-1} & \alpha_N - \lambda \end{vmatrix}$$

$$= (-1)^N \lambda^N + \sum_{m=0}^{N-1} q_{N,m} \lambda^m \tag{5.162}$$

To determine the coefficients $q_{N,m}$, we define

$$P_0(\lambda) \equiv 1 \tag{5.163a}$$

and

$$P_1(\lambda) \equiv \alpha_1 - \lambda \tag{5.163b}$$

From these we consider

$$P_2(\lambda) \equiv \begin{vmatrix} \alpha_1 - \lambda & \beta_1 \\ \beta_1 & \alpha_2 - \lambda \end{vmatrix} = (\alpha_2 - \lambda)(\alpha_1 - \lambda) - \beta_1^2$$

$$= (\alpha_2 - \lambda)P_1(\lambda) - \beta_1^2 P_0(\lambda) \tag{5.164a}$$

and

$$P_3(\lambda) \equiv \begin{vmatrix} \alpha_1 - \lambda & \beta_1 & 0 \\ \beta_1 & \alpha_2 - \lambda & \beta_2 \\ 0 & \beta_2 & \alpha_3 - \lambda \end{vmatrix}$$

$$= (\alpha_3 - \lambda)(\alpha_2 - \lambda)(\alpha_1 - \lambda) - \beta_1^2(\alpha_3 - \lambda) - \beta_2^2(\alpha_1 - \lambda)$$

$$= (\alpha_3 - \lambda)\left[P_2(\lambda) + \beta_1^2\right] - \beta_1^2(\alpha_3 - \lambda) - \beta_2^2(\alpha_1 - \lambda)$$

$$= (\alpha_3 - \lambda)P_2(\lambda) - \beta_2^2 P_1(\lambda) \tag{5.164b}$$

We note that $P_2(\lambda)$ is related to $P_1(\lambda)$ and $P_0(\lambda)$ in the same way that $P_3(\lambda)$ is related to $P_2(\lambda)$ and $P_1(\lambda)$. For $k \leq N$, we define

$$
P_k(\lambda) \equiv
\begin{vmatrix}
\alpha_1 - \lambda & \beta_1 & 0 & 0 & \bullet\bullet & 0 \\
\beta_1 & \alpha_2 - \lambda & \beta_2 & 0 & \bullet\bullet & 0 \\
0 & \beta_2 & \alpha_3 - \lambda & \beta_3 & \bullet\bullet & 0 \\
\bullet & & & & & \\
\bullet & & & & & \\
0 & 0 & \bullet\bullet & \beta_{k-2} & \alpha_{k-1} - \lambda & \beta_{k-1} \\
0 & 0 & 0 & \bullet\bullet & \beta_{k-1} & \alpha_k - \lambda
\end{vmatrix}
\tag{5.165}
$$

Expanding this determinant about the last row (or column), we obtain

$$
P_k(\lambda) \equiv (\alpha_k - \lambda)
\begin{vmatrix}
\alpha_1 - \lambda & \beta_1 & 0 & \bullet\bullet & 0 \\
\beta_1 & \alpha_2 - \lambda & \beta_2 & \bullet\bullet & 0 \\
\bullet & & & & \\
\bullet & & & & \\
0 & \bullet\bullet & \beta_{k-3} & \alpha_{k-2} - \lambda & \beta_{k-2} \\
0 & 0 & \bullet\bullet & \beta_{k-2} & \alpha_{k-1} - \lambda
\end{vmatrix}
$$

$$
- \beta_{k-1}
\begin{vmatrix}
\alpha_1 - \lambda & \beta_1 & 0 & \bullet\bullet & 0 \\
\beta_1 & \alpha_2 - \lambda & \beta_2 & \bullet\bullet & 0 \\
\bullet & & & & \\
\bullet & & & & \\
0 & \bullet\bullet & \beta_{k-3} & \alpha_{k-2} - \lambda & 0 \\
0 & 0 & \bullet\bullet & \beta_{k-2} & \beta_{k-1}
\end{vmatrix}
\tag{5.166a}
$$

Expanding the second determinant about the last column, this becomes

$$
P_k(\lambda) \equiv (\alpha_k - \lambda)
\begin{vmatrix}
\alpha_1 - \lambda & \beta_1 & 0 & \bullet\bullet & 0 \\
\beta_1 & \alpha_2 - \lambda & \beta_2 & \bullet\bullet & 0 \\
\bullet & & & & \\
\bullet & & & & \\
0 & \bullet\bullet & \beta_{k-3} & \alpha_{k-2} - \lambda & \beta_{k-2} \\
0 & 0 & \bullet\bullet & \beta_{k-2} & \alpha_{k-1} - \lambda
\end{vmatrix}
$$

$$
- \beta_{k-1}^2
\begin{vmatrix}
\alpha_1 - \lambda & \beta_1 & \bullet\bullet & 0 \\
\beta_1 & \alpha_2 - \lambda & \beta_2 & 0 \\
\bullet & & & \\
\bullet & & & \\
0 & \bullet\bullet & \beta_{k-3} & \alpha_{k-2} - \lambda
\end{vmatrix}
$$

$$
= (\alpha_k - \lambda)P_{k-1}(\lambda) - \beta_{k-1}^2 P_{k-2}(\lambda)
\tag{5.166b}
$$

We note that if we define $P_{-1}(\lambda)$ to be zero, this recurrence relation is valid for $k = 1$.

To generate $P_N(\lambda)$, we start with $k = 1$ to generate $P_1(\lambda)$ from $P_0(\lambda)$, then $P_2(\lambda)$ from $P_1(\lambda)$ and $P_0(\lambda)$, and so on, until we have obtained $P_N(\lambda)$.

The sequence $\{P_k(\lambda)\}$ is called a *Sturm* sequence. Properties of Sturm sequences are discussed in many places in the literature (see, for example, Patel, V., 1994, pp. 455–457).

Referring to eq. 5.162, the eigenvalues are found using one of the methods developed in ch. 2 for finding the roots of a polynomial. To apply these techniques, we must have the functional form of $P_N(\lambda)$. Thus, we must determine the coefficients $q_{N,m}$. Expressing $P_k(\lambda)$ as

$$P_k(\lambda) = (-1)^k \lambda^k + \sum_{m=0}^{k-1} q_{k,m} \lambda^m \qquad (5.167)$$

the recurrence relation of eq. 5.166b becomes

$$(-1)^k \lambda^k + \sum_{m=0}^{k-1} q_{k,m} \lambda^m = \alpha_k (-1)^{k-1} \lambda^{k-1} + \alpha_k \sum_{m=0}^{k-2} q_{k-1,m} \lambda^m$$

$$- (-1)^{k-1} \lambda^k - \sum_{m=0}^{k-2} q_{k-1,m} \lambda^{m+1} - \beta_{k-1}^2 (-1)^k \lambda^{k-2} - \beta_{k-1}^2 \sum_{m=0}^{k-3} q_{k-2,m} \lambda^m \quad (5.168)$$

We first note that the terms $(-1)^k \lambda^k$ cancel. We also point out that the smallest power of λ in the sum involving λ^{m+1} is λ^1 and the largest power of λ in the last sum is λ^{k-3}. Therefore, we separate out all terms involving λ^0, λ^{k-1}, and λ^{k-2} from the sums, and write

$$\sum_{m=0}^{k-1} q_{k,m} \lambda^m = q_{k,0} \lambda^0 + \sum_{m=1}^{k-3} q_{k,m} \lambda^m + q_{k,k-2} \lambda^{k-2} + q_{k,k-1} \lambda^{k-1} \qquad (5.169a)$$

$$\sum_{m=0}^{k-2} q_{k-1,m} \lambda^m = q_{k-1,0} \lambda^0 + \sum_{m=1}^{k-3} q_{k-1,m} \lambda^m + q_{k-1,k-2} \lambda^{k-2} \qquad (5.169b)$$

$$\sum_{m=0}^{k-2} q_{k-1,m} \lambda^{m+1} = \sum_{m'=1}^{k-1} q_{k-1,m'-1} \lambda^{m'}$$

$$\sum_{m=1}^{k-3} q_{k-1,m-1} \lambda^m + q_{k-1,k-2} \lambda^{k-2} + q_{k-1,k-1} \lambda^{k-1} \qquad (5.169c)$$

(where m' has been renamed as m), and

$$\sum_{m=0}^{k-3} q_{k-2,m}\lambda^m = q_{k-2,0}\lambda^0 + \sum_{m=1}^{k-3} q_{k-2,m}\lambda^m \qquad (5.169\text{d})$$

After canceling the terms $(-1)^k \lambda^k$, eq. 5.168 can be written as

$$q_{k,0}\lambda^0 + \sum_{m=1}^{k-3} q_{k,m}\lambda^m + q_{k,k-2}\lambda^{k-2} + q_{k,k-1}\lambda^{k-1}$$

$$= \alpha_k(-1)^{k-1}\lambda^{k-1} + \alpha_k q_{k-1,0}\lambda^0 + \alpha_k \sum_{m=1}^{k-3} q_{k-1,m}\lambda^m + \alpha_k q_{k-1,k-2}\lambda^{k-2}$$

$$- q_{k-1,k-2}\lambda^{k-1} - q_{k-1,k-3}\lambda^{k-2} - \sum_{m=1}^{k-3} q_{k-1,m-1}\lambda^m$$

$$- \beta_{k-1}^2(-1)^k\lambda^{k-2} - \beta_{k-1}^2 q_{k-2,0}\lambda^0 - \beta_{k-1}^2 \sum_{m=1}^{k-3} q_{k-2,m}\lambda^m \qquad (5.170)$$

With $q_{0,0} = 1$, we equate coefficients of corresponding powers of λ to obtain

$$m = 0,\ k \geq 2 : \quad q_{k,0} = \alpha_k q_{k-1,0} - \beta_{k-1}^2 q_{k-2,0} \qquad (5.171\text{a})$$

$$m = k-2,\ k \geq 2 : \quad q_{k,k-2} = \alpha_k q_{k-1,k-2} - q_{k-1,k-3} - (-1)^k \beta_{k-1}^2 \qquad (5.171\text{b})$$

$$m = k-1,\ k \geq 2 \quad q_{k,k-1} = (-1)^{k-1}\alpha_k - q_{k-1,k-2} \qquad (5.171\text{c})$$

and

$$1 \leq m \leq k-3,\ k \geq 4 : \quad q_{k,m} = \alpha_k q_{k-1,m} - q_{k-1,m-1} - \beta_{k-1}^2 q_{k-2,m} \qquad (5.171\text{d})$$

If we take $0 \leq m \leq k$ and use the conventions that for any indices r and s,

$$q_{r,s} = (-1)^r \quad r = s \qquad (5.172\text{a})$$

$$q_{r,s} = 0 \quad s > r \qquad (5.172\text{b})$$

and

$$q_{r,s} = 0 \quad r < 0 \text{ or } s < 0 \qquad (5.172\text{c})$$

we note that the recurrence relation for the coefficients given in eq. 5.171d reproduces those of eqs. 5.171a, 5.171b and 5.171c.

Therefore, we generate the coefficients for $P_k(\lambda)$ with $0 \leq m \leq k$ using eq. 5.171d only, subject to the constraints of eqs. 5.172. When $k = N$, we obtain the coefficients of $P_N(\lambda)$ from which we can then find the eigenvalues of the tridiagonal matrix T, which are the eigenvalues of the original matrix A.

Unlike the Jacobi approach, the methods of Givens and of Householder transform a symmetric matrix to *tridiagonal* form by similarity transformations that do not destroy zeros obtained in previous iterations.

Givens method

The *Givens* method for tridiagonalizing a real symmetric matrix involves applying a sequence of similarity transformations comprised of orthogonal matrices that first transforms $a_{13} \rightarrow 0$, then a transformation to take $a_{14} \rightarrow 0, \ldots$ then $a_{1N} \rightarrow 0$, then $a_{24} \rightarrow 0$, then $a_{25} \rightarrow 0, \ldots$, then finally a transformation that takes $a_{1(N-2)} \rightarrow 0$. This process results in A being cast into tridiagonal form.

The orthogonal matrix S that transforms a_{pq} and a_{qp} to zero is given by

$$
S = \begin{pmatrix}
1 & \bullet\bullet & 0 & \bullet\bullet & 0 & \bullet\bullet & 0 \\
\bullet & & \bullet & & \bullet & & \bullet \\
\bullet & & \bullet & & \bullet & & \bullet \\
0 & \bullet\bullet & \cos\phi & \bullet\bullet & \sin\phi & \bullet\bullet & 0 \\
\bullet & & \bullet & & \bullet & & \bullet \\
\bullet & & \bullet & & \bullet & & \bullet \\
0 & \bullet\bullet & -\sin\phi & \bullet\bullet & \cos\phi & \bullet\bullet & 0 \\
\bullet & & \bullet & & \bullet & & \bullet \\
\bullet & & \bullet & & \bullet & & \bullet \\
0 & \bullet\bullet & 0 & \bullet\bullet & 0 & \bullet\bullet & 1
\end{pmatrix}
\begin{matrix} \\ \\ \\ \leftarrow row_{p+1} \\ \\ \\ \leftarrow row_q \\ \\ \\ \end{matrix}
\tag{5.173}
$$

$$\uparrow col_{p+1} \qquad \uparrow col_q$$

where

$$\tan\phi = -\frac{a_{pq}}{a_{p(p+1)}} \tag{5.174}$$

Then,

$$A' \equiv S^T A S \tag{5.126a}$$

transforms $a_{qp} = a_{qp}$ to zero.

After M such transformations, the tridiagonal form of A is

$$T = \left(S_M^T S_{M-1}^T ... S_2^T S_1^T\right) A \left(S_1 S_2 ... S_{M-1} S_M\right)$$

$$= \begin{pmatrix} \alpha_1 & \beta_1 & 0 & \bullet\bullet & & 0 \\ \beta_1 & \alpha_2 & \beta_2 & \bullet\bullet & & 0 \\ 0 & \beta_2 & \alpha_3 & \bullet\bullet & & 0 \\ \bullet & & & & & \beta_{N-1} \\ \bullet & & & & & \\ 0 & \bullet\bullet & 0 & \beta_{N-1} & & \alpha_N \end{pmatrix} \tag{5.175}$$

The eigenvalues are then found by the method described above.

Unlike the Jacobi method, since the orthogonal transformation matrix

$$S_f = S_1 S_2 ... S_{M-1} S_M \tag{5.176}$$

does not diagonalize A, the columns of S_f are not the eigenvectors corresponding to the eigenvalues. Each eigenvector X_k corresponding to the eigenvalue λ_k must be found by solving $N - 1$ of the homogeneous linear equations

$$(a_{11} - \lambda_k)x_{k,1} + a_{12}x_{k,2} + ... + a_{1N}x_{k,N} = 0 \tag{5.177a}$$

$$a_{11}x_{k,1} + (a_{12} - \lambda_k)x_{k,2} + ... + a_{1N}x_{k,N} = 0 \tag{5.177b}$$

for $N-1$, x values in terms of one x. A normalization condition imposed by the user determines the value of the remaining x.

Householder method

Tridiagonalizing a real symmetric matrix A by the *Householder* method involves a applying a sequence of similarity transformations with orthogonal matrices, each of which zeros out the required elements in a specific row and column of A.

The first orthogonal transformation matrix, S_1, replaces the $(1,3) = (3,1)$, $(1,4) = (4,1)$, ..., $(1,N) = (N,1)$ elements by zeros. To determine S_1, an $(N-1) \times (N-1)$ matrix Σ_1 is defined such that

$$S_1 \equiv \begin{pmatrix} 1 & 0 & 0 & \bullet\bullet & 0 \\ 0 & & & & \\ 0 & & \Sigma_1 & & \\ \bullet & & & & \\ \bullet & & & & \\ 0 & & & & \end{pmatrix} \tag{5.178}$$

In this form, $S_1{}^T A\, S_1$ does not affect a_{11}.

Tridiagonalization is achieved by constructing Σ_1 so that the (1,2) and (2,1) elements of A are replaced by a non-zero value, and the $(1,3) = (3,1),\ \ldots,$ $(1,N) = (N,1)$, elements in the first row and first column, are zero. To achieve this, we define a column vector

$$V_1 \equiv \begin{pmatrix} a_{21} \\ a_{31} \\ \bullet \\ \bullet \\ a_{N1} \end{pmatrix} \tag{5.179}$$

and determine Σ_1 such that

$$\Sigma_1 V_1 = n_1 \begin{pmatrix} 1 \\ 0 \\ \bullet \\ \bullet \\ 0 \end{pmatrix} \equiv n_1 u_1 \tag{5.180}$$

where u_1 is the *unit column* with $N-1$ elements. With this condition, the similarity transformation by S_1 will replace the (1,2) and (2,1) elements with a non-zero value obtained from n_1 and replace the (1,3) through (1,N) and (3,1) through (N,1) elements with zeros.

Consider

$$(\Sigma_1 V_1)^T (\Sigma_1 V_1) = n_1^2 u_1^T u_1 = n_1^2 = V_1^T \Sigma_1^T \Sigma_1 V_1 \tag{5.181}$$

Since S_1 is an orthogonal matrix, Σ_1 must be orthogonal. Therefore,

$$\Sigma_1^T \Sigma_1 = 1 \tag{5.182}$$

Thus, from eq. 5.181, we have

$$n_1^2 = V_1^T V_1 = a_{12}^2 + a_{13}^2 + \ldots + a_{1N}^2 = \sum_{m=2}^{N} a_{1m}^2 \tag{5.183}$$

To determine Σ_1, we define a column vector

$$W_1 \equiv \begin{pmatrix} w_1 \\ w_2 \\ \bullet \\ \bullet \\ w_{N-1} \end{pmatrix} \tag{5.184}$$

from which we define

$$\Sigma_1 = 1 - 2W_1 W_1^T$$

$$= \begin{pmatrix} 1 & 0 & \bullet & \bullet & 0 \\ 0 & 1 & \bullet & \bullet & 0 \\ \bullet & \bullet & \bullet & & \bullet \\ \bullet & \bullet & \bullet & & \bullet \\ 0 & 0 & \bullet & \bullet & 1 \end{pmatrix} - 2 \begin{pmatrix} w_1 \\ w_2 \\ \bullet \\ \bullet \\ w_{N-1} \end{pmatrix} \begin{pmatrix} w_1 & w_2 & \bullet & \bullet & w_{N-1} \end{pmatrix} \qquad (5.185)$$

Since Σ_1 must be orthogonal,

$$\Sigma_1 \Sigma_1^T = 1 = 1 - \left(1 - 2W_1 W_1^T\right)\left(1 - 2W_1 W_1^T\right)^T$$
$$= 1 - \left(1 - 2W_1 W_1^T\right)\left(1 - 2W_1 W_1^T\right) = 1 - 4W_1 W_1^T + 4W_1 \left(W_1^T W\right)W_1^T \qquad (5.186)$$

To have the last two terms in eq. 5.186 cancel, we must require that

$$W_1^T W_1 = 1 \qquad (5.187)$$

With eq. 5.185, eq. 5.180 becomes

$$\left(1 - 2W_1 W_1^T\right)V_1 = n_1 u_1 \qquad (5.188a)$$

from which

$$2W_1 \left(W_1^T V_1\right) = V_1 - n_1 u_1 \equiv \omega_1 \qquad (5.188b)$$

Defining the number c by

$$2\left(W_1^T V_1\right) \equiv \frac{1}{c} \qquad (5.189)$$

eq. 5.188b becomes

$$W_1 = c\omega_1 \qquad (5.190)$$

and since W_1 is normalized to 1,

$$c^2 = \frac{1}{\omega_1^T \omega_1} \qquad (5.191)$$

Therefore,

$$\Sigma_1 = 1 - 2\frac{\omega_1 \omega_1^T}{\omega_1^T \omega_1} \qquad (5.192)$$

With this form of $\sum_1$, and renaming a_{11} as α_1, we have

$$
S_1^T A S_1 = \begin{pmatrix}
\alpha_1 & \beta_1 & 0 & \bullet\;\bullet & 0 \\
\beta_1 & b_{22} & b_{23} & \bullet\;\bullet & b_{2N} \\
0 & b_{23} & b_{33} & \bullet\;\bullet & b_{3N} \\
\bullet & & & & \bullet \\
\bullet & & & & \bullet \\
0 & b_{2N} & b_{3N} & \bullet\;\bullet & b_{NN}
\end{pmatrix} \equiv A_1 \tag{5.193}
$$

To tridiagonalize the second row and column elements, we define

$$
S_2 \equiv \begin{pmatrix}
1 & 0 & 0 & \bullet\;\bullet & 0 \\
0 & 1 & 0 & \bullet\;\bullet & 0 \\
0 & 0 & & & \\
\bullet & \bullet & & \Sigma_2 & \\
\bullet & \bullet & & & \\
0 & 0 & & &
\end{pmatrix} \tag{5.194}
$$

where $\sum_2$ is an $(N-2) \times (N-2)$ matrix. Then, with the columns containing $(N-2)$ elements

$$
V_2 = \begin{pmatrix}
b_{32} \\
b_{42} \\
\bullet \\
\bullet \\
b_{N2}
\end{pmatrix} \tag{5.195a}
$$

and

$$
u_2 = \begin{pmatrix}
1 \\
0 \\
\bullet \\
\bullet \\
0
\end{pmatrix} \tag{5.195b}
$$

we have

$$
\omega_2 = V_2 - n_2 u_2 \tag{5.195c}
$$

where

$$
n_2^2 = \sum_{m=3}^{N} b_{2m}^2 \tag{5.195d}
$$

Then, with

$$
\Sigma_2 = 1 - 2\frac{\omega_2 \omega_2^T}{\omega_2^T \omega_2} \tag{5.196}
$$

we obtain

$$S_2^T A_1 S_2 = \begin{pmatrix} \alpha_1 & \beta_1 & 0 & \bullet\ \bullet & 0 \\ \beta_1 & \alpha_2 & \beta_2 & \bullet\ \bullet & 0 \\ 0 & \beta_2 & b_{33} & \bullet\ \bullet & b_{3N} \\ \bullet & \bullet & & & \bullet \\ \bullet & \bullet & & & \bullet \\ 0 & 0 & b_{3N} & \bullet\ \bullet & b_{NN} \end{pmatrix} \equiv A_2 \qquad (5.197)$$

We note that the quantities b_{pq} of A_2 do not have the same values as the b_{pq} of A_1. We are simply using the same notation for these elements as place holders for quantities that change value under the transformation.

It is straightforward to generalize this process. The kth orthogonal matrix is given by

$$S_k = \begin{pmatrix} 1 & 0 & 0 & 0 & 0 & \bullet\ \bullet & 0 \\ 0 & 1 & 0 & 0 & 0 & \bullet\ \bullet & 0 \\ & & \bullet & & & & \\ & & \bullet & & & & \\ 0 & \bullet\ \bullet & 0 & 1 & 0 & \bullet\ \bullet & 0 \\ 0 & & & 0 & & & \\ \bullet & & & \bullet & & \Sigma_k & \\ \bullet & & & \bullet & & & \\ 0 & & & 0 & & & \end{pmatrix} \begin{matrix} \\ \\ \\ \\ \leftarrow row_k \\ \\ \\ \\ \\ \end{matrix} \qquad (5.198)$$

where Σ_k is an $(N-k) \times (N\ k)$ matrix. The column vectors with $(N-k)$ elements are

$$V_k = \begin{pmatrix} b_{(k+1)k} \\ b_{(k+2)k} \\ \bullet \\ \bullet \\ b_{Nk} \end{pmatrix} \qquad (5.199a)$$

$$u_k = \begin{pmatrix} 1 \\ 0 \\ \bullet \\ \bullet \\ 0 \end{pmatrix} \qquad (5.199b)$$

and

$$\omega_k = V_k - n_k u_k \qquad (5.199c)$$

where

$$n_k^2 = \sum_{m=k+1}^{N} b_{km}^2 \qquad (5.199d)$$

Then, with

$$\Sigma_k = 1 - 2\frac{\omega_k \omega_k^T}{\omega_k^T \omega_k} \tag{5.200}$$

we have

$$S_k^T A_{k-1} S_k = \begin{pmatrix} \alpha_1 & \beta_1 & 0 & 0 & 0 & \bullet\bullet & 0 \\ \beta_1 & \alpha_2 & \beta_2 & 0 & 0 & \bullet\bullet & 0 \\ \bullet & & \bullet & & & & \bullet \\ \bullet & & & \bullet & & & \bullet \\ & & & & \bullet & & \\ 0 & \bullet\bullet & \beta_{k-1} & \alpha_k & \beta_k & \bullet\bullet & 0 \\ 0 & \bullet\bullet & 0 & \beta_k & b_{(k+1)(k+1)} & \bullet\bullet & b_{(k+1)N} \\ \bullet & & \bullet & \bullet & \bullet & & \bullet \\ 0 & \bullet\bullet & 0 & 0 & b_{N(k+1)} & \bullet\bullet & b_{NN} \end{pmatrix} \equiv A_k \tag{5.201}$$

Example 5.12: Givens and Householder methods for approximating the eigenpairs of a real symmetric matrix

It is straightforward to show that the characteristic equation for the eigenvalues of

$$A = \begin{pmatrix} 2 & 1 & 4 & -1 \\ 1 & 1 & 3 & 2 \\ 4 & 3 & -4 & 1 \\ -1 & 2 & 1 & 3 \end{pmatrix} \tag{5.202}$$

is

$$\lambda^4 - 2\lambda^3 - 45\lambda^2 + 114\lambda + 136 = 0 \tag{5.203}$$

Using one of the techniques for solving such an equation introduced in ch. 2, we find that the eigenvalues are given by

$$\{\lambda_1, \lambda_2, \lambda_3, \lambda_4\} = \{-6.74074, -0.89500, 4.00000, 5.63573\} \tag{5.204}$$

The corresponding eigenvectors

$$X_k \equiv \begin{pmatrix} x_{k,1} \\ x_{k,2} \\ x_{k,3} \\ x_{k,4} \end{pmatrix} \tag{5.205}$$

can be found from the simultaneous equations

$$\begin{pmatrix} 2-\lambda_k & 1 & 4 & -1 \\ 1 & 1-\lambda_k & 3 & 2 \\ 4 & 3 & -4-\lambda_k & 1 \\ -1 & 2 & 1 & 3-\lambda_k \end{pmatrix} \begin{pmatrix} x_{k,1} \\ x_{k,2} \\ x_{k,3} \\ x_{k,4} \end{pmatrix} = 0 \qquad (5.206)$$

which can be expressed in the form

$$(2-\lambda_k)x_{k,1} + x_{k,2} + 4x_{k,3} = x_{k,4} \qquad (5.207a)$$

$$x_{k,1} + (1-\lambda_k)x_{k,2} + 3x_{k,3} = -2x_{k,4} \qquad (5.207b)$$

$$4x_{k,1} + 3x_{k,2} - (4+\lambda_k)x_{k,3} = -x_{k,4} \qquad (5.207c)$$

and

$$-x_{k,1} + 2x_{k,2} + x_{k,3} = (3-\lambda_k)x_{k,4} \qquad (5.207d)$$

Since there are only three independent equations in this set, we solve, for example, eqs. 5.207a, 5.207b, and 5.207c for $x_{k,1}$, $x_{k,2}$, and $x_{k,3}$ in terms of $x_{k,4}$. This set of equations can be expressed as

$$\begin{pmatrix} (2-\lambda_k) & 1 & 4 \\ 1 & (1-\lambda_k) & 3 \\ 4 & 3 & -(4+\lambda_k) \end{pmatrix} \begin{pmatrix} x_{k,1} \\ x_{k,2} \\ x_{k,3} \end{pmatrix} = x_{k,4} \begin{pmatrix} 1 \\ -2 \\ -1 \end{pmatrix} \qquad (5.208)$$

Normalizing the eigenvectors to 1 to determine $x_{k,4}$, we obtain

$$\{X_1, X_2, X_3, X_4\} =$$

$$\left\{ \begin{pmatrix} 0.38012 \\ 0.27324 \\ -0.88061 \\ 0.07333 \end{pmatrix}, \begin{pmatrix} 0.43553 \\ -0.75002 \\ -0.00327 \\ 0.49777 \end{pmatrix}, \begin{pmatrix} -0.60999 \\ 0.15250 \\ -0.15250 \\ 0.76249 \end{pmatrix}, \begin{pmatrix} 0.54197 \\ 0.58272 \\ 0.44862 \\ 0.40675 \end{pmatrix} \right\} \qquad (5.209)$$

To determine the characteristic equation, we first cast the matrix A of eq. 5.202 into tridiagonal form, then use the recurrence relations of eq. 5.171d with the constraints given in eqs. 5.172 to find the coefficients of the various powers of λ.

(a) To tridiagonalize this matrix by the Givens' method, we systematically transform the (1,3) and (3,1) elements, then the (1,4) and (4,1) elements then the (2,4) and (4,2) elements to zero.

   The (1,3) and (3,1) elements are transformed to zero by a similarity transformation via the orthogonal matrix

$$S_1 = \begin{pmatrix} 1 & 0 & 0 & 0 \\ 0 & \cos\phi_1 & \sin\phi_1 & 0 \\ 0 & -\sin\phi_1 & \cos\phi_1 & 0 \\ 0 & 0 & 0 & 1 \end{pmatrix} \tag{5.210}$$

where ϕ_1 is given by

$$\tan\phi_1 = -\frac{a_{13}}{a_{12}} = -\frac{4}{1} \tag{5.211}$$

Then

$$A_1 = S_1^T A S_1 = \begin{pmatrix} 2.00000 & 4.12311 & 0.00000 & -1.00000 \\ 4.12311 & -2.29412 & -3.82353 & 1.45521 \\ 0.00000 & -3.82353 & -0.70588 & -1.69775 \\ -1.00000 & 1.45521 & -1.69775 & 3.00000 \end{pmatrix} \tag{5.212}$$

The (1,4) and (4,1) elements are transformed to zero by

$$S_2 = \begin{pmatrix} 1 & 0 & 0 & 0 \\ 0 & \cos\phi_2 & 0 & \sin\phi_2 \\ 0 & 0 & 1 & 0 \\ 0 & -\sin\phi_2 & 0 & \cos\phi_2 \end{pmatrix} \tag{5.213}$$

where, from eq. 5.212,

$$\tan\phi_2 = -\frac{b_{14}}{b_{12}} = \frac{1.00000}{4.12311} \tag{5.214}$$

Then

$$A_2 = S_2^T A_1 S_2 = \begin{pmatrix} 2.00000 & 4.24264 & 0.00000 & 0.00000 \\ 4.24264 & -2.66667 & -3.31564 & 0.08085 \\ 0.00000 & -3.31564 & -0.70588 & -2.55113 \\ 0.00000 & 0.08085 & -2.55113 & 3.37255 \end{pmatrix} \tag{5.215}$$

To transform the (2,4) and (4,2) elements to zero, we take

$$S_3 = \begin{pmatrix} 1 & 0 & 0 & 0 \\ 0 & 1 & 0 & 0 \\ 0 & 0 & \cos\phi_3 & \sin\phi_3 \\ 0 & 0 & -\sin\phi_3 & \cos\phi_3 \end{pmatrix} \tag{5.216}$$

where ϕ_3 is found from

$$\tan\phi_3 = -\frac{c_{24}}{c_{23}} = \frac{0.08085}{3.31564} \tag{5.217}$$

We obtain the tridiagonal form

$$A_3 = S_3^T A_2 S_3 = \begin{pmatrix} 2.00000 & 4.24264 & 0.00000 & 0.00000 \\ 4.24264 & -2.66667 & -3.31662 & 0.00000 \\ 0.00000 & -3.31662 & -0.57912 & -2.64748 \\ 0.00000 & 0.00000 & -2.64748 & 3.24577 \end{pmatrix} \tag{5.218}$$

(b) To tridiagonalize the matrix of eq. 5.202 by the Householder's method, we refer to eq. 5.179 and define

$$V_1 \equiv \begin{pmatrix} 1 \\ 4 \\ -1 \end{pmatrix} \tag{5.219a}$$

where, from eq. 5.183,

$$n_1 = \sqrt{V_1^T V_1} = 3\sqrt{2} \tag{5.219b}$$

Then, from eq. 5.188b

$$\omega_1 = \begin{pmatrix} 1 \\ 4 \\ -1 \end{pmatrix} - 3\sqrt{2}\begin{pmatrix} 1 \\ 0 \\ 0 \end{pmatrix} = \begin{pmatrix} 1 - 3\sqrt{2} \\ 4 \\ -1 \end{pmatrix} \tag{5.219c}$$

From this, eq. 5.192 becomes

$$\Sigma_1 = \begin{pmatrix} 1 & 0 & 0 \\ 0 & 1 & 0 \\ 0 & 0 & 1 \end{pmatrix} - \frac{2}{(36 - 6\sqrt{2})}\begin{pmatrix} 1 - 3\sqrt{2} \\ 4 \\ -1 \end{pmatrix}\begin{pmatrix} 1 - 3\sqrt{2} & 4 & -1 \end{pmatrix}$$

$$= \begin{pmatrix} 0.23570 & 0.94281 & -0.23570 \\ 0.94281 & -0.16301 & 0.29075 \\ -0.23570 & 0.29075 & 0.92731 \end{pmatrix} \tag{5.219d}$$

Therefore, with

$$S_1 = \begin{pmatrix} 1 & 0 & 0 & 0 \\ 0 & 0.23570 & 0.94281 & -0.23570 \\ 0 & 0.94281 & -0.16301 & 0.29075 \\ 0 & -0.23570 & 0.29075 & 0.92731 \end{pmatrix} = S_1^T \tag{5.220}$$

we obtain

$$A_1 = S_1 A S_1 = \begin{pmatrix} 2 & 4.24264 & 0 & 0 \\ 4.24264 & -2.66667 & 3.18796 & -0.91484 \\ 0 & 3.18796 & 1.11577 & 3.25873 \\ 0 & -0.91484 & 3.25873 & 1.55090 \end{pmatrix} \qquad (5.221)$$

From A_1 we have

$$V_2 = \begin{pmatrix} 3.18796 \\ -0.91484 \end{pmatrix} \qquad (5.222a)$$

from which

$$n_2 = \sqrt{3.18796^2 + 0.91484^2} = 3.31663 \qquad (5.222b)$$

Then

$$\omega_2 = \begin{pmatrix} 3.18796 \\ -0.91484 \end{pmatrix} - 3.31662 \begin{pmatrix} 1 \\ 0 \end{pmatrix} = \begin{pmatrix} -0.12867 \\ -0.91484 \end{pmatrix} \qquad (5.222c)$$

so that

$$\Sigma_2 = \begin{pmatrix} 0.96121 & -0.27583 \\ -0.27583 & -0.96121 \end{pmatrix} \qquad (5.222d)$$

Therefore

$$S_2 = \begin{pmatrix} 1 & 0 & 0 & 0 \\ 0 & 1 & 0 & 0 \\ 0 & 0 & 0.96121 & -0.27583 \\ 0 & 0 & -0.27583 & -0.96121 \end{pmatrix} \qquad (5.223)$$

and we obtain

$$A_2 = \begin{pmatrix} 2.00000 & 4.24264 & 0 & 0 \\ 4.24264 & -2.66667 & 3.31662 & 0 \\ 0 & 3.31662 & -0.57912 & -2.64748 \\ 0 & 0 & -2.64748 & 3.24579 \end{pmatrix} \qquad (5.224)$$

This is the same tridiagonal form of A obtained in eq. 5.218 by the Givens method.

It is straightforward to prove that, except for the signs of the off-diagonal elements, the reduction of a symmetric matrix to tridiagonal form is independent of the method used (see Fox, L., 1965, pp. 251–252).

With

$$\alpha_k = A_2(k, k) \quad k = 1, 2, 3, 4 \tag{5.225a}$$

and

$$\beta_k = A_2(k, k+1) \quad k = 1, 2, 3 \tag{5.225b}$$

eq. 5.171d, constrained by eqs. 5.172, becomes

$$q_{0,0} = 1 \tag{5.226a}$$

$$q_{1,0} = \alpha_1 q_{0,0} - q_{0,-1} - \beta_0^2 q_{-1,0} = \alpha_1 q_{0,0} = 2.00000 \tag{5.226b}$$

$$q_{1,1} = (-1)^1 = -1.00000 \tag{5.226c}$$

$$q_{2,0} = \alpha_2 q_{1,0} - q_{0,-1} - \beta_1^2 q_{0,0} = \alpha_2 q_{1,0} - \beta_1^2 q_{0,0} = -23.33333 \tag{5.226d}$$

$$q_{2,1} = \alpha_2 q_{1,1} - q_{1,0} - \beta_1^2 q_{0,1} = \alpha_2 q_{1,1} - q_{1,0} = 0.66667 \tag{5.226e}$$

$$q_{2,2} = (-1)^2 = 1.00000 \tag{5.226f}$$

$$q_{3,0} = \alpha_3 q_{2,0} - q_{2,-1} - \beta_2^2 q_{1,0} = \alpha_3 q_{2,0} - \beta_2^2 q_{1,0} - 8.48709 \tag{5.226g}$$

$$q_{3,1} = \alpha_3 q_{2,1} - q_{2,0} - \beta_2^2 q_{1,1} = 33.94725 \tag{5.226h}$$

$$q_{3,2} = \alpha_3 q_{2,2} - q_{2,1} - \beta_2^2 q_{1,2} = \alpha_3 q_{2,2} - q_{2,1} = -1.24579 \tag{5.226i}$$

$$q_{3,3} = (-1)^3 = -1.00000 \tag{5.226j}$$

$$q_{4,0} = \alpha_4 q_{3,0} - q_{3,-1} - \beta_3^2 q_{2,0} = \alpha_4 q_{3,0} - \beta_3^2 q_{2,0} = 136.00000 \tag{5.226k}$$

$$q_{4,1} = \alpha_4 q_{3,1} - q_{3,0} - \beta_3^2 q_{2,1} = 114.00000 \tag{5.226\ell}$$

$$q_{4,2} = \alpha_4 q_{3,2} - q_{3,1} - \beta_3^2 q_{2,2} = -45.00000 \tag{5.226m}$$

$$q_{4,3} = \alpha_4 q_{3,3} - q_{3,2} - \beta_3^2 q_{2,3} = \alpha_4 q_{3,3} - q_{3,2} = -2.00000 \tag{5.226n}$$

$$q_{4,4} = (-1)^4 = 1.00000 \tag{5.226p}$$

From these results, we find that the characteristic polynomial is

$$P_4(\lambda) = \lambda^4 + \sum_{m=0}^{3} q_{4,m}\lambda^m = \lambda^4 - 2\lambda^3 - 45\lambda^2 + 114\lambda + 136 \qquad (5.227)$$

which is identical to the polynomial given in eq. 5.202. □

QR method for nonsymmetric matrices

To find the eigenvalues of a real, nonsymmetric matrix A, a commonly used approach is the QR *method*, developed by Francis, J., 1961 and 1962. This method involves *factoring* A as the

$$A = QR \qquad (5.228)$$

where R is an *upper triangular* matrix

$$R = \begin{pmatrix} r_{11} & r_{12} & r_{13} & \bullet & \bullet & r_{1N} \\ & r_{22} & r_{23} & \bullet & \bullet & r_{2N} \\ & & r_{33} & \bullet & \bullet & r_{3N} \\ & \mathbf{0} & & & \bullet & \bullet\bullet \\ & & & & & r_{NN} \end{pmatrix} \qquad (5.229)$$

and Q is the product of orthogonal matrices S.

We choose orthogonal matrices $S_1, S_2, \ldots S_N$ such that

$$S_N^T \bullet \bullet \bullet S_2^T S_1^T A_0 = R \qquad (5.230)$$

The matrix Q is defined as

$$Q \equiv S_1 S_2 \bullet \bullet \bullet S_N \qquad (5.231)$$

so that eq. 5.230 becomes

$$Q^T A = R \qquad (5.232)$$

Since Q is an orthogonal matrix,

$$Q^T = Q^{-1} \qquad (5.233)$$

Therefore, eq. 5.232 can be written as

$$R = Q^{-1}A \qquad (5.234)$$

After determining Q and R, we define a new matrix A' by

$$A' = RQ \qquad (5.235a)$$

Substituting for R from eq. 5.232, this becomes

$$A' = Q^{-1}AQ \qquad (5.235b)$$

Since A and A' are related by this orthogonal similarity transformation, they have the same eigenvalues.

To find the eigenvalues of a real matrix A_0, we first determine Q_1 such that

$$Q_1^T A_0 = R_1 \qquad (5.236)$$

is an upper triangular matrix. We then define

$$A_1 = R_1 Q_1 \qquad (5.237)$$

This process is continued by factoring A_1 as

$$A_1 = Q_2 R_2 \qquad (5.238a)$$

so that R_2 is an upper triangular matrix found from

$$R_2 = Q_2^T A_1 \qquad (5.238b)$$

From Q_2 and R_2 we then define

$$A_2 = R_2 Q_2 \qquad (5.239)$$

This iterative process is continued until

$$A_k = R_k Q_k \qquad (5.240)$$

is an upper triangular matrix to some level of precision. That is, for large enough K

$$A_K \simeq \begin{pmatrix} a_{11}^{(K)} & a_{12}^{(K)} & a_{13}^{(K)} & \bullet \ \bullet & a_{1N}^{(K)} \\ & a_{22}^{(K)} & a_{23}^{(K)} & \bullet \ \bullet & a_{2N}^{(K)} \\ & & a_{33}^{(K)} & \bullet \ \bullet & a_{12}^{(K)} \\ & \mathbf{0} & & \bullet \ \bullet & \bullet \\ & & & & a_{NN}^{(K)} \end{pmatrix} \qquad (5.241)$$

As noted earlier, by expanding

$$|A_K - \lambda 1| = \begin{vmatrix} a_{11}^{(K)} - \lambda & a_{12}^{(K)} & a_{13}^{(K)} & \bullet\ \bullet & a_{1N}^{(K)} \\ & a_{22}^{(K)} - \lambda & a_{23}^{(K)} & \bullet\ \bullet & a_{2N}^{(K)} \\ & & a_{33}^{(K)} - \lambda & \bullet\ \bullet & a_{12}^{(K)} \\ & \mathbf{0} & & \ddots & \vdots \\ & & & & a_{NN}^{(K)} - \lambda \end{vmatrix} \qquad (5.242a)$$

about the elements of the first column of each cofactor, we obtain

$$|A_K - \lambda 1| = \left(a_{11}^{(K)} - \lambda \right)\left(a_{22}^{(K)} - \lambda \right) \bullet\ \bullet\ \bullet \left(a_{NN}^{(K)} - \lambda \right) \qquad (5.242b)$$

Thus, the diagonal elements of an upper (or lower) triangular matrix are its eigenvalues.

As shown in Wilkinsonm, J.H., 1965, pp. 216–219, when the eigenvalues of A_0 are real, this process converges to

$$A_K = \begin{pmatrix} \lambda_1 & a_{12}^{(K)} & a_{13}^{(K)} & & a_{1N}^{(K)} \\ & \lambda_2 & a_{23}^{(K)} & & a_{2N}^{(K)} \\ & & \lambda_3 & & a_{12}^{(K)} \\ & \mathbf{0} & & \ddots & \vdots \\ & & & & \lambda_N \end{pmatrix} \qquad (5.243)$$

with

$$|\lambda_1| > |\lambda_2| > \bullet\ \bullet\ \bullet > |\lambda_N| \qquad (5.244)$$

Jacobi/Givens QR method

One method for determining Q as defined in eq. 5.228 is to create rotation matrices S with elements $\cos\phi$ and $\sin\phi$ such as the matrices used in the Jacobi or the Givens method of tridiagonalization [see eq. 5.140 or 5.173]. This approach for determining Q will be referred to as the *Jacobi/Givens QR* approach. At each step in the process, the angle ϕ is chosen to transform one of the elements below the main diagonal to zero.

Householder QR method

We can also adapt the Householder approach for tridiagonalizing a symmetric matrix to a method of triangularizing a nonsymmetric $N \times N$ matrix A_0.

To do this, we define a vector V_1 containing the N elements in the first column of A_0 (instead of $N-1$ elements as we do when tridiagonalizing a matrix by the Householder method):

$$V_1 = \begin{pmatrix} a_{11} \\ a_{21} \\ \bullet \\ \bullet \\ a_{N1} \end{pmatrix} \tag{5.245}$$

Following the steps outlined in eq. 5.178 through eq. 5.192, we have

$$n_1^2 = \sum_{k=1}^{N} a_{k1}^2 \tag{5.246}$$

from which we define

$$\omega_1 = V_1 - n_1 u_1 = V_1 - n_1 \begin{pmatrix} 1 \\ 0 \\ \bullet \\ \bullet \\ 0 \end{pmatrix} \tag{5.247}$$

With this, we obtain the $N \times N$ orthogonal matrix

$$S_1^T \equiv 1 - 2\frac{\omega_1 \omega_1^T}{\omega_1^T \omega_1} \tag{5.248}$$

where 1 is the $N \times N$ identity matrix. Then

$$S_1^T A_0 = \begin{pmatrix} b_{11} & b_{12} & b_{13} & \bullet\ \bullet & b_{1N} \\ 0 & b_{22} & b_{23} & \bullet\ \bullet & b_{2N} \\ 0 & b_{32} & b_{33} & \bullet\ \bullet & b_{3N} \\ \bullet \\ \bullet \\ 0 & b_{N2} & b_{N3} & \bullet\ \bullet & b_{NN} \end{pmatrix} \equiv B \tag{5.249}$$

We then form a vector with $N-1$ elements

$$V_2 = \begin{pmatrix} b_{22} \\ b_{32} \\ \bullet \\ \bullet \\ b_{N2} \end{pmatrix} \tag{5.250}$$

from which we define the $N - 1$ component vector

$$\omega_2 \equiv V_2 - n_2 u_2 = V_2 - n_2 \begin{pmatrix} 1 \\ 0 \\ 0 \\ \bullet \\ \bullet \\ 0 \end{pmatrix} \qquad (5.251)$$

where

$$n_2^2 = \sum_{k=2}^{N} b_{k2}^2 \qquad (5.252)$$

Then, forming

$$S_2^T = \begin{pmatrix} 1 & 0 & \bullet\bullet & 0 \\ 0 & & & \\ \bullet & & \Sigma_2^T & \\ \bullet & & & \\ 0 & & & \end{pmatrix} \qquad (5.253)$$

with

$$\Sigma_2^T \equiv 1 - 2\frac{\omega_2\omega_2^T}{\omega_2^T\omega_2} \qquad (5.254)$$

such that

$$S_2^T B = \begin{pmatrix} b_{11} & b_{12} & c_{13} & \bullet\bullet & c_{1N} \\ 0 & c_{22} & c_{23} & \bullet\bullet & c_{2N} \\ 0 & 0 & c_{33} & \bullet\bullet & c_{3N} \\ \bullet & \bullet & & & \bullet \\ \bullet & \bullet & & & \bullet \\ 0 & 0 & c_{N3} & \bullet\bullet & c_{NN} \end{pmatrix} \equiv C \qquad (5.255)$$

This process is continued until the resulting matrix has been transformed into

$$Q_1^T A_0 = S_N^T \ldots S_1^T A_0 = R_1 \qquad (5.256)$$

where R_1 is an upper triangular matrix. Then

$$Q_1 = S_1 \ldots S_N \qquad (5.257)$$

and

$$A_1 = R_1 Q_1 \qquad (5.258)$$

This procedure is then applied to A_1 to determine A_2 and so on. After a number of iterations, one obtains A_k that is an upper triangular matrix to an acceptable level of precision.

Example 5.13: Eigenvalues using the QR method

It is easy to verify that the matrix

$$A = \begin{pmatrix} 3 & 0 & -1 \\ 1 & 5 & -2 \\ -2 & 0 & 2 \end{pmatrix} \tag{5.259}$$

has eigenvalues $(5, 4, 1)$.

(a) To determine these eigenvalues using the Jacobi/Givens QR method, we begin by zeroing out the $(2,1)$ element by premultiplying A by

$$S_{21}^T = \begin{pmatrix} \cos\phi_{21} & -\sin\phi_{21} & 0 \\ \sin\phi_{21} & \cos\phi_{21} & 0 \\ 0 & 0 & 1 \end{pmatrix} \tag{5.260}$$

with

$$\phi_{21} = -\tan^{-1}\left(\frac{a_{21}}{a_{11}}\right) = -\tan^{-1}\left(\frac{1}{3}\right) \tag{5.261}$$

This yields

$$S_{21}^T A \equiv B = \begin{pmatrix} 3.16228 & 1.58114 & -1.58114 \\ 0 & 4.74342 & -1.58114 \\ -2.00000 & 0 & 2.00000 \end{pmatrix} \tag{5.262}$$

The $(3,1)$ element of B is transformed to zero by multiplying B by

$$S_{31}^T = \begin{pmatrix} \cos\phi_{31} & 0 & -\sin\phi_{31} \\ 0 & 1 & 0 \\ \sin\phi_{31} & 0 & \cos\phi_{31} \end{pmatrix} \tag{5.263}$$

where

$$\phi_{31} = -\tan^{-1}\left(\frac{b_{31}}{b_{11}}\right) = -\tan^{-1}\left(\frac{-2.00000}{3.16228}\right) \tag{5.264}$$

We obtain

$$S_{31}^T B = S_{31}^T S_{21}^T A \equiv C = \begin{pmatrix} 3.74166 & 1.33631 & -2.40535 \\ 0 & 4.74342 & -1.58114 \\ 0 & 0.84515 & 0.84515 \end{pmatrix} \tag{5.265}$$

The (3,2) element is transformed to zero by multiplying C by

$$S_{32}^T = \begin{pmatrix} 1 & 0 & 0 \\ 0 & \cos\phi_{32} & -\sin\phi_{32} \\ 0 & \sin\phi_{32} & \cos\phi_{32} \end{pmatrix} \qquad (5.266)$$

with

$$\phi_{32} = -\tan^{-1}\left(\frac{c_{32}}{c_{22}}\right) = -\tan^{-1}\left(\frac{0.84515}{-1.58114}\right) \qquad (5.267)$$

From these computations, we find

$$R_1 = \begin{pmatrix} 3.74166 & 1.33631 & -2.40535 \\ 0 & 4.81812 & -1.40837 \\ 0 & 0 & 1.10940 \end{pmatrix} \qquad (5.268)$$

and

$$Q_1 = \left(S_{32}^T S_{31}^T S_{21}^T\right)^T = S_{12}S_{13}S_{23} = \begin{pmatrix} 0.80178 & -0.22237 & 0.55470 \\ 0.26726 & 0.96362 & 0 \\ -0.53452 & 0.14825 & 0.83205 \end{pmatrix}$$

$$(5.269)$$

From these results, we have

$$A_1 = R_1 Q_1 = \begin{pmatrix} 4.64286 & 0.09905 & 0.07412 \\ 2.04050 & 4.43407 & -1.17184 \\ -0.59300 & 0.16447 & 0.92308 \end{pmatrix} \qquad (5.270)$$

We repeat this process until $A_K = R_K Q_K$ is an upper triangular matrix. We find that the QR process converges quite slowly for the matrix of eq. 5.259. After 60 iterations, we obtain

$$A_{60} = \begin{pmatrix} 5.00010 & -2.12126 & -0.70709 \\ 4.58382 \times 10^{-5} & 3.99990 & -1.00001 \\ -1.59750 \times 10^{-5} & 3.86748 \times 10^{-5} & 0.99999 \end{pmatrix} \qquad (5.271)$$

If we accept these (2,1), (3,1), and (3,2) elements as zero, this can be taken to be in lower triangular form, and the eigenvalues, in descending magnitude order, are (5.00010, 3.99990, 0.99999).

(b) Using the Householder QR approach, we define

$$V_1 = \begin{pmatrix} 3 \\ 1 \\ -2 \end{pmatrix} \qquad (5.272)$$

With

$$n_1 = \sqrt{3^2 + 1^2 + 2^2} = \sqrt{14} \qquad (5.273)$$

we construct

$$\omega_1 = V_1 - n_1 u_1 = \begin{pmatrix} 3 \\ 1 \\ -2 \end{pmatrix} - \sqrt{14} \begin{pmatrix} 1 \\ 0 \\ 0 \end{pmatrix} = \begin{pmatrix} 3 - \sqrt{14} \\ 1 \\ -2 \end{pmatrix} \qquad (5.274)$$

from which

$$S_1^T = 1 - 2\frac{\omega_1 \omega_1^T}{\omega_1^T \omega_1} =$$

$$\begin{pmatrix} 1 & 0 & 0 \\ 0 & 1 & 0 \\ 0 & 0 & 1 \end{pmatrix} - \frac{2}{\left(3 - \sqrt{14}\right)^2 + 1^2 + 2^2} \begin{pmatrix} 3 - \sqrt{14} \\ 1 \\ -2 \end{pmatrix} \begin{pmatrix} 3 - \sqrt{14} & 1 & -2 \end{pmatrix}$$

$$= \begin{pmatrix} 0.80178 & 0.26726 & -0.53452 \\ 0.26726 & 0.63964 & 0.72071 \\ -0.53452 & 0.72013 & -0.44143 \end{pmatrix} \qquad (5.275)$$

Thus,

$$S_1^T A = B = \begin{pmatrix} 3.74166 & 1.33631 & -2.40535 \\ 0 & 3.19822 & -0.10512 \\ 0 & 3.60357 & -1.78976 \end{pmatrix} \qquad (5.276)$$

From the second column of B we form

$$V_2 = \begin{pmatrix} 3.19822 \\ 3.60357 \end{pmatrix} \qquad (5.277)$$

so that

$$n_2 = \sqrt{3.19822^2 + 3.60357^2} = 4.81812 \qquad (5.278)$$

This yields

$$\omega_2 = V_2 - n_2 \begin{pmatrix} 1 \\ 0 \end{pmatrix} = \begin{pmatrix} -1.61991 \\ 3.60357 \end{pmatrix} \qquad (5.279)$$

From this,

$$S_2^T = \begin{pmatrix} 1 & 0 & 0 \\ 0 & \Sigma_2 & \\ 0 & & \end{pmatrix} \qquad (5.280)$$

where

$$\Sigma_2 = 1 - 2\frac{\omega_2 \omega_2^T}{\omega_2^T \omega_2} = \begin{pmatrix} 0.66379 & 0.74792 \\ 0.74792 & -0.66379 \end{pmatrix} \qquad (5.281)$$

Thus,

$$S_2^T = \begin{pmatrix} 1 & 0 & 0 \\ 0 & 0.66379 & 0.74792 \\ 0 & 0.74792 & -0.66379 \end{pmatrix} \qquad (5.282)$$

and

$$A_1 = S_2^T B = S_2^T S_1^T A_0$$
$$= \begin{pmatrix} 4.64286 & 9.90536 \times 10^{-2} & 7.41249 \times 10^{-2} \\ 2.04050 & 4.43407 & -1.17184 \\ -0.59300 & 0.16447 & 0.92308 \end{pmatrix} \qquad (5.283)$$

After 60 iterations of this process, we find

$$A_{60} = \begin{pmatrix} 5.00000 & -2.12132 & -0.70711 \\ 8.25632 \times 10^{-7} & 4.00000 & -1.00000 \\ 0 & 0 & 1.00000 \end{pmatrix} \qquad (5.284)$$

Comparing this matrix to A_{60} found by the Jacobi/Givens QR technique [see eq. 5.271], we see that the Householder QR method yields somewhat more accurate results for a given number of iterations. $\square$

Problems

1. Evaluate

$$\text{(a) } A = \begin{vmatrix} 2 & 1 & 0 & 1 & 2 \\ 2 & 2 & 1 & 1 & 0 \\ 2 & 0 & 1 & 1 & 2 \\ 1 & 2 & 1 & 0 & 2 \\ 0 & 1 & 2 & 1 & 2 \end{vmatrix} \qquad \text{(b) } A = \begin{vmatrix} 2 & 4 & 1 & 2 & 1 \\ 1 & 2 & 0 & 1 & 1 \\ 0 & 2 & 3 & 1 & 3 \\ 1 & 1 & 3 & 1 & 1 \\ 3 & 1 & 2 & 2 & 1 \end{vmatrix}$$

   by the method of pivotal condensation.

2. Manipulate the determinant of Problem 1(a) into upper triangular form and the determinant of Problem 1(b) into lower triangular form.

3. Use Cramer's rule to find the solutions to

$$x - y - 2z + 3t = 3$$
$$2x + y + 3z + t = -2$$

$$-x + y + z - t = 6$$
$$3x + y - z + t = 0$$

4. Find the cofactor matrix of $A = \begin{pmatrix} 3 & 5 & 1 \\ 2 & 0 & 3 \\ 7 & 1 & 4 \end{pmatrix}$ and use this cofactor matrix to

   determine A^{-1}. Confirm that the matrix you determined to be A^{-1} is the inverse of A.

5. (a) Use the Gauss–Jordan method to determine the inverse of $A = \begin{pmatrix} 3 & 5 & 1 \\ 2 & 0 & 3 \\ 7 & 1 & 4 \end{pmatrix}$
       Compare your result to that found in Problem 4.
   (b) Use this inverse matrix to find the solution to the set of equations

$$3x + 5y + z = 0$$
$$2x + 3z = -9$$
$$7x + y + 4z = -23$$

6. Solve the set of equations given in Problem 5(b) using the Choleski-Turing method by

   (a) upper triangularizing the coefficient matrix
   (b) lower triangularizing the coefficient matrix

7. Find the three eigenpairs of $A = \begin{pmatrix} 4 & 3 & -2 \\ 3 & 2 & 1 \\ -2 & 1 & 1 \end{pmatrix}$ by the power method.

8. The matrix $A = \begin{pmatrix} 6 & 4 & 3 \\ 4 & 6 & 3 \\ 3 & 3 & 7 \end{pmatrix}$ has eigenvalues 13, 4, and 2.

   (a) Find the eigenvectors normalized to 1.
   (b) Use three sets of iterations of the Jacobi method as in example 5.11 (9 transformations in total) to diagonalize A. Estimate the eigenvalues and the orthonormal eigenvectors of A by the Jacobi method.

9. Cast the matrix $A = \begin{pmatrix} 2 & -1 & 1 & 4 \\ -1 & 3 & 1 & 2 \\ 1 & 1 & 5 & -3 \\ 4 & 2 & -3 & 6 \end{pmatrix}$ into tridiagonal form by

   (a) the Givens method
   (b) the Householder method
      Use the method of example 5.12 to find the characteristic equation for the eigenvalues of this matrix.

10. Find an approximation to the real eigenvalues, accurate to three decimal places, of $A = \begin{pmatrix} 12 & 3 & 1 \\ -9 & -2 & -3 \\ 14 & 6 & 2 \end{pmatrix}$ by transforming A into an upper triangular form by

   (a) the Jacobi/Givens QR method
   (b) the Householder QR method

11. If a matrix that has real eigenvalues is transformed by an orthogonal transformation into lower triangular form, its eigenvalues are given by its diagonal elements. Develop the

   (a) Jacobi/Givens QR method
   (b) Householder QR method for transforming a matrix into lower triangular form.

Chapter 6
ORDINARY FIRST ORDER DIFFERENTIAL EQUATIONS

The general form of an ordinary first order differential equation is

$$F\left(x, y, \frac{dy}{dx}\right) = F(x, y, y') = 0 \tag{6.1}$$

Differential equations that can be written in the form

$$\frac{dy}{dx} = f(x, y) \tag{6.2}$$

have solution given by

$$y(x) = \int f[x, y(x)]dx + C \tag{6.3}$$

The integral is called the *particular solution*. This particular solution plus the constant of integration is the *complete solution* to the differential equation.

To determine the value of the constant of integration C, a value y_0 at a specified x_0 must be given in the form

$$y(x_0) = y_0 \tag{6.4}$$

This constraint is given as part of the problem. If a condition like that given in eq. 6.4 is not given, C is arbitrary.

If eq. 6.2 is of the form

$$\frac{dy}{dx} + P(x)y = f(x) \tag{6.5}$$

H. Cohen, *Numerical Approximation Methods*, DOI 10.1007/978-1-4419-9837-8_6,
© Springer Science+Business Media, LLC 2011

the solution in a closed form can be achieved by multiplying this differential equation by

$$\mu(x) = e^{\int P(x)dx} \qquad (6.6)$$

which is called an *integrating factor*. Then

$$e^{\int P(x)dx}\frac{dy}{dx} + e^{\int P(x)dx}P(x)y = e^{\int P(x)dx}f(x) \qquad (6.7a)$$

can be written as

$$\frac{d\left(ye^{\int P(x)dx}\right)}{dx} = e^{\int P(x)dx}f(x) \qquad (6.7b)$$

the solution to which is

$$y(x) = e^{-\int P(x)dx}\int e^{\int P(x)dx}f(x)dx + C \qquad (6.8)$$

with the constant of integration given by

$$C = y(x_0) - \left[e^{-\int P(x)dx}\int e^{\int P(x)dx}f(x)dx\right]_{x=x_0} \qquad (6.9)$$

Therefore, in principle, the solution to eq. 6.5 can be obtained in closed form. In practice, it might be necessary to approximate

$$\int P(x)dx \quad \text{and/or} \quad \int f(x)e^{\int P(x)dx}dx$$

This could be achieved, for example, by expanding the integrands in Taylor series about x_0, truncated at some power of $(x - x_0)$ that will yield acceptable accuracy.

6.1   Taylor Series Approximation

Single step Taylor series method

Since $y(x_0)$ is specified, we know that $y(x)$ is analytic at x_0. Therefore, it can be expanded in a Taylor series about this point. The Taylor series method involves approximating $y(x)$ at some value of x by a truncated Taylor series in the form

$$y(x) \simeq$$
$$y(x_0) + y'(x_0)(x - x_0) + \frac{y''(x_0)}{2!}(x - x_0)^2 + .. + \frac{y^{(N)}(x_0)}{N!}(x - x_0)^N \qquad (6.10)$$

where $y^{(k)}(x_0)$ represents the kth derivative of y evaluated at x_0.

With $y(x_0)$ given as in eq. 6.4, $y'(x_0)$ is obtained from the differential equation and the higher derivatives are found from derivatives of the differential equation. For example, setting $x = x_0$, we have

$$y'(x_0) = f(x_0, y_0) \qquad (6.11a)$$

$$y''(x_0) = \left[\frac{\partial f}{\partial x} + \frac{\partial f}{\partial y} \frac{dy}{dx} \right]_{\substack{x = x_0 \\ y = y_0}} = \left[\frac{\partial f}{\partial x} + \frac{\partial f}{\partial y} f(x, y) \right]_{\substack{x = x_0 \\ y = y_0}} \qquad (6.11b)$$

and

$$y'''(x_0) = \left\{ \frac{\partial}{\partial x} \left[\frac{\partial f}{\partial x} + \frac{\partial f}{\partial y} f(x, y) \right] \right\}_{\substack{x = x_0 \\ y = y_0}} + \left\{ f(x, y) \frac{\partial}{\partial y} \left[\frac{\partial f}{\partial x} + \frac{\partial f}{\partial y} y'(x) \right] \right\}_{\substack{x = x_0 \\ y = y_0}} \quad (6.11c)$$

As can be seen, finding these higher derivatives can become cumbersome very quickly.

Example 6.1: Single step Taylor series method

The solution to

$$\frac{dy}{dx} = x^2 y^2 \qquad |x| < 1 \qquad (6.12a)$$

subject to

$$y(0) = 3 \qquad (6.12b)$$

can be obtained straightforwardly. The result is

$$y(x) = \frac{3}{(1 - x^3)} \qquad (6.13)$$

Therefore, the exact solution at $x = 0.3$ is

$$y(0.3) = \frac{3}{0.97300} \simeq 3.08325 \qquad (6.14)$$

Approximating $y(x)$ by the first four terms of a Taylor series expansion about $x = 0$, we have

$$y(x) \simeq y(0) + y'(0)x + \frac{y''(0)}{2!}x^2 + \frac{y'''(0)}{3!}x^3 \qquad (6.15)$$

With

$$y(0) = 3 \tag{6.12b}$$

$$y'(0) = \left[x^2 y^2\right]_{\substack{x=0 \\ y=3}} = 0 \tag{6.16a}$$

$$y''(0) = 2\left[xy^2 + x^2 yy'\right]_{\substack{x=0 \\ y=3}} = 2\left[xy^2 + x^4 y^3\right]_{\substack{x=0 \\ y=3}} = 0 \tag{6.16b}$$

and

$$\begin{aligned} y'''(0) &= 2\left[y^2 + 4xyy' + x^2 y'^2 + x^2 yy''\right]_{\substack{x=0 \\ y=3}} \\ &= 2\left[y^2 + 6x^3 y^3 + 3x^6 y^4\right]_{\substack{x=0 \\ y=3}} = 18 \end{aligned} \tag{6.16c}$$

The Taylor series, up to the x^3 term is

$$y(x) \simeq 3 + \frac{18}{3!}x^3 = 3\left(1 + x^3\right) \tag{6.17a}$$

These are the first two terms in the MacLaurin series for the exact solution given in eq. 6.13. From this we obtain

$$y(0.3) \simeq 3.08100 \tag{6.17b}$$

which is accurate to approximately two decimal places. □

Multiple step Taylor series method

If the point x_N at which y is to be evaluated (0.3 in the above example) is not very close to the initial point x_0, a higher degree of accuracy can be obtained by taking more terms in the Taylor series. But as noted above, computing higher derivatives can be a cumbersome task.

To avoid having to compute these higher derivatives, we can improve the accuracy of the results by using a multistep process with a small number of terms in the Taylor series.

The first expansion about x_0 is used to determine a value of y at a point x_1 close to x_0. For example, with a Taylor series of three terms, we take

$$y(x_1) \simeq y(x_0) + y'(x_0)(x_1 - x_0) + \frac{y''(x_0)}{2!}(x_1 - x_0)^2 \tag{6.18a}$$

We then expand $y(x)$ about x_1 to obtain an accurate approximation to $y(x_2)$ as

$$y(x_2) \simeq y(x_1) + y'(x_1)(x_2 - x_1) + \frac{y''(x_1)}{2!}(x_2 - x_1)^2 \qquad (6.18b)$$

The values of $y'(x_1)$ and $y''(x_1)$ are obtained from the differential equation and its derivative. This process is repeated until we obtain an approximation to $y(x_N)$.

Example 6.2: Multiple step Taylor series method

It was shown in example 6.1 that for

$$\frac{dy}{dx} = x^2 y^2 \quad |x| < 1 \qquad (6.12a)$$

with

$$y(0) = 3 \qquad (6.12b)$$

first four terms in the Taylor expansion about $x = 0$ yields

$$y(x) = 3(1 + x^3) \qquad (6.17a)$$

To solve eq. 6.12a at $x = 0.3$ using a two step Taylor expansion, we first expand about $x = 0$ to obtain $y(0.15)$. In example 6.1, it was shown that expansion about $x = 0$ yields eq. 6.17a. Thus, the first iteration yields

$$y(0.15) \simeq 3[1 + (.15^3)] \simeq 3.01013 \qquad (6.19)$$

This result is then used to obtain $y(0.30)$. From the differential equation, we have

$$y'(0.15) = [x^2 y^2]_{\substack{x = 0.15 \\ y = 3.01013}} = 0.20387 \qquad (6.20a)$$

Derivatives of the differential equation yields

$$y''(0.15) = 2[xy^2 + x^4 y^3]_{\substack{x = 0.15 \\ y = 3.01013}} \simeq 2.74587 \qquad (6.20b)$$

and

$$y'''(0.15) = 2\left[y^2 + 6x^3y^3 + 3x^6y^4\right]_{\substack{x = 0.15 \\ y = 3.01013}} \simeq 19.23193 \qquad (6.20c)$$

Then

$$y(0.30) \simeq$$
$$y(0.15) + y'(0.15) * (.15) + \frac{y''(0.1)}{2!} * (.15)^2 + \frac{y'''(0.1)}{3!} * (.15)^3$$
$$\simeq 3.08241 \qquad (6.21)$$

We note that by using the multiple step method, we obtain the solution to the differential equation at more than one point. In this example, we have determined values of $y(x)$ at three values of x;

$$y(x) = \begin{pmatrix} y(0.00) \\ y(0.15) \\ y(0.30) \end{pmatrix} = \begin{pmatrix} 3.00000 \\ 3.01013 \\ 3.08241 \end{pmatrix} \qquad (6.22a)$$

which is a fairly good approximation to

$$y_{exact}(x) = \begin{pmatrix} y_{exact}(0.00) \\ y_{exact}(0.15) \\ y_{exact}(0.30) \end{pmatrix} = \begin{pmatrix} 3.00000 \\ 3.01016 \\ 3.08325 \end{pmatrix} \qquad (6.22b)$$

Because we have values of y at more than one value of x, we can construct an interpolation representation of $y(x)$. With a polynomial interpolation, we have

$$y(x) \simeq \frac{(x - .15)(x - .3)}{(0 - .15)(0 - .3)}y(0) + \frac{(x - 0)(x - .3)}{(.15 - 0)(.15 - .3)}y(.15)$$
$$+ \frac{(x - 0)(x - .15)}{(.3 - 0)(.3 - .15)}y(.30) \qquad (6.23)$$

With the values given in eq. 6.22a, we obtain

$$y(0.25) \simeq 3.05141 \qquad (6.24)$$

which is a reasonable estimate of the exact value of 3.04762.

The results can be improved by a multi-step Taylor series of more than two steps. Doing so results in a larger number of y-values from which a higher order interpolation polynomial can be constructed. □

6.2 Picard Method

Single step Picard method

The solution to the first order ordinary differential equation can be written

$$y(x) = y(x_0) + \int_{x_0}^{x} f[x', y(x')]dx' \qquad (6.25)$$

Since the x dependence of the integrand arising from $y(x)$ is not known, the integral cannot, in general, be evaluated exactly. The *Picard method* involves approximating $y(x')$ in $f[x', y(x')]$ by a constant y_c. Then, at x_N, we have

$$y(x_N) \simeq y_0 + \int_{x_0}^{x_N} f(x', y_c)dx' \qquad (6.26)$$

This resulting integral involves integration of the explicit x dependence of $f(x,y)$ only. If the integral cannot be evaluated exactly, one of the numerical techniques presented in ch. 4 can be used to approximate it.

Since y_0 is the only value of y that is known, the most reasonable choice for y_c is y_0. Then, to determine y at x_N, eq. 6.25 becomes

$$y(x_N) \simeq y_0 + \int_{x_0}^{x_N} f(x', y_0)dx' \qquad (6.27)$$

Example 6.3: Single step Picard method

We again consider

$$\frac{dy}{dx} = x^2 y^2 \quad |x| < 1 \qquad (6.12a)$$

subject to

$$y(0) = 3 \qquad (6.12b)$$

With $x_0 = 0$ and $x_N = 0.3$, we have

$$y(0.30) \simeq y(0) + \int_{0}^{0.3} x^2 3^2 dx = 3 + 9\frac{(0.3)^3}{3} = 3.08100 \qquad (6.28)$$

which is not a very accurate approximation to the exact value of 3.08325. $\square$

Multiple step Picard method

The accuracy of the results obtained by the Picard method can be improved by using multiple steps of the method. We divide $[x_0, x_N]$ into several smaller intervals defined by $\{x_0, x_1,\ldots, x_{N-1}, x_N\}$. It is not necessary for these points to be equally spaced, but most often

$$h = x_{k+1} - x_k \qquad (1.35)$$

is taken to be the same value for all k.

For each interval $[x_k, x_{k+1}]$ we take y_c in $f(x,y_c)$ to be y_k and determine y_{k+1} from

$$y_{k+1} \simeq y_k + \int_{x_k}^{x_{k+1}} f(x, y_k)dx \qquad (6.29)$$

We begin by taking $k = 0$ in eq. 6.29 and obtain y_1 as

$$y_1 \simeq y_0 + \int_{x_0}^{x_1} f(x, y_0)dx \qquad (6.30a)$$

With y_0 given in eq. 6.4, we set $k = 1$ in eq. 6.29 to determine

$$y_2 = y_1 + \int_{x_1}^{x_2} f(x, y_1)dx \qquad (6.30b)$$

and so on. This process is continued until we obtain y_N from

$$y_N = y_{N-1} + \int_{x_{N-1}}^{x_N} f(x, y_{N-1})dx \qquad (6.30c)$$

Since the Picard method produces values of $y(x)$ at more than one value of x, then as we noted in the discussion of the multiple step Taylor series method, we can use interpolation techniques to obtain values of $y(x)$ at points not in the set $\{x_0, x_1, \ldots, x_N\}$.

Example 6.4: Multiple step Picard method

We again find an approximate solution to

$$\frac{dy}{dx} = x^2 y^2 \qquad |x| < 1 \qquad (6.12a)$$

subject to

$$y(0) = 3 \qquad (6.12b)$$

With $x_0 = 0$ and $x_N = 0.3$, we take $h = 0.15$. Then

$$y(0.15) = y(0) + \int_0^{0.15} x^2 3^2 dx = 3 + 9\frac{(0.15)^3}{3} = 3.01013 \qquad (6.31a)$$

which compares well with the exact value of 3.01016. From this

$$y(0.3) = y(0.15) + \int_{0.15}^{0.30} x^2 (3.01013)^2 dx =$$

$$3.01013 + (3.01013)^2 \frac{\left[(0.3)^3 - (0.15)^3\right]}{3} = 3.08148 \qquad (6.31b)$$

Thus, the solution we obtain is

$$y(x) = \begin{pmatrix} y(0.00) \\ y(0.15) \\ y(0.30) \end{pmatrix} \simeq \begin{pmatrix} 3.00000 \\ 3.01013 \\ 3.08148 \end{pmatrix} \qquad (6.32)$$

which is a reasonable approximation to

$$y_{exact}(x) = \begin{pmatrix} y_{exact}(0.00) \\ y_{exact}(0.15) \\ y_{exact}(0.30) \end{pmatrix} = \begin{pmatrix} 3.00000 \\ 3.01016 \\ 3.08325 \end{pmatrix} \qquad (6.22b)$$

As noted above, the values given in eq. 6.32 can be used to construct an interpolation representation of $y(x)$ to obtain an approximate solution at any value of x. $\square$

6.3 Runge–Kutta Method

Single step Runge–Kutta method

To solve

$$\frac{dy}{dx} = f(x, y) \qquad (6.2)$$

subject to

$$y(x_0) = y_0 \qquad (6.4)$$

at x_N, by the single step *Runge–Kutta method*, we compute the values of the four Runge–Kutta parameters

$$R_1 \equiv f(x_0, y_0)h \tag{6.33a}$$

$$R_2 \equiv f\left(x_0 + \frac{1}{2}h, y_0 + \frac{1}{2}R_1\right)h \tag{6.33b}$$

$$R_3 \equiv f\left(x_0 + \frac{1}{2}h, y_0 + \frac{1}{2}R_2\right)h \tag{6.33c}$$

and

$$R_4 \equiv f(x_0 + h, y_0 + R_3)h \tag{6.33d}$$

where

$$h \equiv x_N - x_0 \tag{6.34}$$

$y(x_N)$ is then given by the Runge–Kutta rule

$$y(x_N) = y_0 + \frac{1}{6}[R_1 + 2R_2 + 2R_3 + R_4] \tag{6.35}$$

Example 6.5: Single step Runge–Kutta method

As in previous examples, we consider the differential equation

$$\frac{dy}{dx} = x^2 y^2 \quad |x| < 1 \tag{6.12a}$$

with

$$y(0) = 3 \tag{6.12b}$$

To determine $y(0.3)$ using a single step Runge–Kutta method, we take $h = 0.3$, and define

$$R_1 = [x_0 y_0]^2 h = 0 \tag{6.36a}$$

$$R_2 = \left[\left(x_0 + \frac{1}{2}h\right)\left(y_0 + \frac{1}{2}R_1\right)\right]^2 h = [(0.15)(3)]^2 0.3 = 0.06075 \tag{6.36b}$$

$$R_3 = \left[\left(x_0 + \frac{1}{2}h\right)\left(y_0 + \frac{1}{2}R_2\right)\right]^2 h = [(0.15)(3.06075)]^2 0.3 = 0.06324 \tag{6.36c}$$

and

$$R_4 = [(x_0 + h)(y_0 + R_3)]^2 h = [(0.3)(3.06324)]^2 0.3 = 0.25335 \qquad (6.36d)$$

Then, referring to eq. 6.34,

$$y(0.3) \simeq y_0 + \frac{1}{6}[R_1 + 2R_2 + 2R_3 + R_4] = 3.08356 \qquad (6.37)$$

which is reasonably close to the exact value of 3.08325. □

Multiple step Runge-Kutta method

As with the Taylor and Picard methods, the multiple step Runge–Kutta method involves dividing the interval $[x_0, x_N]$ into several smaller intervals. We then apply the Runge–Kutta method over the interval $[x_0, x_1]$ to find y_1. From this, the Runge–Kutta scheme applied to the interval $[x_1, x_2]$ yields y_2, and so on.

Example 6.6: Multiple step Runge–Kutta method

As before, we consider

$$\frac{dy}{dx} = x^2 y^2 \qquad |x| < 1 \qquad (6.12a)$$

subject to

$$y(0) = 3 \qquad (6.12b)$$

with $h = 0.15$. The value of $y(0.15)$ is determined by calculating

$$R_1 = [x_0 y_0]^2 h = 0 \qquad (6.38a)$$

$$R_2 = \left[(x_0 + \frac{1}{2}h)(y_0 + \frac{1}{2}R_1) \right]^2 h = [(0.075)(3)]^2 (.15) = 0.00759 \qquad (6.38b)$$

$$R_3 = \left[(x_0 + \frac{1}{2}h)(y_0 + \frac{1}{2}R_2) \right]^2 h = [(.075)(3.00380)]^2 (.15) = 0.00761 \qquad (6.38c)$$

and

$$R_4 = [(x_0 + h)(y_0 + R_3)]^2 h = [(.15)(3.00761)]^2 (.15) = 0.03053 \qquad (6.38d)$$

from which

$$y(0.15) = y_0 + \frac{1}{6}[R_1 + 2R_2 + 2R_3 + R_4] = 3.01016 \qquad (6.39)$$

This result is then used to determine $y(0.3)$ by a second application of the Runge–Kutta method. With $x_1 = 0.15$, we generate the set of Runge–Kutta parameters

$$R_1 = [x_1 y_1]^2 h = [(.15)(3.01016)]^2(.15) = 0.03058 \qquad (6.40a)$$

$$R_2 = \left[(x_1 + \frac{1}{2}h)(y_1 + \frac{1}{2}R_1)\right]^2 h = [(.225)(3.0545)]^2(.15) = 0.06951 \qquad (6.40b)$$

$$R_3 = \left[(x_1 + \frac{1}{2}h)(y_1 + \frac{1}{2}R_2)\right]^2 h = [(.225)(3.04492)]^2(.15) = 0.07041 \qquad (6.40c)$$

and

$$R_4 = [(x_1 + h)(y_1 + R_3)]^2 h = [(.3)(3.08057)]^2(.15) = 0.12811 \qquad (6.40d)$$

Then

$$y(0.3) = y_1 + \frac{1}{6}[R_1 + 2R_2 + 2R_3 + R_4] = 3.08324 \qquad (6.41)$$

which is quite accurate. $\square$

Runge–Kutta method and numerical integration

To solve eq. 6.2 by the single step Picard method, we approximate the value of y in the integrand $f(x, y)$ by y_0 and evaluate the integral of $f(x', y_0)$ to obtain

$$y(x_N) \simeq y_0 + \int_{x_0}^{x_N} f(x', y_0)dx' \qquad (6.27)$$

We evaluate this integral by the Simpson's 1/3 rule given in eq. 4.56 with two intervals to make one segment of width $h = x_N - x_0$. Therefore, the width of one interval is $h/2$ and the approximation to eq. 6.27 by Simpson's 1/3 rule becomes

$$y(x_N) = y(x_0) + \frac{\frac{1}{2}h}{3}\left[f(x_0, y_0) + 4f(x_0 + \frac{1}{2}h, y_0) + f(x_0 + h, y_0)\right] \qquad (6.42)$$

To relate this to the Runge–Kutta method, we take y in $f(x, y)$ be the constant y_0. Then, the Runge–Kutta parameters are

$$R_1 = f(x_0, y_0)h \tag{6.43a}$$

$$R_2 = f(x_0 + \frac{1}{2}h, y_0)h \tag{6.43b}$$

$$R_3 = f(x_0 + \frac{1}{2}h, y_0)h = R_2 \tag{6.43c}$$

and

$$R_4 = f(x + h, y_0)h \tag{6.43d}$$

and the Runge–Kutta solution is

$$y(x) = y_0 + \frac{h}{6}\left[f(x_0, y_0) + 4f(x_0 + \frac{1}{2}h, y_0) + f(x_0 + h, y_0)\right] \tag{6.44}$$

This is identical to the Simpson's 1/3 rule approximation given in eq. 6.42.

6.4 Finite Difference Methods

To solve differential equations using the method of finite differences, we divide the domain $[x_0, x_N]$ into N equally spaced intervals and take

$$h = x_{k+1} - x_k \tag{1.35}$$

to have a fixed value independent of k.

Euler finite difference approximation to $y'(x_k)$

In chapter 1, we defined a difference operator Δ such that

$$\Delta y(x_k) = y(x_{k+1}) - y(x_k) \equiv y_{k+1} - y_k \tag{1.36b}$$

from which, the derivative operator was found to be

$$D \equiv \frac{d}{dx} = \frac{1}{h}\ln(1 + \Delta) = \frac{1}{h}\sum_{n=1}^{\infty}(-1)^{n+1}\frac{\Delta^n}{n} \tag{1.55b}$$

Approximating this series by the $n = 1$ term, we obtain

$$y'(x_k) \equiv y'_k \simeq \frac{1}{h}\Delta y_k = \frac{y_{k+1} - y_k}{h} \tag{6.45}$$

which is the *Euler finite difference approximation* of $y'(x_k)$.

The Taylor series for $y(x_{k+1})$ is given by

$$y_{k+1} = y(x_k + h) = y_k + hy'_k + \frac{h^2}{2!}y''_k + \frac{h^3}{3!}y'''_k + \dots \tag{6.46a}$$

from which

$$\frac{y_{k+1} - y_k}{h} = y'_k + \frac{h}{2!}y''_k + \frac{h^2}{3!}y'''_k + \dots \tag{6.46b}$$

Ignoring the terms containing h on the right side of eq. 6.46b, we see that approximating $y'(x_k)$ by the Euler finite difference $(y_{k+1} - y_k)/h$ is equivalent to ignoring terms of order h^2 and higher in the Taylor series for y_{k+1}.

Milne finite difference approximation to $y'(x_k)$

The Taylor series for $y(x_{k-1})$ is

$$y_{k-1} = y(x_k - h) = y_k - hy'_k + \frac{h^2}{2!}y''_k - \frac{h^3}{3!}y'''_k + \dots \tag{6.47}$$

Subtracting eqs. 6.46a and 6.47 we have

$$y_{k+1} = y_{k-1} + 2hy'_k + 2\frac{h^3}{3!}y'''_k + \dots \tag{6.48a}$$

Ignoring terms of order h^3 and higher, we obtain

$$y'_k \simeq \frac{y_{k+1} - y_{k-1}}{2h} \tag{6.48b}$$

This is the *Milne finite difference approximation* of $y'(x_k)$. It is as accurate as the Taylor series for y_{k+1} up to order h^2 and is therefore more accurate than the Euler finite difference approximation.

Euler finite difference solution

By approximating $y'(x_k)$ by the Euler finite difference, the Euler finite difference equation at x_k is given by

$$\frac{y_{k+1} - y_k}{h} \simeq f(x_k, y_k) \tag{6.49a}$$

which we write as

$$y_{k+1} \simeq y_k + hf(x_k, y_k) \tag{6.49b}$$

where

$$h = x_{k+1} - x_k \tag{1.35}$$

To determine $y(x_N)$, we divide $[x_0, x_N]$ into several small intervals defined by $\{x_0, x_1, \ldots, x_N\}$. We then take $k = 0, 1, \ldots, N - 1$ to obtain

$$y_1 \simeq y_0 + hf(x_0, y_0) \tag{6.50a}$$

$$y_2 \simeq y_1 + hf(x_1, y_1) \tag{6.50b}$$

$$y_3 \simeq y_2 + hf(x_2, y_2) \tag{6.50c}$$

$$y_N \simeq y_{N-1} + hf(x_{N-1}, y_{N-1}) \tag{6.50d}$$

Example 6.7: Euler finite difference solution

To solve

$$\frac{dy}{dx} = x^2 y^2 \qquad |x| < 1 \tag{6.12a}$$

with

$$y(0) = 3 \tag{6.12b}$$

we take $h = 0.15$, to obtain

$$y(0.15) = y_1 = y_0 + h(x_0 y_0)^2 = 3 + 0.15(0 * 3)^2 = 3 \tag{6.51a}$$

Then

$$y(0.3) = y_2 = y_1 + h(x_1 y_1)^2 = 3 + 0.15(.15 * 3)^2 = 3.03038 \tag{6.51b}$$

which is not very accurate.

As discussed earlier, with these three values of y; we can approximate $y(x)$ at any value of x ε $[0.0, 0.3]$ by an interpolation scheme. □

Milne finite difference solution

To apply the Milne finite difference method, we approximate the differential equation as

$$y_{k+1} \simeq y_{k-1} + 2hy'_k = y_{k-1} + 2hf(x_k, y_k) \qquad (6.52)$$

Setting $k = 0$, eq. 6.52 becomes

$$y_1 \simeq y_{-1} + 2hf(x_0, y_0) \qquad (6.53a)$$

Since $x_{-1} = x_0 - h$ is outside $[x_0, x_N]$, any approximation to y_{-1} yields an extrapolated value of y, which, as noted in ch. 1, is unreliable. It is preferable to use an approximation of y at a value of x within $[x_0, x_N]$.

For $k = 1$, eq. 6.52 becomes

$$y_2 \simeq y_0 + 2hf(x_1, y_1) \qquad (6.53b)$$

The value of y_1 is unknown and must be determined independently of the finite difference method. For example, since the Milne finite difference approximation is identical to the Taylor series expansion up to h^2, a reasonable approach is to approximate y_1 by the truncated Taylor series

$$y_1 = y(x_0 + h) \simeq y_0 + y'_0 h + \frac{1}{2} y''_0 h^2 \qquad (6.54)$$

where y_0' and y_0'' are obtained from the differential equation and its derivative.

The Picard approach yields a value of y_1 as

$$y_1 = y_0 + \int_{x_0}^{x_1} f(x, y_0) dx \qquad (6.30a)$$

which requires that the integral be evaluated exactly or approximated accurately by some numerical method.

The Runge–Kutta scheme yields a value of y_1 as

$$y_1 \simeq y_0 + \frac{1}{6}[R_1 + 2R_2 + 2R_3 + R_4] \qquad (6.34)$$

and the Euler finite difference method approximates y_1 as

$$y_1 \simeq y_0 + hf(x_0, y_0) \tag{6.50a}$$

However, since this is not as accurate as the Taylor series estimate given in eq. 6.54, we will not consider it in approximating y_1.

Example 6.8: Milne finite difference solution

As above, we solve the differential equation

$$\frac{dy}{dx} = x^2 y^2 \quad |x| < 1 \tag{6.12a}$$

subject to

$$y(0) = 3 \tag{6.12b}$$

at $x = 0.3$. With $h = 0.1$ (and thus, $x_1 = 0.1$), we find the value of $y(0.3)$ in two steps, given by

$$y(0.2) = y_2 = y_0 + 2h(x_1 y_1)^2 \tag{6.55a}$$

from which

$$y(0.3) = y_3 = y_1 + 2h(x_2 y_2)^2 \tag{6.55b}$$

Approximating y_1 by a Taylor series truncated at h^2, we obtain

$$y_1 = 3 + 0.1(x_0 y_0)^2 + \frac{(0.1)^2}{2}\left(2x_0 y_0^2 + 2x_0^2 y_0^3\right) = 3 \tag{6.56}$$

With the Picard approximation, we have

$$y_1 \simeq y_0 + \int_0^{0.1} x^2 \times y_0^2 dx = 3.00300 \tag{6.57}$$

and with the Runge–Kutta method, we find

$$R_1 = 0.1 \times (x_0 y_0)^2 = 0 \tag{6.58a}$$

$$R_2 = 0.1 \times [(x_0 + 0.05)y_0]^2 = 0.00225 \tag{6.58b}$$

$$R_3 = 0.1 \times [(x_0 + 0.05)(y_0 + 0.001125)]^2 = 0.00225 \tag{6.58c}$$

and

$$R_4 = 0.1 \times [(x_0 + 0.1)(y_0 + 0.00225)]^2 = 0.00901 \qquad (6.58d)$$

from which

$$y_1 \simeq y_0 + \frac{1}{6}[0 + 2 \times 0.00225 + 2 \times 0.00225 + 0.00901] = 3.00301 \qquad (6.59)$$

With the value of y_1 obtained from the Runge–Kutta method, we have

$$y(0.2) = y_2 = 3 + 2(.1)(.1 \times 3.00301)^2 = 3.01804 \qquad (6.60)$$

Then, from eq. 6.55b, we obtain

$$y(0.3) = y_3 = 3.00301 + 2(0.1)(0.2 \times 3.01804)^2 = 3.07588 \qquad (6.61)$$

which is in fair agreement with the exact result, 3.08325.

The accuracy of the solution is improved by applying the multistep Milne finite difference approximation for smaller values of h and thus a larger number of steps. □

6.5   Predictor–Corrector Methods

Predictors–corrector methods consists of a pair of equations with which the accuracy of numerical solutions to

$$\frac{dy}{dx} = f(x, y) \qquad (6.2)$$

subject to

$$y(x_0) = y_0 \qquad (6.4)$$

can be improved.

Let $y(x_{k+1}) = y_{k+1}$ be approximated by some multistep approximation (such as the Picard method). In general, y_{k+1} is found from some expression that depends on the values of $\{x_1, \ldots, x_{k+1}\}$ and $\{y_1, \ldots, y_k\}$ and is of the form

$$y_{k+1} = P(x_1, x_2, ..., x_{k+1}; y_1, y_2, ..., y_k) \qquad (6.62a)$$

$P(x_1, \ldots, x_{k+1}; y_1, \ldots, y_k)$ is called the *predictor* of y_{k+1}. The *corrector* equation for y_{k+1} is of the form

$$y_{k+1}^{(m+1)} = C(x_1, x_2, ..., x_k, x_{k+1}; y_1, y_2, ..., y_k, y_{k+1}^{(m)}) \qquad (6.62b)$$

We point out that the predictor P must not depend on y_{k+1} and the corrector C must depend on y_{k+1}.

The predictor equation is usually one that is well established [such as the Picard estimate of eq. 6.29]. The corrector C is a function developed by the user.

Corrections to each predicted value are achieved by iterations of eq. 6.62b in the index m. The predicted value of y_{k+1} given in eq. 6.62a serves as $y^{(0)}{}_{k+1}$ in the argument of the corrector function to generate

$$y_{k+1}^{(1)} = C(x_1, x_2, ..., x_k, x_{k+1}; y_1, y_2, ..., y_k, y_{k+1}^{(0)}) \qquad (6.63a)$$

The second correction to y_{k+1} is then obtained from

$$y_{k+1}^{(2)} = C(x_1, x_2, ..., x_k, x_{k+1}; y_1, y_2, ..., y_k, y_{k+1}^{(1)}) \qquad (6.63b)$$

and so on. This process is repeated until

$$\left| \frac{y_{k+1}^{(m+1)} - y_{k+1}^{(m)}}{y_{k+1}^{(m+1)}} \right| < \varepsilon \qquad (6.64)$$

where ε is a small number chosen by the user that defines the required level of accuracy.

Picard predictor and corrector

For each value of the index $k \geq 0$, the *Picard predictor* is obtained by replacing y_c in eq. 6.26 by y_k. Then

$$y_{k+1} \simeq y_k + \int_{x_k}^{x_{k+1}} f(x, y_k)\, dx \qquad (6.29)$$

with the value of y_0, given by the constraint of eq. 6.4.

A reasonable *Picard corrector* is one developed by choosing some expression for y_c in $f(x, y_c)$ that involves y_{k+1}. A simple example of such a corrector is one in which y_c is replaced by a linear combination of y_k and y_{k+1} to obtain

$$y_{k+1}^{(m+1)} \simeq y_k + \int_{x_k}^{x_{k+1}} f\left[x, (\alpha y_k + \beta y_{k+1}^{(m)})\right] dx \qquad (6.65)$$

A reasonable choice is to take

$$\alpha = \beta = \frac{1}{2} \tag{6.66}$$

so that y_c is approximated by the average of $y(x)$ over the interval $[x_k, x_{k+1}]$. Then this Picard corrector equation becomes

$$y_{k+1}^{(m+1)} \simeq y_k + \int_{x_k}^{x_{k+1}} f\left[x, \frac{1}{2}\left(y_k + y_{k+1}^{(m)}\right)\right] dx \tag{6.67}$$

where y_k is the most accurate value of $y(x_k)$ obtained from the previous prediction and correction, and the value of $y^{(0)}{}_{k+1}$ is the value of y_{k+1} obtained from predictor. Thus, for example, with the value of $y^{(0)}{}_{k+1}$ obtained from eq. 6.29, the first correction is given by

$$y_{k+1}^{(1)} \simeq y_k + \int_{x_k}^{x_{k+1}} f\left[x, \frac{1}{2}\left(y_k + y_{k+1}^{(0)}\right)\right] dx \tag{6.68a}$$

Then

$$y_{k+1}^{(2)} \simeq y_k + \int_{x_k}^{x_{k+1}} f\left[x, \frac{1}{2}\left(y_k + y_{k+1}^{(1)}\right)\right] dx \tag{6.68b}$$

and so on. This correction process is terminated when

$$\left|\frac{y_{k+1}^{(m+1)} - y_{k+1}^{(m)}}{y_{k+1}^{(m+1)}}\right| < \varepsilon \tag{6.64}$$

where the small value of ε is chosen by the user.

Example 6.9: Picard predictor–corrector method

We again consider

$$\frac{dy}{dx} = x^2 y^2 \quad |x| < 1 \tag{6.12a}$$

constrained by

$$y(0) = 3 \tag{6.12b}$$

The Picard predictor equation for this example is

$$y_{k+1} = y_k + \int_{x_k}^{x_{k+1}} x^2 y_k^2 dx = y_k + \frac{\left(x_{k+1}^3 - x_k^3\right) y_k^2}{3} = y_{k+1}^{(0)} \tag{6.69}$$

and, using the corrector defined by eq. 6.68, the Picard corrector is

$$y_{k+1}^{(m+1)} = y_k + \frac{\left(x_{k+1}^3 - x_k^3\right)\left(y_k + y_{k+1}^{(m)}\right)^2}{12} \tag{6.70}$$

Taking $h = 0.15$, we predict the value of y_1 from eq. 6.29 as

$$y(.15) = y_1 = y_1^{(0)} = y_0 + \int_0^{.15} (xy_0)^2 dx = 3 + 9\frac{(.15)^3}{3} = 3.01013 \tag{6.71}$$

Corrections to this value are obtained from iterations of eq. 6.70:

$$y_1^{(m+1)} = y_0 + \int_0^{.15} x^2 \left[\frac{1}{2}\left(y_0 + y_1^{(m)}\right)\right]^2 dx = 3 + \frac{(.15)^3}{12}\left(3 + y_1^{(m)}\right)^2 \tag{6.72}$$

With $m = 0$, we obtain

$$y_1^{(1)} = 3 + \frac{(.15)^3}{12}\left(3 + y_1^{(0)}\right)^2 = 3.01016 \tag{6.73a}$$

from which

$$y_1^{(2)} = 3 + \frac{(.15)^3}{12}\left(3 + y_1^{(1)}\right)^2 = 3.01016 \tag{6.73b}$$

Comparing $y_1^{(2)}$ to $y_1^{(1)}$, we find that

$$\left|\frac{y_1^{(2)} - y_1^{(1)}}{y_1^{(2)}}\right| \simeq 6.6 * 10^{-14} \tag{6.74}$$

which indicates that the result of eq. 6.73b is an extremely precise corrected value of $y(0.15)$.

Using this most precise value of y_1, we predict

$$y_2^{(0)} = y_1 + \int_{.15}^{.30} (xy_1)^2 dx$$

$$= 3.01016 + 3.01016^2 \left(\frac{.30^3 - .15^3}{3}\right) = 3.08152 \tag{6.75}$$

The corrected values of y_2 are obtained from

$$y_2^{(m+1)} = y_1 + \int_{.15}^{.30} x^2 \left[\frac{1}{2} \left(y_1 + y_2^{(m)} \right) \right]^2 dx =$$

$$3.01016 + \frac{\left[(.30)^3 - (.15)^3 \right]}{12} \left(3.01016 + y_2^{(m)} \right)^2 \tag{6.76}$$

With four iterations, we obtain

$$y_2^{(4)} = y(0.30) = 3.08326 \tag{6.77}$$

which compares very well with the exact value of 3.08325. $\square$

Picard Rectangular and Trapezoidal Predictor–Correctors

When the Picard integral

$$\int_{x_k}^{x_{k+1}} f(x, y_c) dx$$

cannot be evaluated exactly, predictor–corrector methods can be developed using one of the quadrature rules developed in ch. 4. Then the predictor equation becomes

$$y_{k+1} = y_k + \sum_{n=k}^{k+1} w_n f(x_n, y_c) \tag{6.78}$$

and, taking y_c to be the average of y_k and y_{k+1}, the corrector equation is

$$y_{k+1}^{(m+1)} = y_k + \sum_{n=k}^{k+1} w_n f \left[x_n, \frac{1}{2} \left(y_k + y_{k+1}^{(m)} \right) \right] \tag{6.79}$$

We note that, in general, the predictor and corrector of eqs. 6.78 and 6.79, involve sums over points in the interval $[x_k, x_{k+1}]$. Unless the quadrature rule accesses only x_k and x_{k+1}, it will be necessary to obtain approximations to y at points within the interval $[x_k, x_{k+1}]$. The only quadrature rules we have discussed that access only x_k and x_{k+1} are the two simplest Newton-Cotes rules, the two rectangular rules and the trapezoidal rule given in eqs. 4.48. For that reason, we approximate the integral by one of these quadrature sums.

Using the midpoint rectangular rule of eq. 4.48a, eqs. 6.78, and 6.79 become

$$y_{k+1} = y_k + hf\left[\frac{1}{2}(x_k + x_{k+1}), y_k\right] \qquad (6.80a)$$

and

$$y_{k+1}^{(m+1)} = y_k + hf\left[\frac{1}{2}(x_k + x_{k+1}), \tfrac{1}{2}(y_k + y_{k+1}^{(m)})\right] \qquad (6.80b)$$

These equations comprise the *rectangular predictor–correctors.*

As noted, the trapezoidal rule of eq. 4.48b is identical to the average function rectangular quadrature. This quadrature rule yields the predictor and correction equations

$$y_{k+1} = y_k + \frac{1}{2}h[f(x_k, y_k) + f(x_{k+1}, y_k)] \qquad (6.81a)$$

and

$$y_{k+1}^{(m+1)} = y_k + \frac{1}{2}h\left[f\left(x_k, \frac{1}{2}(y_k + y_{k+1}^{(m)})\right) + f\left(x_{k+1}, \frac{1}{2}(y_k + y_{k+1}^{(m)})\right)\right] \qquad (6.81b)$$

We refer to this as the *trapezoidal predictor–correctors.*

Example 6.10: Rectangular and trapezoidal predictor–correctors

We apply the rectangular and trapezoidal predictor–correctors to

$$\frac{dy}{dx} = x^2 y^2 \quad |x| < 1 \qquad (6.12a)$$

subject to

$$y(0) = 3 \qquad (6.12b)$$

In the following, we define the level of accuracy by taking $\varepsilon = 10^{-6}$.

(a) From eqs. 6.80, the predictor obtained for the midpoint rectangular rule is

$$y_{k+1} = y_k + \frac{h}{4}(x_k + x_{k+1})^2 y_k^2 \qquad (6.82a)$$

and the rectangular corrector is given by

$$y_{k+1}^{(m+1)} = y_k + \frac{h}{16}(x_k + x_{k+1})^2 \left(y_k + y_{k+1}^{(m)}\right)^2 \qquad (6.82b)$$

With $h = 0.15$ and $x_N = 0.30$, the predicted values of y are

$$y = \begin{pmatrix} y(0.00) \\ y(0.15) \\ y(0.30) \end{pmatrix} = \begin{pmatrix} 3.00000 \\ 3.00759 \\ 3.07630 \end{pmatrix} \qquad (6.83)$$

After four iterations, the corrector of eq. 6.82b yields

$$y = \begin{pmatrix} 3.00000 \\ 3.00761 \\ 3.07792 \end{pmatrix} \qquad (6.84)$$

which are not very accurate.
(b) From eqs. 6.81, the predictor for the trapezoidal rule is

$$y_{k+1} = y_k + \frac{h}{2}\left[(x_k)^2 + (x_{k+1})^2\right] y_k^2 \qquad (6.85a)$$

and the corrector is

$$y_{k+1}^{(m+1)} = y_k + \frac{h}{8}\left(x_k^2 + x_{k+1}^2\right)\left(y_k + y_{k+1}^{(m)}\right)^2 \qquad (6.85b)$$

The values, given by the trapezoidal predictor, are

$$y = \begin{pmatrix} y(0.00) \\ y(0.15) \\ y(0.30) \end{pmatrix} = \begin{pmatrix} 3.00000 \\ 3.01519 \\ 3.09198 \end{pmatrix} \qquad (6.86)$$

and four iterations of the trapezoidal corrector yields

$$y = \begin{pmatrix} 3.00000 \\ 3.01526 \\ 3.09399 \end{pmatrix} \qquad (6.87)$$

which are only slightly more accurate than the values obtained with the rectangular predictor–corrector. These inaccuracies arise from the use of a rudimentary quadrature rule to approximate the integral.

Predictors and correctors using different algorithms

It is not necessary to use a corrector equation based on the same algorithm as the predictor. By using an algorithm for the corrector that is different from the predictor algorithm, we get a sense of the importance of using an accurate predictor and the importance of the accuracy of the corrector.

Example 6.11: Picard, rectangular, and trapezoidal predictors and correctors

Referring to examples 6.11 and 6.12, we apply correctors that are based on algorithms, some of which are different from the predictor algorithm. We obtain the results shown below.

$$
y_{\substack{Picard \\ predicted}} = \begin{pmatrix} 3.00000 \\ 3.01013 \\ 3.08152 \end{pmatrix} \quad y_{\substack{Picard \\ corrected}} = \begin{pmatrix} 3.00000 \\ 3.01016 \\ 3.08326 \end{pmatrix} \tag{6.88a}
$$

$$
y_{\substack{rectangular \\ predicted}} = \begin{pmatrix} 3.00000 \\ 3.00759 \\ 3.07897 \end{pmatrix} \quad y_{\substack{Picard \\ corrected}} = \begin{pmatrix} 3.00000 \\ 3.01016 \\ 3.08326 \end{pmatrix} \tag{6.88b}
$$

$$
y_{\substack{trapezoidal \\ predicted}} = \begin{pmatrix} 3.00000 \\ 3.01519 \\ 3.08661 \end{pmatrix} \quad y_{\substack{Picard \\ corrected}} - \begin{pmatrix} 3.00000 \\ 3.01016 \\ 3.08326 \end{pmatrix} \tag{6.88c}
$$

$$
y_{\substack{Picard \\ predicted}} = \begin{pmatrix} 3.00000 \\ 3.01013 \\ 3.07885 \end{pmatrix} \quad y_{\substack{rectangular \\ corrected}} = \begin{pmatrix} 3.00000 \\ 3.00761 \\ 3.07792 \end{pmatrix} \tag{6.89a}
$$

$$
y_{\substack{rectangular \\ predicted}} = \begin{pmatrix} 3.00000 \\ 3.00759 \\ 3.07630 \end{pmatrix} \quad y_{\substack{rectangular \\ corrected}} = \begin{pmatrix} 3.00000 \\ 3.00761 \\ 3.07792 \end{pmatrix} \tag{6.89b}
$$

$$
y_{\substack{trapezoidal \\ predicted}} = \begin{pmatrix} 3.00000 \\ 3.01519 \\ 3.08394 \end{pmatrix} \quad y_{\substack{rectangular \\ corrected}} = \begin{pmatrix} 3.00000 \\ 3.00761 \\ 3.07792 \end{pmatrix} \tag{6.89c}
$$

$$
y_{\substack{Picard \\ predicted}} = \begin{pmatrix} 3.00000 \\ 3.01013 \\ 3.08686 \end{pmatrix} \quad y_{\substack{trapezoidal \\ corrected}} = \begin{pmatrix} 3.00000 \\ 3.01526 \\ 3.09399 \end{pmatrix} \tag{6.90a}
$$

$$y_{\substack{rectangular \\ predicted}} = \begin{pmatrix} 3.00000 \\ 3.00759 \\ 3.08686 \end{pmatrix} \quad y_{\substack{trapezoidal \\ corrected}} = \begin{pmatrix} 3.00000 \\ 3.01526 \\ 3.09399 \end{pmatrix} \qquad (6.90b)$$

$$y_{\substack{trapezoidal \\ predicted}} = \begin{pmatrix} 3.00000 \\ 3.01519 \\ 3.09198 \end{pmatrix} \quad y_{\substack{trapezoidal \\ corrected}} = \begin{pmatrix} 3.00000 \\ 3.01526 \\ 3.09399 \end{pmatrix} \qquad (6.90c)$$

Comparing the results given in eqs. 6.88–6.90 to

$$y_{exact} = \begin{pmatrix} 3.00000 \\ 3.01016 \\ 3.08325 \end{pmatrix} \qquad (\textbf{6.22b})$$

we see that the Picard corrector yields more accurate results than either the rectangular or trapezoidal correctors, no matter which predicted value is used.

This is more easily illustrated in terms of the fractional errors defined by

$$E \equiv \left| \frac{y_{corrected} - y_{exact}}{y_{exact}} \right| \qquad (6.91)$$

at each value of x (0.15 and 0.30) for each of these predictor–corrector pairs. For each of the three predictors with the Picard corrector, the largest fractional error is 3.5×10^{-4}. This maximum fractional error for the rectangular corrector and the trapezoidal corrector are 1.7×10^{-1} and 3.5×10^{-1}, respectively, independent of which predictor is used. □

Adams–Bashforth–Moulton predictors and correctors

The Adams–Bashforth–Moulton methods are developed by approximating $f(x,y)$ by a Lagrange interpolation over the points $x_k, x_{k-1}, \ldots, x_{k-p}$ to obtain the predictor of y_{k+1}. The corrector is then obtained by representing $f(x,y)$ by a second Lagrange interpolation over the points $x_{k+1}, x_k, \ldots, x_{k-q}$. The values of p and q, which determine the order of the interpolating polynomial, are chosen by the user. By approximating $f(x,y)$ by polynomials, it is straightforward to evaluate the integrals of these polynomials exactly.

Although it is not required, for most predictors and correctors of this type, the order of the interpolating polynomial for the predictor and the one for the corrector are the same. In this way, the levels of precision of the predictor are the same as the order of the corrector polynomial. We will use such interpolations in our discussion.

The predictor is defined by interpolating $f(x,y)$ over $\{x_{k-p}, x_{k-p+1}, \ldots, x_k\}$ in the form

$$f(x, y) = \mu_k(x)f_k + \mu_{k-1}(x)f_{k-1} + \ldots + \mu_{k-p}(x)f_{k-p}$$

$$= \sum_{n=0}^{p} \mu_{k-n}(x)f_{k-n} \tag{6.92a}$$

To obtain the corrector, we interpolate over $\{x_{k-p+1}, x_{k-p+2}, \ldots, x_{k+1}\}$ as

$$f(x, y) = \mu_{k+1}(x)f_{k+1}^{(m)} + \mu_k(x)f_k + \ldots + \mu_{k-p+1}(x)f_{k-p+1}$$

$$= \mu_{k+1}(x)f_{k+1}^{(m)} + \sum_{n=1}^{p} \mu_{k+1-n}(x)f_{k+1-n} \tag{6.92b}$$

where the functions $\mu(x)$ are polynomials expressed in the form given in eq. 1.7.

The predicted values, y_{k+1}, are then given by the integral of these interpolated forms of $f(x,y)$ in eq. 6.92a:

$$y_{k+1} = y_k + \sum_{n=0}^{p} f_{k-n} \int_{x_k}^{x_{k+1}} \mu_{k-n}(x)dx \tag{6.93a}$$

The corrected values are found by integrating the interpolated forms of $f(x,y)$ in eq. 6.92b;

$$y_{k+1}^{(m+1)} = y_k + f_{k+1}^{(m)} \int_{x_k}^{x_{k+1}} \mu_{k+1}(x)dx + \sum_{n=1}^{p} f_{k+1-n} \int_{x_k}^{x_{k+1}} \mu_{k+1-n}(x)dx \tag{6.93b}$$

For example, for $p = 2$, the interpolating polynomials are second order, and the representation of $f(x,y)$ has three terms. The predictor is obtained by integrating

$$f(x, y(x)) \simeq \frac{(x - x_{k-2})(x - x_{k-1})}{(x_k - x_{k-2})(x_k - x_{k-1})}f_k +$$

$$\frac{(x - x_{k-2})(x - x_k)}{(x_{k-1} - x_{k-2})(x_{k-1} - x_k)}f_{k-1} + \frac{(x - x_{k-1})(x - x_k)}{(x_{k-2} - x_{k-1})(x_{k-2} - x_k)}f_{k-2} \tag{6.94a}$$

and the corrector is found from

$$f(x, y) \simeq \frac{(x - x_{k-1})(x - x_k)}{(x_{k+1} - x_{k-1})(x_{k+1} - x_k)}f_{k+1}^{(m)} +$$

$$\frac{(x - x_{k-1})(x - x_{k+1})}{(x_k - x_{k-1})(x_k - x_{k+1})}f_k + \frac{(x - x_k)(x - x_{k+1})}{(x_{k-1} - x_k)(x_{k-1} - x_{k+1})}f_{k-1} \tag{6.94b}$$

Since only y_{k+1} (and therefore f_{k+1}) is corrected by iterating the corrector equation in m, only f_{k+1} has a superscript (m).

Writing each factor in the denominator of eq. 6.94b in the form

$$(x_{k-\ell} - x_{k-n}) = (n - \ell)h \tag{6.95}$$

and integrating $f(x,y)$ given in eq. 6.94a, the *Adams–Bashforth–Moulton predictor* is

$$y_{k+1} \simeq y_k + \sum_{n=0}^{2} f_{k-n} \int_{x_k}^{x_{k+1}} \mu_{k-n}(x)dx = y_k + \frac{h}{12}(23f_k - 16f_{k-1} + 5f_{k-2}) \tag{6.96a}$$

From the integral of eq. 6.89b, the *Adams–Bashforth–Moulton corrector* is

$$y_{k+1}^{(m+1)} \simeq y_k + \sum_{n=0}^{2} f_{k+1-n} \int_{x_k}^{x_{k+1}} \mu_{k+1-n}(x)dx$$

$$= y_k + \frac{h}{12}\left(5f_{k+1}^{(m)} + 8f_k - f_{k-1}\right) \tag{6.96b}$$

We note that the predictor of y_{k+1} given in eq. 6.96a contains f_{k-2}, f_{k-1}, and f_k which depend on x_{k-2}, x_{k-1}, and x_k and on y_{k-2}, y_{k-1}, and y_k. Therefore, to predict a value for y_{k+1}, we must know the values of y_k, y_{k-1}, and y_{k-2}. We see that for $k = 0$ and $k = 1$, some of these y values are outside the range $[y_0, y_N]$. To avoid using these extrapolated values, we begin with $k = 2$. Then, the value of y with the smallest index that we can determine from eq. 6.96a is

$$y_3 \simeq y_2 + \frac{h}{12}(23f_2 - 16f_1 + 5f_0) \tag{6.97a}$$

We find y_0 and therefore f_0 from the initial condition. But we must approximate y_1 and y_2 by some other method such as a truncated Taylor series. Once we have values for y_1 and y_2, and we then predict y_3, we can then predict y_4 as

$$y_4 \simeq y_3 + \frac{h}{12}(23f_3 - 16f_2 + 5f_1) \tag{6.97b}$$

and so on.

Using a similar argument, if an Adams–Bashforth–Moulton corrector is developed using an interpolating polynomial of order 2, the corrector depends on f_{k-1}, f_k, and f_{k+1}. (and thus y_{k-1}, y_k, and y_{k+1}). Again, to insure that we only use y values at points within the domain of y, we must start with $k = 1$. Thus, from eq. 6.96b, we have correctors of the form

$$y_2^{(m+1)} \simeq y_1 + \frac{h}{12}\left(5f_2^{(m)} + 8f_1 - f_0\right) \qquad (6.98a)$$

$$y_3^{(m+1)} \simeq y_2 + \frac{h}{12}\left(5f_3^{(m)} + 8f_2 - f_1\right) \qquad (6.98b)$$

etc. We note that since the value of y_1 cannot be corrected using this corrector, we must determine its value by a method other than the predictor.

Source code for this and other types of Adams–Bashforth–Moulton methods, written in C, can be found at http://mymathlib.webtrellis.net/diffeq/adams_top.html

Example 6.12: Adams–Bashforth–Moulton predictor–corrector

We apply the Adams–Bashforth–Moulton method to the differential equation

$$\frac{dy}{dx} = x^2y^2 \qquad (6.12a)$$

subject to

$$y(0) = 3 \qquad (6.12b)$$

with $p = 2$. The predictor of eq. 6.96a is

$$y_{k+1} = y_k + \frac{h}{12}\left[23x_k^2y_k^2 - 16x_{k-1}^2y_{k-1}^2 + 5x_{k-2}^2y_{k-2}^2\right] \qquad (6.99)$$

Starting with $k = 2$, we obtain y_3 from

$$y_3 = y_2 + \frac{h}{12}\left[23x_2^2y_2^2 - 16x_1^2y_1^2 + 5x_0^2y_0^2\right] \qquad (6.100)$$

where the values of y_1 and y_2 must be obtained by some method such as a truncated Taylor series, Runge–Kutta approach, etc. Once we have found these values, we can determine $y_3, y_4, \ldots$

For example, with example 6.6 as our guide, we use a multistep Runge–Kutta scheme with $h = 0.1$ to predict the values of y_1 and y_2, then eq. 6.100 to predict a value of y_3. We find

$$y = \begin{pmatrix} y_0 \\ y_1 \\ y_2 \\ y_3 \end{pmatrix} = \begin{pmatrix} y(0.0) \\ y(0.1) \\ y(0.2) \\ y(0.3) \end{pmatrix} = \begin{pmatrix} 3.00000 \\ 3.00300 \\ 3.02419 \\ 3.08229 \end{pmatrix} \qquad (6.101)$$

The Adams–Bashforth–Moulton corrector for these values of y is given by approximating $f(x,y)$ by

$$f\left(x, y_{k-1}, y_k, y_{k+1}^{(m)}\right) \simeq \frac{(x - x_{k-1})(x - x_k)}{(x_{k+1} - x_{k-1})(x_{k+1} - x_k)} f_{k+1}^{(m)}$$
$$+ \frac{(x - x_{k-1})(x - x_{k+1})}{(x_k - x_{k-1})(x_k - x_{k+1})} f_k + \frac{(x - x_k)(x - x_{k+1})}{(x_{k-1} - x_k)(x_{k-1} - x_{k+1})} f_{k-1} \qquad (6.102)$$

from which

$$y_{k+1}^{(m+1)} \simeq y_k + \int_{x_k}^{x_{k+1}} f\left(x, y_{k-1}, y_k, y_{k+1}^{(m)}\right) dx$$

$$= y_k + \frac{h}{12}\left(5 f_{k+1}^{(m)} + 8 f_k - f_{k-1}\right) \qquad (6.103)$$

Starting with $k = 1$, the corrector equation can be applied to the predicted values of y_2 and y_3, but not to the predicted value of y_1. We have

$$y_2^{(m+1)} \simeq y_1 + \frac{h}{12}\left(5(x_2 y_2^{(m)})^2 + 8(x_1 y_1)^2 - (x_0 y_0)^2\right) \qquad (6.104a)$$

and

$$y_3^{(m+1)} \simeq y_2 + \frac{h}{12}\left(5(x_3 y_3^{(m)})^2 + 8(x_2 y_2)^2 - (x_1 y_1)^2\right) \qquad (6.104b)$$

After several iterations, we obtain

$$y = \begin{pmatrix} 3.00000 \\ 3.00300 \\ 3.02426 \\ 3.08355 \end{pmatrix} \qquad (6.105)$$

to a level of precision defined by $\varepsilon = 10^{-6}$. $\square$

Problems

The analytic solution to

$$\frac{dy}{dx} = x + y$$

subject to $y(0) = 1$ is $y(x) = 2e^x - x - 1$

From this solution, we see that

$$y(0.5) = 1.79744$$

As specified in each problem below, find an approximation to this solution.

1. Use a single step Taylor series expansion about $x = 0$ to find $y(0.5)$, keeping terms up to $h^2 = (0.5-0)^2$.
2. Use a two step iteration of a Taylor series expansion. Use the first iteration to determine $y(0.25)$ expanding about $x = 0$, and a second iteration expanding about $x = 0.25$ to determine $y(0.5)$. For each iteration, keep terms in each Taylor series up to h^2.
3. Use a single step Picard method to find $y(0.50)$.
4. Use a two step iteration of the Picard method to find $y(0.50)$, using the first iteration to determine a value of $y(0.25)$.
5. Use a single step Runge–Kutta method to find $y(0.50)$.
6. Use a two-step iteration of the Runge–Kutta method to find $y(0.50)$, using the first iteration to determine a value of $y(0.25)$.
7. Use the Euler finite difference approximation to find $y(0.50)$ taking $h = 0.25$.
8. Use the Milne finite difference approximation to find $y(0.5)$ taking $h = 0.25$.
9. Use the Milne finite difference approximation to determine $y(0.3)$ taking $= 0.1$.

10. (a) Develop a corrector for the Picard method by replacing y_c by $y^{(m)}{}_{k+1}$ in the integral of eq. 6.67. That is, in eq. 6.67, take $\alpha = 0$, $\beta = 1$.
    (b) Determine corrected values of $y(0.25)$ and $y(0.50)$ to a level of precision of 10^{-3}.

11. (a) Develop a predictor–corrector method based on the Euler finite difference approximation to a first order differential equation.
    (b) Use this predictor–corrector method to find the solution to the differential equation above at $x = 0.25$ and $x = 0.50$ to a level of precision of 10^{-3}.

12. (a) Develop a predictor–corrector method based on the Milne finite difference approximation to a first order differential equation.
    (b) Use this predictor–corrector method to find the solution to the differential equation above. Use the Runge–Kutta approximation to predict $y(0.25)$, and from this, find a corrected value of $y(0.50)$, to a level of precision of 10^{-3}.

13. (a) Develop a predictor for the Adams–Bashforth–Moulton method with third order (four term) polynomial interpolation using a Taylor series truncated at the h^2 term to find y_1 and y_2.
    (b) To a level of precision of 10^{-3}, find a solution to the differential equation above, at $x = 0.25$, and at $x = 0.50$, using a third order interpolated Adams–Bashforth–Moulton corrector.

14. (a) Develop a predictor for the Adams–Bashforth–Moulton method using a second order (three term) interpolation over e^x [see eq. 1.18 with $q(x) = e^x$]. Use a Taylor series truncated at the h^2 term to find y_1 and y_2.

    (b) To a level of precision of 10^{-3}, find a solution to the differential equation above, at $x = 0.25$ and $x = 0.50$, using a three term Adams–Bashforth–Moulton corrector, interpolated over e^x.

Chapter 7
ORDINARY SECOND ORDER DIFFERENTIAL EQUATIONS

The general form of an ordinary second order differential equation is

$$F(x, y, y', y'') = 0 \qquad (7.1)$$

Many physical systems are described by second order differential equations of the form

$$y'' + P(x)y' + Q(x)y = f(x, y, y') \qquad (7.2)$$

We will restrict our discussion to numerical solutions to this type of equation.

7.1 Initial and Boundary Conditions

To find the complete solution to a first-order differential equation, it was necessary to specify one value of y at a specific value of x. For a second order differential equation, there are two constants of integration, requiring two independent conditions to determine the complete solution. For many differentials that describe physical phenomena, those are in the form of either *initial conditions* or *boundary conditions*.

Initial conditions

Initial conditions are expressed in the form

$$\begin{cases} y(x_0) = y_0 = constant \\ y'(x_0) = y'_0 = constant \end{cases} \qquad (7.3)$$

H. Cohen, *Numerical Approximation Methods*, DOI 10.1007/978-1-4419-9837-8_7,
© Springer Science+Business Media, LLC 2011

These conditions are so named because they specify the values of $y(x)$ and $y'(x)$ at x_0, the starting or initial point of the problem. In applied problems described by differential equations constrained by initial conditions, x often represents an instant of time, with x_0 defining the instant at which the values of y are first measured.

Boundary conditions

In many applied problems, the values of $y(x)$, $y'(x)$, or a combination of $y(x)$ and $y'(x)$ are specified at the spatial boundaries of the problem. When the domain of y is defined by $a \leq x \leq b$, then a and b are the boundaries of the physical system. For differential equations that describe such problems, $y(x)$ and/or $y'(x)$ satisfy one of three types of *boundary conditions*:

- *Dirichlet* boundary conditions specify the values of $y(x)$ at the boundaries. These conditions are given by

$$\begin{cases} y(a) = y_a = constant \\ y(b) = y_b = constant \end{cases} \tag{7.4a}$$

- *Neumann* boundary conditions specify the values of $y'(x)$ at the boundaries. They are of the form

$$\begin{cases} y'(a) = y'_a = constant \\ y'(b) = y'_b = constant \end{cases} \tag{7.4b}$$

- *Gauss* or *mixed* boundary conditions specify the values of a linear combination of $y(x)$ and $y'(x)$ at the boundaries. They are given in the form

$$\begin{cases} \lambda y(a) + \mu y'(a) = v_a = constant \\ \lambda y(b) + \mu y'(b) = v_b = constant \end{cases} \tag{7.4c}$$

When the constants λ and μ are non-zero, we divide by λ, and rename the constants μ, v_a, and v_b to write these conditions in the form

$$\begin{cases} y(a) + \mu y'(a) = v_a = constant \\ y(b) + \mu y'(b) = v_b = constant \end{cases} \tag{7.4d}$$

7.2   Taylor Series Approximations for Equations Subject to Initial Conditions

If a second order differential equation is subject to the initial conditions of eq. 7.3, $y(x)$ is analytic at x_0. Then one can approximate $y(x)$ by a truncated Taylor series expanded about x_0:

$$y(x) \simeq$$
$$y(x_0) + y'(x_0)(x - x_0) + \ldots + \frac{y^{(n)}(x_0)}{n!}(x - x_0)^n = \sum_{k=0}^{n} \frac{y^{(k)}(x_0)}{k!}(x - x_0)^k \quad (6.5)$$

The initial conditions specify the values of $y(x_0)$ and $y'(x_0)$. From the differential equation given in eq. 7.2, we have

$$y''(x_0) = f(x_0, y_0, y_0') - P(x_0)y_0' - Q(x_0)y_0 \quad (7.5a)$$

Higher derivatives are obtained from derivatives of the differential equation. For example,

$$y'''(x_0) = \left(\frac{\partial f}{\partial x} + \frac{\partial f}{\partial y}y' + \frac{\partial f}{\partial y'}y'' \right)_{x_0}$$
$$- P(x_0)y''(x_0) - [P'(x_0) + Q(x_0)] y_0' - Q'(x_0)y_0 \quad (7.5b)$$

Example 7.1: Taylor series approximations for an equation subject to initial conditions

The solution to

$$y'' - 2xy' - y = y^2 e^{-x^2} \quad (7.6a)$$

subject to the initial conditions

$$\begin{cases} y(0) = 1 \\ y'(0) = 0 \end{cases} \quad (7.6b)$$

is

$$y(x) = e^{x^2} \quad (7.7)$$

Approximating the solution by a Taylor series expanded about $x_0 = 0$ keeping terms up to x^4, we have

$$y(x) \simeq y(0) + y'(0)x + \frac{1}{2!}y''(0)x^2 + \frac{1}{3!}y'''(0)x^3 + \frac{1}{4!}y''''(0)x^4 \quad (7.8)$$

The coefficients of the first two terms are given by the initial conditions. From the differential equation we have

$$y''(0) = \left[y^2 e^{-x^2} + 2xy' + y\right]_{x=0} = 2 \tag{7.9a}$$

$$y'''(0) = \frac{d}{dx}\left[y^2 e^{-x^2} + 2xy' + y\right]\Big|_{x=0}$$
$$= \left[-2xy^2 e^{-x^2} + 2yy' e^{-x^2} + 3y' + 2xy''\right]_{x=0} = 0 \tag{7.9b}$$

and

$$y''''(0) = = \frac{d}{dx}\left[-2xy^2 e^{-x^2} + 2yy' e^{-x^2} + 3y' + 2xy''\right]\Big|_{x=0}$$
$$\left[\left(-2y^2 - 8xyy' + 4x^2 y^2 + 2y'^2 + 2yy''\right)e^{-x^2} + 5y'' + 2xy'''\right]_{x=0} = 12 \tag{7.9c}$$

so that

$$y(x) \simeq 1 + x^2 + \frac{x^4}{2} \tag{7.10}$$

These are, of course, the first three non-zero terms in the MacLaurin series for the exponential solution. From this, we obtain

$$y(0.50) \simeq 1.28125 \tag{7.11a}$$

which is a reasonable estimate of

$$y(0.50) = e^{(0.50)^2} = 1.28403 \tag{7.11b}$$

A more accurate approximation is obtained if a multistep Taylor series is used. For example, to generate a two step Taylor series at $x = 0.50$, we use the truncated Taylor series about $x = 0$ to determine the values of $y(0.25)$ and $y'(0.25)$. From eq. 7.10, we have

$$y(0.25) \simeq \left[1 + x^2 + \frac{x^4}{2}\right]_{x=0.25} = 1.06445 \tag{7.12a}$$

and

$$y'(0.25) \simeq \left[2x + 2x^3\right]_{x=0.25} = 0.53125 \tag{7.12b}$$

Then, from the differential equation and its derivatives (see eqs. 7.9), we have

$$y''(0.25) = \left[y^2 e^{-x^2} + 2xy' + y \right]_{x=0.25} = 2.39449 \qquad (7.12c)$$

$$y'''(0.25) = \left[2\left(-xy^2 + yy' \right)e^{-x^2} + 3y' + 2xy'' \right]_{x=0.25} = 3.32125 \qquad (7.12d)$$

and

$$y''''(0.25) = \left[\left(-2y^2 - 8xyy' + 4x^2 y^2 + 2y'^2 + 2yy'' \right)e^{-x^2} + 5y'' + 2xy''' \right]_{x=0.25}$$

$$= 16.02694 \qquad (7.12e)$$

With such values, we can now express the Taylor series about $x = 0.25$, truncated at the $(x - 0.25)^4$ term as

$$y(x) \simeq \left[y(0.25) + y'(0.25)(x - 0.25) + \frac{y''(0.25)}{2!}(x - 0.25)^2 \right.$$
$$\left. + \frac{y'''(0.25)}{3!}(x - 0.25)^3 + \frac{y''''(0.25)}{4!}(x - 0.25)^4 \right] \qquad (7.13)$$

from which we obtain

$$y(0.50) \simeq 1.28335 \qquad (7.14)$$

which is more accurate than the result obtained using the single step Taylor series approach. □

7.3   Runge–Kutta Approximations for Equations Subject to Initial Conditions

The Runge–Kutta method was introduced in ch. 6 to approximate the solution to a first-order differential equation. It can also be used to estimate the solution of a second order differential equation by expressing it as two first-order equations.

For

$$y'' + P(x)y' + Q(x)y = f(x, y, y') \qquad (7.2)$$

constrained by the initial conditions

$$\begin{cases} y(x_0) = y_0 \\ y'(x_0) = y'_0 \end{cases} \qquad (7.3)$$

we define the first-order differential equation for $y(x)$,

$$y'(x) \equiv w(x) \qquad (7.15a)$$

With this definition, the second order differential equation for $y(x)$ given in eq. 7.2 can be written as a first order equation for $w(x)$:

$$y''(x) = w'(x) = f(x, y, y') - P(x)w - Q(x)y \equiv G(x, y, w) \qquad (7.15b)$$

The initial conditions for $y(x)$ become

$$\begin{cases} y(x_0) = y_0 \\ w(x_0) = w_0 = y'_0 \end{cases} \qquad (7.16)$$

The Runge–Kutta method for solving a second order differential equation involves defining two sets of Runge–Kutta parameters, R and S. With the values of y_0 and w_0 given by initial conditions, we determine the value of R_1. With R_1, we find S_1. Then $S_1 \Rightarrow R_2 \Rightarrow S_2 \Rightarrow \ldots \Rightarrow R_4 \Rightarrow S_4$.

Referring to eq. 7.15a, we begin with

$$R_1 = y'(x_0)h = w(x_0)h \qquad (7.17a)$$

S_1, the Runge–Kutta parameter for the first-order equation for $w(x)$ given in eq. 7.15b is

$$S_1 = G(x_0, y_0, w_0)h = [f(x_0, y_0, w_0) - P(x_0)w_0 - Q(x_0)y_0]h \qquad (7.17b)$$

From these, we determine

$$R_2 = \left(w_0 + \frac{1}{2}S_1\right)h \qquad (7.17c)$$

and

$$S_2 = G\left(x_0 + \frac{1}{2}h, y_0 + \frac{1}{2}R_1, w_0 + \frac{1}{2}S_1\right)h =$$
$$\left[f\left(x_0 + \frac{1}{2}h, y_0 + \frac{1}{2}R_1, w_0 + \frac{1}{2}S_1\right) - P(x_0 + \frac{1}{2}h)(w_0 + \frac{1}{2}S_1)\right.$$
$$\left. -Q(x_0 + \frac{1}{2}h)(y_0 + \frac{1}{2}R_1)\right]h \qquad (7.17d)$$

Then

$$R_3 = \left(w_0 + \frac{1}{2}S_2\right)h \tag{7.17e}$$

and

$$S_3 = G\left(x_0 + \frac{1}{2}h, y_0 + \frac{1}{2}R_2, w_0 + \frac{1}{2}S_2\right)h =$$
$$\left[f\left(x_0 + \frac{1}{2}h, y_0 + \frac{1}{2}R_2, w_0 + \frac{1}{2}S_2\right) - P\left(x_0 + \frac{1}{2}h\right)\left(w_0 + \frac{1}{2}S_2\right)\right.$$
$$\left. -Q\left(x_0 + \frac{1}{2}h\right)\left(y_0 + \frac{1}{2}R_2\right)\right]h \tag{7.17f}$$

Then

$$R_4 = (w_0 + S_3)h \tag{7.17g}$$

and

$$S_4 = G(x_0 + h, y_0 + R_3, w_0 + S_3)h =$$
$$[f(x_0 + h, y_0 + R_3, w_0 + S_3) - P(x_0 + h)(w_0 + S_3)$$
$$-Q(x_0 + h)(y_0 + R_3)]h \tag{7.17h}$$

From these parameters, we obtain

$$y(x) = y(x_0 + h) = y_0 + \frac{1}{6}[R_1 + 2R_2 + 2R_3 + R_4] \tag{7.18a}$$

and

$$w(x) = y'(x_0 + h) = w_0 + \frac{1}{6}[S_1 + 2S_2 + 2S_3 + S_4] \tag{7.18b}$$

Example 7.2: Runge–Kutta approximations for an equation subject to initial conditions

We again consider

$$y'' - 2xy' - y = y^2 e^{-x^2} \tag{7.6a}$$

subject to

$$\begin{cases} y(0) = y_0 = 1 \\ y'(0) = w_0 = 0 \end{cases} \tag{7.6b}$$

the solution to which is

$$y(x) = e^{x^2} \tag{7.7}$$

From this solution, we have the exact values

$$y(0.50) = e^{(0.50)^2} = 1.28403 \tag{7.11b}$$

and

$$w(0.50) = y'(0.50) = 2(0.50)e^{(0.50)^2} = 1.28403 \tag{7.19}$$

The corresponding pair of first-order equations for eq. 7.6a are

$$y'(x) = w(x) \tag{7.20a}$$

and

$$w' = y^2 e^{-x^2} + 2xw + y \tag{7.20b}$$

With $x_0 = 0$ and $h = 0.5$, the R and S parameters are

$$R_1 = w_0 h = 0 \tag{7.21a}$$

$$S_1 = \left[y_0^2 e^{-x_0^2} + 2x_0 w_0 + y_0 \right] h = 1.00000 \tag{7.21b}$$

$$R_2 = \left[w_0 + \frac{1}{2} S_1 \right] h = 0.25000 \tag{7.21c}$$

$$S_2 = \left[\left(y_0 + \frac{1}{2} R_1 \right)^2 e^{-\left(x_0 + \frac{1}{2}h\right)^2} + 2\left(x_0 + \frac{1}{2}h\right)\left(w_0 + \frac{1}{2} S_1\right) + \left(y_0 + \frac{1}{2} R_1\right) \right] h$$
$$= 1.09471 \tag{7.21d}$$

$$R_3 = \left[w_0 + \frac{1}{2} S_2 \right] h = 0.27368 \tag{7.21e}$$

$$S_3 = \left[\left(y_0 + \frac{1}{2} R_2 \right)^2 e^{-\left(x_0 + \frac{1}{2}h\right)^2} + 2\left(x_0 + \frac{1}{2}h\right)\left(w_0 + \frac{1}{2} S_2\right) + \left(y_0 + \frac{1}{2} R_2\right) \right] h$$
$$= 1.29381 \tag{7.21f}$$

$$R_4 = [w_0 + S_3]h = 0.64691 \qquad (7.21\text{g})$$

and

$$S_4 = \left[(y_0 + R_3)^2 e^{-(x_0+h)^2} + 2(x_0 + h)(w_0 + S_3) + (y_0 + R_3)\right]h$$
$$= 1.91545 \qquad (7.21\text{h})$$

Then,

$$y(0.50) \simeq y_0 + \frac{1}{6}[R_1 + 2R_2 + 2R_3 + R_4] = 1.28238 \qquad (7.22\text{a})$$

and

$$w(0.50) \simeq w_0 + \frac{1}{6}[S_1 + 2S_2 + 2S_3 + S_4] = 1.28208 \qquad (7.22\text{b})$$

To develop a multiple step Runge–Kutta process, we take $h = 0.25$ and apply the above process to find $y(0.25)$ and $w(0.25)$. Then setting $x_0 = 0.25$, $y_0 = y(0.25)$, and $w_0 = w(0.25)$, and applying the Runge–Kutta method to find $y(0.50)$ and $w(0.50)$, we obtain

$$y(0.50) \simeq 1.28382 \qquad (7.23\text{a})$$

and

$$w(0.50) \simeq 1.28378 \qquad (7.23\text{b})$$

Comparing eqs. 7.14 and 7.23a to the exact value given in eq. 7.11b, we see that the two step Runge–Kutta approach is slightly more accurate than the two step truncated Taylor series of example 7.1. $\square$

7.4   Finite Difference Approximate Solutions for Equations Subject to Initial Conditions

Euler finite differences

As developed in ch. 6, the Euler finite difference approximation to y' is given by

$$y'_k \simeq \frac{y_{k+1} - y_k}{h} \qquad (6.39)$$

From this, the Euler approximation to $y''(x)$ is obtained by

$$\left.\frac{d^2 y}{dx^2}\right|_{x_k} = \left.\frac{dy'}{dx}\right|_{x_k} \equiv y''_k \simeq \frac{1}{h}\left(y'_{k+1} - y'_k\right)$$

$$\simeq \frac{1}{h^2}\left[(y_{k+2} - y_{k+1}) - (y_{k+1} - y_k)\right] = \frac{y_{k+2} - 2y_{k+1} + y_k}{h^2} \quad (7.24)$$

Since the accuracy of the Euler approximation to y'_k is the same as the accuracy of the Taylor series for $y(x)$ expanded about x_k up to order h, the approximation to y''_k given in eq. 7.24 is also accurate up to order h.

Milne finite differences

The Milne finite difference approximation to $y'(x)$ is

$$y'(x_k) \simeq \frac{y_{k+1} - y_{k-1}}{2h} \quad (6.49)$$

To obtain the Milne approximation to y''_k, we consider the Taylor series

$$y_{k+1} = y(x_k + h) = y_k + y'_k h + \frac{1}{2}y''_k h^2 + \dots \quad (6.40)$$

and

$$y_{k-1} = y(x_k - h) = y_k - y'_k h + \frac{1}{2}y''_k h^2 - \dots \quad (6.42)$$

Adding these equations and truncating these Taylor series at the term containing h^2, we have

$$y_{k+1} + y_{k-1} = 2y_k + y''_k h^2 \quad (7.25)$$

from which, the Milne approximation to y''_k is

$$y''_k \simeq \frac{y_{k+1} - 2y_k + y_{k-1}}{h^2} \quad (7.26)$$

Since the Taylor series is truncated at the h^2 term, eq. 7.26 is as accurate as the Taylor series up to order h^2.

One might consider approximating a Milne approximation to y''_k from

$$y''_k \simeq \frac{y'_{k+1} - y'_{k-1}}{2h} \quad (7.27)$$

Referring to

$$y'_k \simeq \frac{y_{k+1} - y_{k-1}}{2h} \qquad (6.44)$$

the Milne approximation to y_k'' can be expressed as

$$y_k'' \simeq \frac{1}{2h}\left[\left(\frac{y_{k+2} - y_k}{2h}\right) - \left(\frac{y_k - y_{k-2}}{2h}\right)\right] = \frac{y_{k+2} - 2y_k + y_{k-2}}{4h^2} \qquad (7.28)$$

This approximation requires determining values of y at x_{k+2}, x_k, and x_{k-2}, which is a wider range of x than is accessed in eq. 7.26, with no improvement in the accuracy of the approximated y_k''. As such, eq. 7.28 is never used to approximate y_k''.

To determine the solution to

$$y'' + P(x)y' + Q(x)y = f(x, y, y') \qquad (7.2)$$

with given initial conditions

$$\begin{cases} y(x_0) = y_0 \\ y'(x_0) = y'_0 \end{cases} \qquad (7.3)$$

we substitute the finite difference approximations to y_k' and y_k'' into eq. 7.2. We use the notation

$$P_k \equiv P(x_k) \qquad (7.29a)$$

$$Q_k \equiv Q(x_k) \qquad (7.29b)$$

and

$$f_k \equiv f(x_k, y_k, y'_k) \qquad (7.29c)$$

Since it is specified by the initial conditions, we do not approximate y_0' by finite differences. y_k' is only approximated by finite differences for $k > 0$.

Euler and Milne finite difference equations

For $k = 0$, with y_0' given by initial conditions, the *Euler finite difference equation* for eq. 7.2 is given by

$$\frac{y_2 - 2y_1 + y_0}{h^2} + P_0 y'_0 + Q_0 y_0 = f(x_0, y_0, y'_0) \equiv f_0 \qquad (7.30a)$$

For $k > 0$, y_k' must be approximated by finite differences. Then the Euler approximation to eq. 7.2 is

$$\frac{y_{k+2} - 2y_{k+1} + y_k}{h^2} + P_k \frac{y_{k+1} - y_k}{h} + Q_k y_k = f_k \tag{7.30b}$$

Eq. 7.30a can be written

$$\left(\frac{1}{h^2} + Q_0\right) y_0 - \frac{2}{h^2} y_1 + \frac{1}{h^2} y_2 = f_0 - P_0 y_0' \tag{7.31a}$$

and eq. 7.30b can be rearranged as

$$\left(\frac{1}{h^2} - \frac{P_k}{h} + Q_k\right) y_k + \left(\frac{P_k}{h} - \frac{2}{h^2}\right) y_{k+1} + \frac{1}{h^2} y_{k+2}$$
$$\simeq f\left(x_k, y_k, \frac{y_{k+1} - y_k}{h}\right) \equiv f_k \tag{7.31b}$$

To determine y_2 from eq. 7.31a, it is necessary to obtain an approximate value of y_1 from the initial conditions. Referring to eq. 6.44,

$$y_1 \simeq y_0 + h y_0' \tag{7.32}$$

Then eq. 7.31a becomes

$$\left(Q_0 - \frac{1}{h^2}\right) y_0 + \frac{1}{h^2} y_2 = f_0 - \left(P_0 - \frac{2}{h}\right) y_0' \tag{7.33}$$

from which we obtain y_2. With the values of y_1 and y_2 we find y_3 from eq. 7.31b, then use that result to determine y_4 and so on. This process is repeated until we obtain a value for $y_N = y(x_N)$.

Using the Milne approximations to y_k' and y_k'', the *Milne finite difference equation* to eq. 7.2 for $k = 1$ is given by

$$\frac{y_2 - 2y_1 + y_0}{h^2} + Q_1 y_1 = f(x_1, y_1, y_0') - P_1 y_0' \tag{7.34a}$$

and for $k > 1$, the Milne finite difference equation is

$$\frac{y_{k+1} - 2y_k + y_{k-1}}{h^2} + P_k \frac{y_{k+1} - y_{k-1}}{2h} + Q_k y_k = f\left(x_k, y_k, \frac{y_{k+1} - y_{k-1}}{2h}\right) \tag{7.34b}$$

We write these equations as

$$\frac{1}{h^2}y_0 + \left(Q_1 - \frac{2}{h^2}\right)y_1 + \frac{1}{h^2}y_2 = f(x_1, y_1, y_0') - P_1 y_0' \qquad (7.35a)$$

and

$$\left(\frac{1}{h^2} - \frac{P_k}{2h}\right)y_{k-1} + \left(Q_k - \frac{2}{h^2}\right)y_k + \left(\frac{1}{h^2} + \frac{P_k}{2h}\right)y_{k+1}$$
$$= f\left(x_k, y_k, \frac{y_{k+1} - y_{k-1}}{2h}\right) \qquad (7.35b)$$

To obtain y_2 from eq. 7.35a, we must have an estimate of the value of y_1. Since the Milne approximation is as accurate as a Taylor series truncated at h^2, we use the differential equation at x_0 to determine y_0'' and approximate y_1 by

$$y_1 \simeq y_0 + hy_0' + \frac{1}{2}h^2 y_0'' = y_0 + y_0'h + \frac{1}{2}h^2(f_0 - P_0 y_0' - Q_0 y_0) \qquad (7.36)$$

Then, from eq. 7.35a, the equation for y_2 becomes

$$y_2\left(\frac{1}{h^2} + \frac{P_1}{2h}\right) = f\left(x_1, y_1, \frac{y_2 - y_0}{2h}\right) + y_0\left(\frac{P_1}{2h} - \frac{1}{h^2}\right) + y_1\left(\frac{2}{h^2} - Q_1\right) \qquad (7.37)$$

Example 7.3: Finite difference approximations for an equation subject to initial conditions

We again consider

$$y'' - 2xy' - y = y^2 e^{-x^2} \qquad (7.6a)$$

subject to

$$\begin{cases} y_0 = 1 \\ y_0' = 0 \end{cases} \qquad (7.6b)$$

the solution to which is

$$y(x) = e^{x^2} \qquad (7.7)$$

(a) For the Euler approximation, we find y_1 from the initial conditions to be

$$y_1 \simeq y_0 + hy_0' = y_0 = 1 \qquad (7.38)$$

The Euler finite difference approximation to eq. 7.6a is given by

$$\frac{y_{k+2} - 2y_{k+1} + y_k}{h^2} - 2x_k \frac{y_{k+1} - y_k}{h} - y_k = y_k^2 e^{-x_k^2} \qquad (7.39a)$$

from which

$$y_{k+2} = h^2 y_k^2 e^{-x_k^2} + 2(1 + hx_k)y_{k+1} + \left(h^2 - 2hx_k - 1\right)y_k \qquad (7.39b)$$

To determine $y(0.50)$, with $h = 0.10$, we set $k = 0$ in eq. 7.39b. With y_1 given by eq. 7.38, we obtain

$$y_2 = y(0.2) \simeq h^2 y_0^2 e^{-x_0^2} + 2(1 + hx_0)y_1 + \left(h^2 - 2hx_0 - 1\right)y_0 = 1.02000 \quad (7.40a)$$

Then, for $k = 1$, we have

$$
\begin{aligned}
y_3 = y(0.3) &\simeq h^2 y_1^2 e^{-x_1^2} + 2(1 + hx_1)y_2 + \left(h^2 - 2hx_1 - 1\right)y_1 \\
&= (.1)^2 e^{-(.1)^2} + 2\left(1 + (.1)^2\right)(1.02000) + \left((.1)^2 - 2(.1)(.1) - 1\right) \\
&= 1.06030
\end{aligned}
\qquad (7.40b)
$$

With $k = 2$, we find

$$y_4 = 1.12241 \qquad (7.40c)$$

and finally, setting $k = 3$, we obtain

$$y_5 = y(0.50) = 1.20912 \qquad (7.40d)$$

Comparing this to the exact value of 1.28403, we see that this result is not very accurate.
The accuracy of the result can be improved by taking smaller values of h. For $h = 0.01$, we obtain

$$y(0.50) \simeq 1.27542 \qquad (7.41a)$$

and with $h = 0.001$, we find

$$y(0.50) \simeq 1.28315 \qquad (7.41b)$$

(b) The Milne finite difference equation for eq. 7.6a is

$$\frac{y_{k+1} - 2y_k + y_{k-1}}{h^2} - 2x_k \frac{y_{k+1} - y_{k-1}}{2h} - y_k = y_k^2 e^{-x_k^2} \qquad (7.42a)$$

from which

$$y_{k+1} = \frac{h^2 y_k^2 e^{-x_k^2} + (2 + h^2)y_k - (1 + hx_k)y_{k-1}}{(1 - hx_k)} \tag{7.42b}$$

We immediately see that we must choose each x_k and an interval size h such that the denominator is never zero for any $x \,\varepsilon\, [x_0, x_N]$.
We begin the process with $k = 1$ so that we do not include the extrapolated value y_{-1}. This yields

$$y_2 = \frac{h^2 y_1^2 e^{-x_1^2} + (2 + h^2)y_1 - (1 + hx_1)y_0}{(1 - hx_1)} \tag{7.43}$$

To obtain an independent value of y_1, we use the Taylor series truncated at h^2. We obtain

$$y_1 \simeq y_0 + hy_0' + \frac{1}{2}h^2 y_0'' = 1 + h^2 \tag{7.44}$$

With this analysis, for $h = 0.10$, we obtain

$$y(0.50) \simeq 1.28253 \tag{7.45a}$$

With $h = 0.01$, we find

$$y(0.50) \simeq 1.28401 \tag{7.45b}$$

which is a fairly accurate result. $\square$

7.5 Finite Difference Approximations for Equations Subject to Boundary Conditions

All linear terms in y and y' in

$$y'' + P(x)y' + Q(x)y = f(x, y, y') \tag{7.2}$$

are included in $Q(x)y$ and $P(x)y'$, respectively. Therefore, when eq. 7.2 is approximated by a finite difference equation, $f(x, y, y')$ contains only nonlinear terms in y and y'.
Solving simultaneous equations obtained from finite difference approximations that contain nonlinear terms in the unknowns is always a difficult task. The only way to avoid this nonlinearity in the unknowns is to impose the restriction that

$$f(x, y, y') = f(x) \tag{7.46}$$

Therefore, our discussion and examples will be restricted to differential equations of the form

$$y'' + P(x)y' + Q(x)y = f(x) \tag{7.47}$$

Referring to eq. 7.4a, boundary conditions are given by specifying the values of $y(x_0)$ and $y(x_N)$, or $y'(x_0)$ and $y'(x_N)$, or a linear combination of $y(x_0)$ $y'(x_0)$ and the same linear combination of $y(x_N)$ and $y'(x_N)$, where x_0 and x_N define the boundaries of the problem and $y(x)$ is defined for $x \ \varepsilon \ [x_0, x_N]$.

Finite difference solution to equations constrained by Dirichlet boundary conditions

Let eq. 7.47 be constrained by Dirichlet boundary conditions of the form

$$\begin{cases} y(x_0) = y_0 = constant \\ y(x_N) = y_N = constant \end{cases} \tag{7.4a}$$

With y_0 and y_N given, we must solve for the $N - 1$ unknowns $\{y_1, \ldots, y_{N-1}\}$.

Referring to eq. 7.30b, the Euler finite difference approximation to eq. 7.47 can be written as

$$\left(\frac{1}{h^2} - \frac{P_k}{h} + Q_k\right)y_k + \left(\frac{P_k}{h} - \frac{2}{h^2}\right)y_{k+1} + \frac{1}{h^2}y_{k+2} = f(x_k) \equiv f_k \tag{7.31b}$$

With y_0 and y_N given by the boundary conditions, we take $0 \le k \le N - 2$ to generate the following set of linear equations for $\{y_1, \ldots, y_{N-1}\}$:

$$\left(\frac{P_0}{h} - \frac{2}{h^2}\right)y_1 + \frac{1}{h^2}y_2 = f_0 - \left(\frac{1}{h^2} - \frac{P_0}{h} + Q_0\right)y_0 \tag{7.48a}$$

$$\left(\frac{1}{h^2} - \frac{P_1}{h} + Q_1\right)y_1 + \left(\frac{P_1}{h} - \frac{2}{h^2}\right)y_2 + \frac{1}{h^2}y_3 = f_1 \tag{7.48b}$$

$$\left(\frac{1}{h^2} - \frac{P_{N-3}}{h} + Q_{N-3}\right)y_{N-3} + \left(\frac{P_{N-3}}{h} - \frac{2}{h^2}\right)y_{N-2} + \frac{1}{h^2}y_{N-1} = f_{N-3} \tag{7.48c}$$

$$\left(\frac{1}{h^2} - \frac{P_{N-2}}{h} + Q_{N-2}\right)y_{N-2} + \left(\frac{P_{N-2}}{h} - \frac{2}{h^2}\right)y_{N-1} = f_{N-2} - \frac{1}{h^2}y_N \tag{7.48d}$$

The Milne finite difference equation for eq. 7.47 is

$$\frac{y_{k+1} - 2y_k + y_{k-1}}{h^2} + P_k \frac{y_{k+1} - y_{k-1}}{2h} + Q_k y_k = f_k \qquad (7.49a)$$

which we write as

$$\left(\frac{1}{h^2} - \frac{P_k}{2h}\right) y_{k-1} + \left(Q_k - \frac{2}{h^2}\right) y_k + \left(\frac{1}{h^2} + \frac{P_k}{2h}\right) y_{k+1} = f_k \qquad (7.49b)$$

Taking $1 \le k \le N-1$, we obtain the following $N-1$ linear equations for $\{y_1, \ldots, y_{N-1}\}$:

$$\left(Q_1 - \frac{2}{h^2}\right) y_1 + \left(\frac{1}{h^2} + \frac{P_1}{2h}\right) y_2 = f_1 - \left(\frac{1}{h^2} - \frac{P_1}{2h}\right) y_0 \qquad (7.50a)$$

$$\left(\frac{1}{h^2} - \frac{P_2}{2h}\right) y_1 + \left(Q_2 - \frac{2}{h^2}\right) y_2 + \left(\frac{1}{h^2} + \frac{P_2}{2h}\right) y_3 = f_2 \qquad (7.50b)$$

$$\left(\frac{1}{h^2} - \frac{P_{N-2}}{2h}\right) y_{N-3} + \left(Q_{N-2} - \frac{2}{h^2}\right) y_{N-2} + \left(\frac{1}{h^2} + \frac{P_{N-2}}{2h}\right) y_{N-1} = f_{N-2} \qquad (7.50c)$$

$$\left(\frac{1}{h^2} - \frac{P_{N-1}}{2h}\right) y_{N-2} + \left(Q_{N-1} - \frac{2}{h^2}\right) y_{N-1} + = f_{N-1} - \left(\frac{1}{h^2} + \frac{P_{N-1}}{2h}\right) y_N \qquad (7.50d)$$

Example 7.4: Finite difference approximations for an equation subject to Dirichlet boundary conditions

The solution to

$$y'' - 2xy' - 2y = 0 \qquad (7.51a)$$

constrained by the Dirichlet boundary conditions

$$\begin{cases} y(0) = 1 \\ y(1) = e \end{cases} \qquad (7.51b)$$

is

$$y(x) = e^{x^2} \quad 0 \le x \le 1 \qquad (7.52)$$

With $h = 0.25$, $x_0 = 0$, and $x_4 = 1$. We therefore solve for $\{y_1, y_2, y_3\}$ by finite difference methods, and compare the solutions to

$$Y_{exact} = \begin{pmatrix} y(0.25) \\ y(0.50) \\ y(0.75) \end{pmatrix} = \begin{pmatrix} y_1 \\ y_2 \\ y_3 \end{pmatrix} = \begin{pmatrix} 1.06449 \\ 1.28403 \\ 1.75505 \end{pmatrix} \qquad (7.53a)$$

If we take $h = 0.10$, then $x_0 = 0$ and $x_{10} = 1$ and we solve for $\{y_1, \ldots, y_9\}$. We compare those results to

$$Y_{exact} = \begin{pmatrix} y(0.10) \\ y(0.20) \\ y(0.30) \\ y(0.40) \\ y(0.50) \\ y(0.60) \\ y(0.70) \\ y(0.80) \\ y(0.90) \end{pmatrix} = \begin{pmatrix} y_1 \\ y_2 \\ y_3 \\ y_4 \\ y_5 \\ y_6 \\ y_7 \\ y_8 \\ y_9 \end{pmatrix} = \begin{pmatrix} 1.01001 \\ 1.04081 \\ 1.09417 \\ 1.17351 \\ 1.28403 \\ 1.43333 \\ 1.63232 \\ 1.89648 \\ 2.24791 \end{pmatrix} \qquad (7.53b)$$

(a) Using the Euler finite difference approximations with $h = 0.25$, eqs. 7.48 become

$$-32y_1 + 16y_2 = -14 \quad (k = 0) \qquad (7.54a)$$

$$16y_1 - 34y_2 + 16y_3 = 0 \quad (k = 1) \qquad (7.54b)$$

and

$$18y_2 - 36y_3 == -16e = -43.49251 \quad (k = 2) \qquad (7.54c)$$

The solution to this set of equations is

$$Y = \begin{pmatrix} y_1 \\ y_2 \\ y_3 \end{pmatrix} = \begin{pmatrix} y(0.25) \\ y(0.50) \\ y(0.75) \end{pmatrix} = \begin{pmatrix} 1.16889 \\ 1.46278 \\ 1.93951 \end{pmatrix} \qquad (7.55)$$

Referring to eq. 7.53a we see that the Euler finite difference solutions are not very accurate.

With $h = 0.10$, we obtain

$$Y = \begin{pmatrix} y_1 \\ y_2 \\ y_3 \\ y_4 \\ y_5 \\ y_6 \\ y_7 \\ y_8 \\ y_9 \end{pmatrix} = \begin{pmatrix} y(0.1) \\ y(0.2) \\ y(0.3) \\ y(0.4) \\ y(0.5) \\ y(0.6) \\ y(0.7) \\ y(0.8) \\ y(0.9) \end{pmatrix} = \begin{pmatrix} 1.02836 \\ 1.07673 \\ 1.14663 \\ 1.24086 \\ 1.36367 \\ 1.52113 \\ 1.72161 \\ 1.97657 \\ 2.30165 \end{pmatrix} \qquad (7.56)$$

Comparing these to the exact values given in eq. 7.53b, as expected, a smaller step size h yields a more accurate solution.

(b) For $h = 0.25$, the Milne finite difference equations for $\{y_1, y_2, y_3\}$ are

$$-34y_1 + 15y_2 = -17 \quad (k = 1) \tag{7.57a}$$

$$18y_1 - 34y_2 + 14y_3 = 0 \quad (k = 2) \tag{7.57b}$$

and

$$19y_2 - 34y_3 = -13e = -35.33766 \quad (k = 3) \tag{7.57c}$$

from which we obtain

$$Y = \begin{pmatrix} y_1 \\ y_2 \\ y_3 \end{pmatrix} = \begin{pmatrix} y(0.25) \\ y(0.50) \\ y(0.75) \end{pmatrix} = \begin{pmatrix} 1.06978 \\ 1.29150 \\ 1.76106 \end{pmatrix} \tag{7.58}$$

Comparing this to the exact values given in eq. 7.53a, we see that is a more accurate result than that obtained using the Euler finite difference with the same step size h. $\square$

When boundary conditions other than Dirichlet conditions are imposed, in addition to y_k with $1 \le k \le N{-}1$, y_0 and y_N are unknown. Therefore, we must solve for $N + 1$ unknown quantities $\{y_0, \ldots, y_N\}$ and thus, must generate $N + 1$ equations.

Finite difference solution to equations constrained by Neumann boundary conditions

Let

$$y'' + P(x)y' + Q(x)y = f(x) \tag{7.47}$$

be subject to Neumann boundary conditions

$$\begin{cases} y'(x_0) = y'_0 = constant \\ y'(x_N) = y'_N = constant \end{cases} \tag{7.59}$$

With Neumann conditions, the Euler finite difference equations for eq. 7.47 are

$$\frac{y_2 - 2y_1 + y_0}{h^2} + Q_0 y_0 = f_0 - P_0 y'_0 \tag{7.60a}$$

$$\frac{y_{k+2} - 2y_{k+1} + y_k}{h^2} + P_k \frac{y_{k+1} - y_k}{h} + Q_k y_k = f_k \quad (1 \le k \le N - 2) \qquad (7.60\text{b})$$

$$\frac{y_{N+1} - 2y_N + y_{N-1}}{h^2} + P_{N-1} \frac{y_N - y_{N-1}}{h} + Q_{N-1} y_{N-1} = f_{N-1} \qquad (7.60\text{c})$$

and

$$\frac{y_{N+2} - 2y_{N+1} + y_N}{h^2} + Q_N y_N = f_N - P_N y'_N \qquad (7.60\text{d})$$

Eqs. 7.60c and 7.60d contain y_{N+1} and y_{N+2} which are values outside the domain of y. We can approximate y_{N+1} from the Euler finite difference approximation to the Neumann boundary condition at x_N. With

$$y'_N \simeq \frac{y_{N+1} - y_N}{h} \qquad (7.61\text{a})$$

we obtain

$$y_{N+1} \simeq y_N + h y'_N \qquad (7.61\text{b})$$

Substituting this into eq. 7.60c we have

$$\frac{-y_N + y_{N-1}}{h^2} + P_{N-1} \frac{y_N - y_{N-1}}{h} + Q_{N-1} y_{N-1} = f_{N-1} - \frac{y'_N}{h} \qquad (7.62)$$

We cannot use the boundary conditions to approximate a value of y_{N+2}. We could approximate y_{N+2} by a Taylor series truncated at the term in the first power of h

$$y_{N+2} = y(x_N + 2h) \simeq y_N + 2h y'_N \qquad (7.63)$$

Substituting this and eq. 7.61b into the Euler finite difference approximation to y_N'', this approximation results in

$$y_N'' \simeq 0 \qquad (7.64)$$

which, in general, is not a valid approximation to y_N''. Thus, there is no reliable estimate of y_{N+2} and eq. 7.60d cannot be used.

This leaves us with N equations to determine $N + 1$ unknown y values. We therefore, must have one more independent equation. We obtain this from the Neumann boundary condition at x_0:

$$y'_0 \simeq \frac{y_1 - y_0}{h} \qquad (7.65)$$

Then, with eqs. 7.65, 7.60b, and 7.62,

$$- y_0 + y_1 = hy_0' \tag{7.66a}$$

$$\left(\frac{1}{h^2} - \frac{P_k}{h} + Q_k\right) y_k - \left(\frac{2}{h^2} - P_k\right) y_{k+1} + \frac{1}{h^2} y_{k+2} = f_k \quad 0 \le k \le N - 2 \tag{7.66b}$$

and

$$\left(Q_{N-1} - \frac{P_{N-1}}{h} + \frac{1}{h^2}\right) y_{N-1} + \left(\frac{P_{N-1}}{h} - \frac{1}{h^2}\right) y_N = f_{N-1} - \frac{y_N'}{h} \tag{7.66c}$$

are the $N + 1$ independent equations needed to determine $\{y_0, \ldots, y_N\}$ using Euler finite differences.

The Milne finite difference equations for eq. 7.47 are given by

$$\frac{y_1 - 2y_0 + y_{-1}}{h^2} + Q_0 y_0 = f_0 - P_0 y_0' \tag{7.67a}$$

$$\frac{y_{k+1} - 2y_k + y_{k-1}}{h^2} + P_k \frac{y_{k+1} - y_{k-1}}{2h} + Q_k y_k = f_k \quad 1 \le k \le N - 1 \tag{7.67b}$$

and

$$\frac{y_{N+1} - 2y_N + y_{N-1}}{h^2} + Q_N y_N = f_N - P_N y_N' \tag{7.67c}$$

We see that eqs. 7.67a and 7.67c contain y_{-1} and y_{N+1} values outside the domain of y. Reliable approximations of these y values are found from the Milne finite difference approximations to the Neumann boundary conditions by writing

$$y_0' \simeq \frac{y_1 - y_{-1}}{2h} \tag{7.68a}$$

from which

$$y_{-1} \simeq y_1 - 2hy_0' \tag{7.68b}$$

and

$$y_N' \simeq \frac{y_{N+1} - y_{N-1}}{2h} \tag{7.68c}$$

which yields

$$y_{N+1} \simeq y_{N-1} + 2hy_N' \tag{7.68d}$$

Then, eq. 7.67a can be written as

$$\left(Q_0 - \frac{2}{h^2}\right)y_0 + \frac{2}{h^2}y_1 = f_0 - \left(P_0 - \frac{2}{h}\right)y_0' \qquad (7.69a)$$

and eq. 7.67c becomes

$$\frac{2}{h^2}y_{N-1} + \left(Q_N - \frac{2}{h^2}\right)y_N = f_N - \left(P_N + \frac{2}{h}\right)y_N' \qquad (7.69b)$$

For $1 \leq k \leq N-1$, the Milne finite difference approximations to eq. 7.67b, written in the form

$$\left(\frac{1}{h^2} - \frac{P_k}{2h}\right)y_{k-1} + \left(Q_k - \frac{2}{h^2}\right)y_k + \left(\frac{1}{h^2} + \frac{P_k}{2h}\right)y_{k+1} = f_k \qquad (7.70)$$

which, along with eqs. 7.69, are the $N+1$ equations needed to solve for $\{y_0, \ldots, y_N\}$.

Example 7.5: Finite difference approximations for an equation subject to Neumann boundary conditions

We again consider the differential equation

$$y'' - 2xy' - 2y = 0 \qquad (7.51a)$$

which is satisfied by

$$y(x) = e^{x^2} \quad 0 \leq x \leq 1 \qquad (7.52)$$

The Neumann boundary conditions that yield this solution are

$$\begin{cases} y'(0) = y_0' = 0 \\ y'(1) = y_N' = 2e \simeq 5.43656 \end{cases} \qquad (7.71)$$

We take $h = 0.25$, and solve for the set $\{y_0, \ldots, y_4\}$. The exact solution for this set, given by eq. 7.52, is

$$Y_{exact} = \begin{pmatrix} 1.00000 \\ 1.06449 \\ 1.28403 \\ 1.75505 \\ 2.71828 \end{pmatrix} \qquad (7.72)$$

(a) Referring to eqs. 7.66, the Euler finite difference equations are

$$- y_0 + y_1 = 0 \qquad (7.73a)$$

$$7y_0 - 16y_1 + 8y_2 = 0 \qquad (7.73b)$$

$$8y_1 - 17y_2 + 8y_3 = 0 \qquad (7.73c)$$

$$9y_2 - 18y_3 + 8y_4 = 0 \qquad (7.73d)$$

and

$$10y_3 - 11y_4 = -4e = -10.07313 \qquad (7.73e)$$

The solution to this set of equations is

$$Y = \begin{pmatrix} y_0 \\ y_1 \\ y_2 \\ y_3 \\ y_4 \end{pmatrix} = \begin{pmatrix} y(0.00) \\ y(0.25) \\ y(0.50) \\ y(0.75) \\ y(1.00) \end{pmatrix} = \begin{pmatrix} 1.64998 \\ 1.64998 \\ 1.85623 \\ 2.29451 \\ 3.07438 \end{pmatrix} \qquad (7.74)$$

Comparing these to the exact values, it is clear that they are quite inaccurate, particularly the value of y_0. Dirichlet conditions specify the exact values of y_0 and y_N but for Neumann (and later Gauss) conditions, these values of y are obtained from finite difference of the boundary conditions. The inaccuracy in these approximations adds to the inaccuracy inherent in the finite difference approximations to the differential equation.

Substituting the exact values for y_0 and y_1 into eq. 7.73a we obtain

$$- y_0 + y_1 = 0.06449 \qquad (7.75a)$$

Thus, eq. 7.73a is a poor approximation to the exact value of $-y_0 + y_1$. Similarly, substituting the exact values of y_3 and y_4 into eq. 7.73e, we have

$$10y_3 - 11y_4 = -12.35058 \qquad (7.75b)$$

indicating that $10y_3 - 11y_4$ is inaccurately approximated by eq. 7.73e.

(b) Using Milne finite differences with $h = 0.25$, we refer to eqs. 7.69 and 7.70 to obtain

$$- 17y_0 + 16y_1 = 0 \qquad (7.76a)$$

$$17y_0 - 34y_1 + 15y_2 = 0 \qquad (7.76b)$$

$$9y_1 - 17y_2 + 7y_3 = 0 \tag{7.76c}$$

$$19y_2 - 34y_3 + 13y_4 = 0 \tag{7.76d}$$

and

$$16y_3 - 17y_4 = -6e = -16.30969 \tag{7.76e}$$

The solution to this set of equations is found to be

$$Y = \begin{pmatrix} y(0.00) \\ y(0.25) \\ y(0.50) \\ y(0.75) \\ y(1.00) \end{pmatrix} = \begin{pmatrix} 0.92828 \\ 0.98630 \\ 1.18356 \\ 1.60626 \\ 2.47117 \end{pmatrix} \tag{7.77}$$

which compares reasonably well with the exact values given in eq. 7.72.

Thus, as expected, the Milne finite difference methods are more accurate than Euler schemes for solving second order differential equations with Neumann boundary conditions. To get a sense of how much better the Milne approach is, we compare the solutions obtained by these methods to the exact y values at the points specified in eq. 7.72. The percent errors from the exact values by the two finite difference methods for each y are

$$\%Error_{Euler} = \begin{pmatrix} 65\% \\ 55\% \\ 45\% \\ 31\% \\ 13\% \end{pmatrix} \quad \%Error_{Milne} = \begin{pmatrix} 7\% \\ 7\% \\ 8\% \\ 8\% \\ 9\% \end{pmatrix} \tag{7.78}$$

From these results, it is clear that one should use Milne rather than Euler finite differences when solving an equation subject to Neumann boundary conditions. □

Finite difference solution to equations constrained by Gauss boundary conditions

The finite difference techniques used to find the solution to

$$y'' + P(x)y' + Q(x)y = f(x) \tag{7.47}$$

subject to Gauss (mixed) boundary conditions

$$\begin{cases} y(x_0) + \mu y'(x_0) = y_0 + \mu y'_0 = v_0 \\ y(x_N) + \mu y'(x_N) = y_N + \mu y'_N = v_N \end{cases} \tag{7.79}$$

are the same as those developed for equations subject to Neumann conditions with y_0' replaced by $(v_0 - y_0)/\mu$ and y_N' replaced by $(v_N - y_N)/\mu$.

As with equations subject to Neumann boundary conditions, we do not use the Euler finite difference equation for $k = N$ because it contains y_{N+2} which cannot be estimated from the boundary condition at x_N. In addition, as discussed above [see eq. 7.64], a Taylor series truncated at h yields an unrealistic result for y_N'. Thus, we discard the Euler finite difference equation for $k = N$ and are left with N equations to determine the set of $N + 1$ y values. The Euler approximation to the Gauss boundary condition at x_0, written in the form

$$y_0\left(1 - \frac{\mu}{h}\right) + y_1\frac{\mu}{h} = v_0 \tag{7.80}$$

augments this set of equations, giving us the necessary $N + 1$ equations.

The Euler approximation to y_{N+1} is found from the Gauss boundary condition at x_N:

$$y_N + \mu\frac{y_{N+1} - y_N}{h} = v_N \tag{7.81a}$$

from which

$$y_{N+1} = y_N\left(1 - \frac{h}{\mu}\right) + \frac{hv_N}{\mu} \tag{7.81b}$$

With this, the Euler finite difference equation for $k = N{-}1$ becomes

$$y_{N-1}\left(\frac{1}{h^2} - \frac{P_{N-1}}{h} + Q_{N-1}\right) + y_N\left(-\frac{1}{h^2} - \frac{1}{\mu h} + \frac{P_{N-1}}{h}\right)$$
$$= f_{N-1} - \frac{v_N}{\mu h} \tag{7.82}$$

The Euler finite difference equations for $0 \le k \le N{-}2$

$$\frac{y_{k+2} - 2y_{k+1} + y_k}{h^2} + P_k\frac{y_{k+1} - y_k}{h} + Q_k y_k = f_k \tag{7.83}$$

along with eqs. 7.80 and 7.83 are the $N + 1$ independent equations used to solve for $\{y_0, \ldots, y_N\}$.

The Milne finite difference approximations to eq. 7.47 for $0 \leq k \leq N$ are given by

$$\frac{y_1 - 2y_0 + y_{-1}}{h^2} + y_0 \left(Q_0 - \frac{P_0}{\mu} \right) = f_0 - \frac{P_0}{\mu} v_0 \tag{7.84a}$$

$$\frac{y_{k+1} - 2y_k + y_{k-1}}{h^2} + P_k \frac{y_{k+1} - y_{k-1}}{2h} + Q_k y_k = f_k \quad 1 \leq k \leq N - 1 \tag{7.84b}$$

and

$$\frac{y_{N+1} - 2y_N + y_{N-1}}{h^2} + \left(Q_N - \frac{P_N}{\mu} \right) y_N = f_N - \frac{P_N v_N}{\mu} \tag{7.84c}$$

The approximation to y_{-1} is obtained from the Milne finite difference form of the Gauss boundary condition at x_0,

$$y_0 + \mu \frac{y_1 - y_{-1}}{2h} = v_0 \tag{7.85a}$$

which yields

$$y_{-1} = y_1 + \frac{2h}{\mu} (y_0 - v_0) \tag{7.85b}$$

Substituting this into eq. 7.84a, we obtain

$$y_0 \left(\frac{2}{\mu h} - \frac{2}{h^2} + Q_0 - \frac{P_0}{\mu} \right) + y_1 \frac{2}{h^2} = f_0 - \frac{P_0}{\mu} + \frac{2v_0}{\mu h} \tag{7.86}$$

y_{N+1} is obtained from the Milne approximation to the Gauss boundary condition at x_N:

$$y_N + \mu \frac{y_{N+1} - y_{N-1}}{2h} = v_N \tag{7.87a}$$

from which

$$y_{N+1} = y_{N-1} + \frac{2h}{\mu} (v_N - y_N) \tag{7.87b}$$

With this, eq. 7.84c becomes

$$y_{N-1} \frac{2}{h^2} + y_N \left(Q_N - \frac{P_N}{\mu} - \frac{2}{\mu h} - \frac{2}{h^2} \right) = f_N - \frac{P_N v_N}{\mu} - \frac{2v_N}{\mu h} \tag{7.88}$$

Eqs. 7.86, 7.84b, and 7.88 are the $N + 1$ equations needed to solve for the set $\{y_0, \ldots, y_N\}$.

Example 7.6: Finite difference approximations for an equation subject to Gauss boundary conditions

The equation

$$y'' - 2xy' - 2y = 0 \qquad (7.51a)$$

with

$$y(x) = e^{x^2} \quad 0 \le x \le 1 \qquad (7.52)$$

can be subjected to the Gauss boundary conditions

$$\begin{cases} y_0 + y'_0 = 1 \\ y_N + y'_N = 3e = 8.15485 \end{cases} \qquad (7.89)$$

As noted in example 7.5, the exact values obtained using $h = 0.25$ are given by

$$Y_{exact} = \begin{pmatrix} 1.00000 \\ 1.06449 \\ 1.28403 \\ 1.75505 \\ 2.71828 \end{pmatrix} \qquad (7.72)$$

(a) The Euler finite difference approximations to eqs. 7.86, 7.84b, and 7.88 are

$$-3y_0 + 4y_1 = 1 \qquad (7.90a)$$

$$7y_0 - 16y_1 + 8y_2 = 0 \qquad (7.90b)$$

$$8y_1 - 17y_2 + 8y_3 = 0 \qquad (7.90c)$$

$$9y_2 - 18y_3 + 8y_4 = 0 \qquad (7.90d)$$

and

$$10y_3 - 13y_4 = -6e = -16.30969 \qquad (7.90e)$$

The solution to this set of equations is

$$Y = \begin{pmatrix} y_0 \\ y_1 \\ y_2 \\ y_3 \\ y_4 \end{pmatrix} = \begin{pmatrix} y(0.00) \\ y(0.25) \\ y(0.50) \\ y(0.75) \\ y(1.00) \end{pmatrix} = \begin{pmatrix} 4.01430 \\ 3.26073 \\ 3.00894 \\ 3.13327 \\ 3.66480 \end{pmatrix} \tag{7.91}$$

which is highly inaccurate.

As discussed for the equation subject to Neumann conditions, the cause of this inaccuracy is the failure of the Euler approximation to yield acceptable estimates of the boundary conditions. Substituting the exact values of y_0 and y_1 from eq. 7.72 into eq. 7.90a, we have

$$-3y_0 + 4y_1 = 1.25796 \tag{7.92a}$$

Therefore, eq. 7.90a is an inaccurate approximation to the exact value of $-3y_0 + 4y_1$. Likewise, substituting the exact values of y_3 and y_4 into eq. 7.90e, we obtain

$$10y_3 - 13y_4 = -17.78714 \tag{7.92b}$$

Thus, eq. 7.90e is not a very accurate estimate of $10y_3 - 13y_4$.

(b) The Milne finite difference approximations to eqs. 7.86, 7.84b, and 7.88 are

$$-13y_0 + 16y_1 = 4 \tag{7.93a}$$

$$17y_0 - 34y_1 + 15y_2 = 0 \tag{7.93b}$$

$$9y_1 - 17y_2 + 7y_3 = 0 \tag{7.93c}$$

$$19y_2 - 34y_3 + 13y_4 = 0 \tag{7.93d}$$

and

$$16y_3 - 20y_4 = -9e = -24.46454 \tag{7.93e}$$

The solution to these equations is found to be

$$Y = \begin{pmatrix} y_0 \\ y_1 \\ y_2 \\ y_3 \\ y_4 \end{pmatrix} = \begin{pmatrix} 0.71458 \\ 0.83059 \\ 1.07283 \\ 1.53753 \\ 2.45325 \end{pmatrix} \tag{7.94}$$

Although these results are not very accurate, it is clear that unlike the results obtained using Euler finite differences, they are reasonable approximations to the solution and can be improved by using a smaller value of h. $\square$

Once again, this example illustrates the fact that the Milne finite difference approximation yields more reliable and more accurate results for second order differential equations constrained by boundary conditions and should always be used rather than Euler methods.

7.6 Shooting Method for Equations Subject to Dirichlet Boundary Conditions

The *shooting method* is used to solve

$$y'' + P(x)y' + Q(x)y = f(x, y, y') \tag{7.2}$$

constrained by Dirichlet boundary conditions

$$\begin{cases} y(x_0) = y_0 \\ y(x_N) = y_N \end{cases} \tag{7.4a}$$

Let $u(x)$ be the particular solution and $v(x)$ be the homogeneous solution to eq. 7.2. That is, $u(x)$ satisfies

$$u'' + P(x)u' + Q(x)u = f(x, y, y') \tag{7.95a}$$

and $v(x)$ is the solution to

$$v'' + P(x)v' + Q(x)v = 0 \tag{7.95b}$$

Subtracting eq. 7.95a from eq. 7.2, we obtain

$$(y - u)'' + P(x)(y - u)' + Q(x)(y - u) = 0 \tag{7.96}$$

Comparing this to eq. 7.95b, we see that since $y(x) - u(x)$ is a solution to the homogeneous equation, it must be proportional to $v(x)$. Therefore,

$$y(x) = u(x) + \alpha v(x) \tag{7.97}$$

With this relation, we see from eq. 7.95a that $u(x)$ satisfies

$$u'' + P(x)u' + Q(x)u = f(x, u + \alpha v, u' + \alpha v') \tag{7.98}$$

But $u(x)$ cannot depend on the proportionality constant α. The only way to eliminate dependence on α is to require that $f(x, y, y')$ be independent of y and y'. Thus, as with finite difference methods, the shooting method is applicable to equations of the form

$$y'' + P(x)y' + Q(x)y = f(x) \tag{7.47}$$

If $y'(x_0) \equiv y_0'$ were known, we could also solve this differential equation using the methods described earlier for equations subject to initial conditions. The goal of the shooting method is to solve the boundary value problem by solving it as initial value problem.

The initial conditions for $u(x)$ and $v(x)$ are obtained from eq. 7.97 by setting

$$y_0 = u(x_0) + \alpha v(x_0) = u_0 + \alpha v_0 \tag{7.99a}$$

and

$$y_0' = u'(x_0) + \alpha v'(x_0) = u_0' + \alpha v_0' \tag{7.99b}$$

Since α and y_0' are undetermined, we choose $v(x)$ such that

$$y_0 = u_0 \tag{7.100a}$$

and

$$y_0' = \alpha \tag{7.100b}$$

This results in

$$\begin{cases} u_0 = y_0 \\ v_0 = 0 \end{cases} \tag{7.101a}$$

and

$$\begin{cases} u'_0 = 0 \\ v'_0 = 1 \end{cases} \tag{7.101b}$$

The equations for $u(x)$ and $v(x)$, constrained by these initial conditions, can be solved by techniques such as the truncated Taylor series, Runge–Kutta or finite difference method, from which we determine $\{u_1, \ldots, u_N\}$ and $\{v_1, \ldots, v_N\}$. Then, if $v_N \neq 0$, the Dirichlet boundary condition for $y(x_N)$ becomes

$$u_N + \alpha v_N = y_N \tag{7.102a}$$

Since u_N and v_N are known and y_N is specified by the boundary condition, we have

$$\alpha = \frac{y_N - u_N}{v_N} \tag{7.102b}$$

With this value of α, we determine all y values from

$$y_k = u_k + \alpha v_k \quad 1 \le k \le N - 1 \tag{7.103}$$

Example 7.7: Shooting method for an equation subject to Dirichlet boundary conditions

We consider the equation

$$y'' - xy' - xy = -4e^{-x} \tag{7.104}$$

constrained by the Dirichlet boundary conditions

$$\begin{cases} y(1) = 2e^{-1} \\ y(3) = 6e^{-2} \end{cases} \tag{7.105}$$

the solution to which is

$$y(x) = 2xe^{-x} \quad 1 \le x \le 3 \tag{7.106}$$

For this example, eq. 7.95a is

$$u'' - xu' - xu = -4e^{-x} \tag{7.107}$$

subject to

$$\begin{cases} u(1) = 2e^{-1} \\ u'(1) = 0 \end{cases} \tag{7.108}$$

and eq. 7.95b is

$$v'' - xv' - xv = 0 \tag{7.109}$$

subject to

$$\begin{cases} v(1) = 0 \\ v'(1) = 1 \end{cases} \tag{7.110}$$

We solve for as $u(x)$ and $v(x)$ by using a four step Runge–Kutta scheme with $h = 0.5$ (as in Sect. 7.3). We obtain

$$U \equiv \begin{pmatrix} u(1.5) \\ u(2.0) \\ u(2.5) \\ u(3.0) \end{pmatrix} = \begin{pmatrix} 0.66833 \\ 0.53788 \\ 0.39752 \\ 0.23582 \end{pmatrix} \qquad (7.111a)$$

and

$$V \equiv \begin{pmatrix} v(1.5) \\ v(2.0) \\ v(2.5) \\ v(3.0) \end{pmatrix} = \begin{pmatrix} 0.51869 \\ 1.84930 \\ 6.70887 \\ 29.84624 \end{pmatrix} \qquad (7.111b)$$

from which we find

$$y_0' = \alpha = 2.10757 \times 10^{-3} \qquad (7.112)$$

Then, from eq. 7.103, we have

$$Y \equiv \begin{pmatrix} y(1.5) \\ y(2.0) \\ y(2.5) \\ y(3.0) \end{pmatrix} = \begin{pmatrix} 0.66942 \\ 0.54178 \\ 0.41166 \\ 0.29872 \end{pmatrix} \qquad (7.113)$$

This compares quite well with the exact values obtained from eq. 7.106:

$$Y_{exact} = \begin{pmatrix} 0.66940 \\ 0.54134 \\ 0.41042 \\ 0.29872 \end{pmatrix} \qquad (7.114)\square$$

In examples 7.4–7.7, we presented finite difference methods for finding $y(x)$ at $\{x_0, \ldots, x_N\}$ when $y(x)$ is constrained by boundary conditions. To determine approximate values of y at points not in the set $\{x_0, \ldots, x_N\}$, one can use interpolation methods over $y(x_0), \ldots, y(x_N)$.

7.7 Frobenius Approximations

Ordinary, regular singular, and irregular singular points

There are two independent solutions, $y_1(x)$ and $y_2(x)$, to equations of the form

$$y'' + P(x)y' + Q(x)y = f(x) \qquad (7.47)$$

The complete solution to such an equation is

$$y(x) = c_1 y_1(x) + c_2 y_2(x) \qquad (7.115)$$

where c_1 and c_2 are determined by either initial or boundary conditions.

The *Frobenius solution* is in the form of an infinite series

$$y(x) = \sum_{k=0}^{\infty} c_k (x - x_0)^{k+s} \qquad (7.116)$$

The parameter s, which allows for the possibility that $y(x)$ may be singular at x_0, is called the *index of the solution*.

If $P(x)$ and $Q(x)$ can be written as the following series

$$P(x) = \frac{p_{-1}}{(x - x_0)} + \sum_{k=0}^{\infty} p_k (x - x_0)^k \qquad (7.117a)$$

and

$$Q(x) = \frac{q_{-2}}{(x - x_0)^2} + \frac{q_{-1}}{(x - x_0)} + \sum_{k=0}^{\infty} q_k (x - x_0)^k \qquad (7.117b)$$

then, if $P(x)$ is singular, its singularity is a *simple pole* and if $Q(x)$ is singular, its singularity is a simple pole or a *second order pole*.

If $f(x)$ is analytic at x_0 and if the three constants, p_{-1}, q_{-1}, and q_{-2} are zero so that $P(x)$ and $Q(x)$ are analytic at x_0, then x_0 is an *ordinary point*. If x_0 is a *regular singular point*, then at least one of these constants is non-zero and $(x-x_0)^2 f(x)$ must be analytic at x_0. Another way to define a regular singularity is that at least one of the functions $P(x)$ and $Q(x)$ is singular at x_0 and $(x-x_0)P(x)$, $(x-x_0)^2 Q(x)$, and $(x-x_0)^2 f(x)$ are analytic at x_0. If $P(x)$, $Q(x)$, and/or $f(x)$ have singularities at x_0 other than as expressed in eqs. 7.117, then x_0 is called an *irregular singular point*.

Fuch's theorem states that if x_0 is an ordinary point, both $y_1(x)$ and $y_2(x)$ given in eq. 7.115 can be found in series form. If x_0 is a regular singular point, then it is guaranteed at least one of the solutions, $y_1(x)$ or $y_2(x)$ can be found in series form.

If x_0 is an irregular singular point, there is no guarantee that a Frobenius solution exists. An expanded discussion of these details about Fuchs' theorem can be found in the literature (see, for example, Cohen, H., 1992, pp. 241–243 and pp. 248–250).

The *Frobenius approximation* is obtained by truncating the Frobenius series of eq. 7.116 to obtain

$$y(x) \simeq \sum_{k=0}^{N} c_k (x - x_0)^{k+s} \tag{7.118}$$

x_0 Is an Ordinary Point

When x_0 is an ordinary point, $y(x)$ is analytic at x_0 and it is always possible to take both independent values of s to be zero. Then, the Frobenius solution is

$$y(x) = \sum_{k=0}^{\infty} c_k (x - x_0)^k \tag{7.119}$$

When x_0 is an ordinary point, eqs. 7.117 become

$$P(x) = \sum_{n=0}^{\infty} p_n (x - x_0)^n \tag{7.120a}$$

$$Q(x) = \sum_{n=0}^{\infty} q_n (x - x_0)^n \tag{7.120b}$$

and

$$f(x) = \sum_{k=0}^{\infty} f_k (x - x_0)^k \tag{7.120c}$$

Then, the series form of the differential equation is

$$\sum_{k=2}^{\infty} c_k k(k - 1)(x - x_0)^{k-2} + \sum_{\substack{k=1 \\ m=0}}^{\infty} c_k p_m (x - x_0)^{m+k-1} + \sum_{\substack{k=0 \\ m=0}}^{\infty} c_k q_m (x - x_0)^{m+k}$$

$$= \sum_{k=0}^{\infty} f_k (x - x_0)^k \tag{7.121}$$

We equate coefficients of corresponding powers of $(x-x_0)$ to obtain an equation relating a coefficient of some larger index to terms involving coefficients of smaller indices. Such an equation is called a *recurrence relation*. To obtain the Frobenius approximation, we generate coefficients up to c_N and truncate the Frobenius series at $k = N$.

Example 7.8: Frobenius approximation with x_0 an ordinary point

The inhomogeneous differential equation

$$y'' - xy' + xy = 2e^x \tag{7.122}$$

constrained by the initial conditions

$$\begin{cases} y(0) = 0 \\ y'(0) = 1 \end{cases} \tag{7.123}$$

has solution

$$y(x) = xe^x \tag{7.124}$$

Since $x = 0$ is an ordinary point for this equation, the Frobenius series is written as

$$y(x) = \sum_{k=0}^{\infty} c_k x^k \tag{7.125}$$

With

$$e^x = \sum_{k=0}^{\infty} \frac{x^k}{k!} \tag{7.126}$$

we substitute eq. 7.125 into eq. 7.122 to obtain

$$\sum_{k=2}^{\infty} c_k k(k-1)x^{k-2} - \sum_{k=1}^{\infty} c_k k x^k + \sum_{k=0}^{\infty} c_k x^{k+1} = 2\sum_{k=0}^{\infty} \frac{x^k}{k!} \tag{7.127a}$$

Writing this as

$$2c_2 + \sum_{k=3}^{\infty} c_k k(k-1)x^{k-2} - \sum_{k=1}^{\infty} c_k k x^k + \sum_{k=0}^{\infty} c_k x^{k+1}$$

$$= 2\left[1 + \sum_{k=1}^{\infty} \frac{x^k}{k!} \right] \tag{7.127b}$$

we equate the coefficients of x^0 to obtain

$$c_2 = 1 \tag{7.128}$$

Then, eq. 7.127b becomes

$$\sum_{k=3}^{\infty} c_k k(k-1) x^{k-2} - \sum_{k=1}^{\infty} c_k k x^k + \sum_{k=0}^{\infty} c_k x^{k+1} = 2 \sum_{k=1}^{\infty} \frac{x^k}{k!} \tag{7.129a}$$

On the left hand side of this equation, we set $k' = k-2$ in the first series and $k' = k+1$ in the third series. Then, after renaming k' as k (ignoring the primes on k'), eq. 7.129a can be written as

$$\sum_{k=1}^{\infty} [(k+2)(k+1)c_{k+2} - kc_k + c_{k-1}] x^k = 2 \sum_{k=1}^{\infty} \frac{x^k}{k!} \tag{7.129b}$$

Thus, the recurrence relation for this example is given by

$$c_{k+2} = \frac{1}{(k+2)(k+1)} \left[\frac{2}{k!} + kc_k - c_{k-1} \right] \quad k \geq 1 \tag{7.130}$$

The 6th order Frobenius approximation to this solution is

$$y(x) \simeq c_0 + c_1 x + c_2 x^2 + c_3 x^3 + c_4 x^4 + c_5 x^5 + c_6 x^6 \tag{7.131}$$

Applying the initial conditions of eq. 7.123 we have

$$c_0 = 0 \tag{7.132a}$$

and

$$c_1 = 1 \tag{7.132b}$$

Then, along with

$$c_2 = 1 \tag{\textbf{7.128}}$$

we set $1 \leq k \leq 4$ in eq. 7.130 to obtain

$$c_3 = \frac{1}{6}[2 + c_1 - c_0] = \frac{1}{2!} \tag{7.133a}$$

$$c_4 = \frac{1 + 2c_2 - c_1}{12} = \frac{1}{3!} \tag{7.133b}$$

$$c_5 = \frac{1}{20}\left[\frac{1}{3} + 3c_3 - c_2\right] = \frac{1}{4!} \qquad (7.133c)$$

and

$$c_6 = \frac{1}{30}\left[\frac{1}{12} + 4c_4 - c_3\right] = \frac{1}{5!} \qquad (7.133d)$$

Therefore, the Frobenius approximation to this solution, up to x^6, is

$$y(x) \simeq x + x^2 + \frac{x^3}{2!} + \frac{x^4}{3!} + \frac{x^5}{4!} + \frac{x^6}{5!} = x\left(1 + x + \frac{x^2}{2!} + \frac{x^3}{3!} + \frac{x^4}{4!} + \frac{x^5}{5!}\right) \qquad (7.134)$$

We recognize this as the first six non-zero terms in the MacLaurin expansion of xe^x.

This Frobenius approximation at $x = (0.25, 0.50, 0.75, 1.00)$ is

$$Y = \begin{pmatrix} y(0.25) \\ y(0.50) \\ y(0.75) \\ y(1.00) \end{pmatrix} = \begin{pmatrix} 0.32101 \\ 0.82435 \\ 1.58754 \\ 2.71667 \end{pmatrix} \qquad (7.135a)$$

which, as expected, agrees very well with

$$Y_{exact} = \begin{pmatrix} 0.32101 \\ 0.82436 \\ 1.58775 \\ 2.71828 \end{pmatrix} \qquad (7.135b)\,\square$$

x_0 is a regular singular point

When x_0 is a regular singular point, $y(x)$ will have a singularity at x_0. Referring to the singularity structure of $P(x)$, $Q(x)$, and $f(x)$ given in eqs. 7.117, we multiply the differential equation of eq. 7.47 by $(x - x_0)^2$. Defining

$$U(x) \equiv (x - x_0)P(x) \qquad (7.136a)$$

$$V(x) \equiv (x - x_0)^2 Q(x) \qquad (7.136b)$$

and

$$F(x) \equiv (x - x_0)^2 f(x) \tag{7.136c}$$

eq. 7.47 is written as

$$(x - x_0)^2 y'' + (x - x_0)U(x)y' + V(x)y = F(x) \tag{7.137}$$

where $U(x)$, $V(x)$, and $F(x)$ are not singular at x_0.

  With

$$y(x) = \sum_{k=0}^{\infty} c_k (x - x_0)^{k+s} \tag{7.118}$$

we write $U(x)$, $V(x)$, and $F(x)$ in series about x_0:

$$U(x) = \sum_{k=0}^{\infty} \mu_k (x - x_0)^k \tag{7.138a}$$

$$V(x) = \sum_{k=0}^{\infty} v_k (x - x_0)^k \tag{7.138b}$$

and

$$F(x) \equiv \sum_{k=0}^{\infty} \phi_k (x - x_0)^{k+s} \tag{7.138c}$$

Substituting eqs. 7.138 into eq. 7.137, the differential equation in series form is

$$\sum_{k=0}^{\infty} c_k(k+s)(k+s-1)(x - x_0)^{k+s} + \sum_{\substack{k=0 \\ m=0}}^{\infty} c_k \mu_m (k+s)(x - x_0)^{k+m+s}$$

$$+ \sum_{\substack{k=0 \\ m=0}}^{\infty} c_k v_m (x - x_0)^{k+m+s} = \sum_{k=0}^{\infty} \phi_k (x - x_0)^{k+s} \tag{7.139}$$

  The lowest power of $(x - x_0)$ in each series is $(x - x_0)^s$ arising from the $k = 0$ and $m = 0$ terms in each series. This yields

$$c_0[s(s-1) + \mu_0 s + v_0] = \phi_0 \qquad (7.140a)$$

We see that since c_0 is not known, we cannot solve this for values of s unless $\phi_0 = 0$. Then

$$c_0[s(s-1) + \mu_0 s + v_0] = 0 \qquad (7.140b)$$

yields

$$s = \frac{1}{2}\left[(1 - \mu_0) \pm \sqrt{(1 - \mu_0)^2 - 4v_0}\right] \qquad (7.141)$$

This equation for s is referred to as the *indicial equation*, at least one solution of which is non-zero.

Recurrence relations for the coefficients c_k are found by equating the coefficients of like powers of $(x - x_0)$ that are greater than s. Following the analysis by Cohen, H., (1992), pp. 248–250,

$$c_k = \frac{\phi_k - \sum\limits_{m=1}^{k} c_{k-m}[\mu_m(k - m + s) + v_m]}{(k + s)(k + s - 1) + \mu_0(k + s) + v_0} \qquad k \geq 1 \qquad (7.142)$$

Let the two independent values of the index of solution s be defined as s_1 and s_2. If $s_1 - s_2$ is not an integer, the two solutions are those obtained from the coefficients of eq. 7.142 with $s = s_1$ and with $s = s_2$. Let

$$s_1 - s_2 \equiv n \qquad (7.143a)$$

with n a positive integer. When

$$s = s_2 = s_1 - n \qquad (7.143b)$$

eq. 7.142 becomes

$$c_k = \frac{\phi_k - \sum\limits_{m=1}^{k} c_{k-m}[\mu_m(k - m + s_1 - n) + v_m]}{(k + s_1 - n)(k + s_1 - n - 1) + \mu_0(k + s_1 - n) + v_0} \qquad (7.144a)$$

Setting $k = n$, this becomes

$$c_n = \frac{\phi_k - \sum\limits_{m=1}^{k} c_{k-m}[\mu_m(s_1 - m) + v_m]}{s_1(s_1 - 1) + \mu_0 s_1 + v_0} \qquad (7.144b)$$

Referring to eq. 7.140b, since s_1 is a solution to the indicial equation, the denominator of this expression is zero. If the numerator is not zero, c_n is undefined and there will be only one Frobenius solution, a series multiplied by an overall constant. We designate this series solution as $y_1(x)$. In that case, the second solution is given by

$$y_2(x) = y_1(x) \int \frac{e^{\int \frac{P(x)}{Q(x)} dx}}{[y_1(x)]^2} dx \qquad (7.145)$$

(see Cohen, H., 1992, pp. 250–252).

If the numerator of eq. 7.144b is zero, c_n is arbitrary, and the second solution in series form, which multiplies c_n, exists.

Example 7.9: Frobenius approximation with x_0 a regular singular point

The *half order Bessel equation* is given by

$$y'' + \frac{1}{x} y' + \left(1 - \frac{1}{4x^2}\right) y = 0 \qquad (7.146a)$$

Referring to eqs. 7.117, we see that $x = 0$ is a regular singular point.

Multiplying eq. 7.146a by x^2, this Bessel equation can be written as

$$x^2 y'' + xy' + \left(x^2 - \frac{1}{4}\right) y = 0 \qquad (7.146b)$$

If we constrain $y(x)$ by

$$\begin{cases} y(0) = 0 \\ y(\pi/6) = 1 \end{cases} \qquad (7.147)$$

the solution to the half order Bessel equation is

$$y(x) = \sqrt{\frac{2\pi}{3}} \frac{\sin x}{\sqrt{x}} \qquad (7.148)$$

With

$$y(x) = \sum_{k=0}^{\infty} c_k x^{k+s} \qquad (7.149)$$

eq. 7.146b becomes

$$\sum_{k=0}^{\infty} c_k(k+s)(k+s-1)x^{k+s} + \sum_{k=0}^{\infty} c_k(k+s)x^{k+s} - \frac{1}{4}\sum_{k=0}^{\infty} c_k x^{k+s}$$

$$+ \sum_{k=0}^{\infty} c_k x^{k+s+2} = \sum_{k=0}^{\infty} c_k\left[(k+s)^2 - \frac{1}{4}\right]x^{k+s} + \sum_{k=0}^{\infty} c_k x^{k+s+2} = 0 \qquad (7.150a)$$

Writing this as

$$c_0\left[s^2 - \frac{1}{4}\right]x^s + c_1\left[(s+1)^2 - \frac{1}{4}\right]x^{s+1}$$

$$+ \sum_{k=2}^{\infty} c_k\left[(k+s)^2 - \frac{1}{4}\right]x^{k+s} + \sum_{k=0}^{\infty} c_k x^{k+s+2} = 0 \qquad (7.150b)$$

we see that the first two terms are the only ones involving x^s and x^{s+1}. Therefore,

$$c_0\left[s^2 - \frac{1}{4}\right] = 0 \qquad (7.151a)$$

and

$$c_1\left[(s+1)^2 - \frac{1}{4}\right] = 0 \qquad (7.151b)$$

from which

$$s = \frac{1}{2} \qquad (7.152a)$$

$$c_0 = \text{indeterminant} \qquad (7.152b)$$

$$c_1 = 0 \qquad (7.152c)$$

or

$$s = -\frac{1}{2} \qquad (7.153a)$$

$$c_0 \text{ and } c_1 \text{ indeterminant} \qquad (7.153b)$$

or

$$s = -\frac{3}{2} \qquad (7.154a)$$

$$c_0 = 0 \qquad\qquad\qquad\qquad (7.154b)$$

$$c_1 = \text{indeterminant} \qquad\qquad\qquad (7.154c)$$

Since any two independent values of s will yield the two solutions to a second-order equation, we take

$$s = \pm \frac{1}{2}, \quad c_1 = 0 \qquad\qquad\qquad (7.155)$$

It is straightforward to demonstrate that the solution obtained for $s = -3/2$ duplicates the $s = -1/2$ solution.

The first two terms in eq. 7.150b are zero and eq. 7.150b becomes

$$\sum_{k=2}^{\infty} c_k \left[(k+s)^2 - \frac{1}{4} \right] x^{k+s} + \sum_{k=0}^{\infty} c_k x^{k+s+2} =$$

$$\sum_{k'=0}^{\infty} c_{k'+2} \left[(k'+s+2)^2 - \frac{1}{4} \right] x^{k'+s+2} + \sum_{k=0}^{\infty} c_k x^{k+s+2} =$$

$$\sum_{k'=0}^{\infty} \left\{ c_{k+2} \left[(k+s+2)^2 - \frac{1}{4} \right] + c_k \right\} x^{k+s+2} = 0 \qquad (7.156)$$

Thus, the recurrence relation for the coefficients c_k are

$$c_{k+2} = \frac{-c_k}{(k+s+2)^2 - \frac{1}{4}} \qquad\qquad\qquad (7.157)$$

For $s = 1/2$, the recurrence relation is

$$c_{k+2} = \frac{-c_k}{\left(k+\frac{5}{2}\right)^2 - \frac{1}{4}} = \frac{-c_k}{(k+2)(k+3)} \qquad\qquad (7.158)$$

from which we obtain

$$c_2 = \frac{-c_0}{3!} \qquad\qquad\qquad\qquad (7.159a)$$

$$c_3 = \frac{-c_1}{4!} = 0 \qquad\qquad\qquad\qquad (7.159b)$$

$$c_4 = \frac{-c_2}{5 \times 4} = \frac{c_0}{5!} \qquad\qquad\qquad (7.159c)$$

$$c_5 = \frac{-c_3}{6 \times 5} = 0 \tag{7.159d}$$

and

$$c_6 = \frac{-c_4}{7 \times 6} = \frac{-c_0}{7!} \tag{7.159e}$$

Therefore, the Frobenius approximation to the $s = 1/2$ solution, up to x^6, is

$$y_{1/2}(x) \simeq c_0 x^{1/2}\left[1 - \frac{x^2}{3!} + \frac{x^4}{5!} - \frac{x^6}{7!}\right] \tag{7.160a}$$

By identical analysis, the $s = -1/2$ solution is approximated by

$$y_{-1/2} \simeq c_0' x^{-1/2}\left[1 - \frac{x^2}{2!} + \frac{x^4}{4!} - \frac{x^6}{6!}\right] \tag{7.160b}$$

The reader may recognize the terms in the $s = 1/2$ approximation as the first four non-zero terms in the series representation of

$$y_{1/2}(x) = c_0 \frac{\sin x}{\sqrt{x}} \tag{7.161a}$$

and the terms in the $s = -1/2$ approximation are the first four non-zero terms of

$$y_{-1/2}(x) = c_0' \frac{\cos x}{\sqrt{x}} \tag{7.161b}$$

The approximation to the complete solution is given by

$$y(x) \simeq c_0 x^{1/2}\left[1 - \frac{x^2}{3!} + \frac{x^4}{5!} - \frac{x^6}{7!}\right] + c_0' x^{-1/2}\left[1 - \frac{x^2}{2!} + \frac{x^4}{4!} - \frac{x^6}{6!}\right] \tag{7.162}$$

We see that the solution multiplying c_0' is not defined at $x = 0$. Therefore, the only way the constraint at $x = 0$ can be satisfied if

$$c_0' = 0 \tag{7.163}$$

Then

$$y(x) \simeq c_0 x^{1/2}\left[1 - \frac{x^2}{3!} + \frac{x^4}{5!} - \frac{x^6}{7!}\right] = \frac{c_0}{\sqrt{x}}\left[x - \frac{x^3}{3!} + \frac{x^5}{5!} - \frac{x^7}{7!}\right] \tag{7.164}$$

c_0 is determined by

$$c_0 \sqrt{\frac{\pi}{6}} \left[1 - \frac{1}{3!} \left(\frac{\pi}{6}\right)^2 + \frac{1}{5!} \left(\frac{\pi}{6}\right)^4 - \frac{1}{7!} \left(\frac{\pi}{6}\right)^6 \right] = 1 \qquad (7.165)$$

from which

$$y(x) \simeq 1.44720 x^{1/2} \left[1 - \frac{x^2}{3!} + \frac{x^4}{5!} - \frac{x^6}{7!} \right] \qquad (7.166)$$

Therefore, for example, we obtain

$$y(\pi/9) \simeq 0.83777 \qquad (7.167)$$

which differs from the exact value by $1.6 \times 10^{-6}\%$. $\square$

Problems

The differential equation $y'' + 2xy' - 6y = 2x$ has solution $y = x + x^3$ when constrained by

(I)   Initial conditions $\begin{cases} y(0) = 0 \\ y'(0) = 1 \end{cases}$

(II) Dirichlet boundary conditions $\begin{cases} y(0) = 0 \\ y(1) = 2 \end{cases}$

(III) Neumann boundary conditions $\begin{cases} y'(0) = 1 \\ y'(1) = 4 \end{cases}$

(IV) Gauss boundary conditions $\begin{cases} y'(0) - 2y(0) = 1 \\ y'(1) - 2y(1) = 0 \end{cases}$

In Problems 1, 2, 3, 5, 6, 7, and 9, solve the above differential equation by the method specified. For each problem, compare your approximate answer to the exact value of y at the specified value of x. To make the comparison, use

$$\%Error \equiv \frac{|estimated\ value - exact\ value|}{exact\ value} * 100$$

1. Find the approximate solution to the above differential equation subject to the initial conditions given above, by

   (a) a single step Taylor series expansion about $x = 0$ to find $y(0.5)$, keeping terms up to the term containing the third derivative of y.
   (b) a two step Taylor series expansion; first about $x = 0$ to find $y(0.25)$, then an expansion about $x = 0.25$ to determine $y(0.5)$. Keep terms in each expansion up to the term containing the third derivative of y.

2. Find the approximate solution to the above differential equation subject to the initial conditions given above, by

   (a) a single step Runge–Kutta technique to find $y(0.5)$.
   (b) a two step Runge–Kutta technique; first apply the Runge–Kutta method to find $y(0.25)$, then a second application of the Runge–Kutta method to determine $y(0.5)$.

3. Find the approximate solution to the above differential equation subject to the initial conditions given above, by finite difference methods. For $h = 0.2$, find $y(0.6)$ using

   (a) Euler finite difference approximations.
   (b) Milne finite difference approximations.

4. Find the approximate solution to the above differential equation subject to the Dirichlet conditions given above, by the shooting method. With $h = 0.1$, find $y(0.2)$ by solving the inhomogeneous equation for $u(x)$ using a two step Runge–Kutta method, and by finding the homogeneous solution $v(x)$ using a two step Taylor series up to h^4. Compare the approximate value of $y'(0)$ found by the shooting method to the exact value $y'(0) = 1$.

5. Find the approximate solution to the above differential equation subject to the Dirichlet conditions given above, by finite difference methods. For $h = 0.2$, find $y(0.6)$ using

   (a) Euler finite difference approximations.
   (b) Milne finite difference approximations.

6. Find the approximate solution to the above differential equation subject to the Neumann conditions given above, by finite difference methods. For $h = 0.2$, find $y(0.6)$ using

   (a) Euler finite difference approximations.
   (b) Milne finite difference approximations.

7. Find the approximate solution to the above differential equation subject to the Gauss conditions given above, by finite difference methods. For $h = 0.2$, find $y(0.6)$ using

   (a) Euler finite difference approximations.
   (b) Milne finite difference approximations.

8. Develop a single step Runge–Kutta method for a third order differential equation $y''' + P(x)y'' + Q(x)y' + R(x)y = g(x, y, y', y'')$ subject to the initial conditions

$$\begin{cases} y(x_0) = y_0 \\ y'(x_0) = y'_0 \\ y''(x_0) = y''_0 \end{cases}$$

9. Taking $h = 0.25$, use the shooting method to find the solution to the second-order differential equation above, subject to the stated Dirichlet conditions

$$\begin{cases} y(0) = 0 \\ y(1) = 2 \end{cases}$$

   In addition to comparing the results to the exact value, compare the value of $y'(0)$ estimated by the shooting method to the exact value given above.

10. Find the terms up to x^6 of the Frobenius approximation to the differential equation $y'' - 2xy' - 2y = 0$ subject to the initial conditions

$$\begin{cases} y(0) = 1 \\ y'(0) = 0 \end{cases}$$

11. Find the terms up to x^6 of the Frobenius approximation to the differential equation of example 7.8, eq. 7.124, constrained by the Dirichlet boundary conditions

$$\begin{cases} y(0) = 0 \\ y(1) = e \end{cases}$$

12. The two differential equations $y'' + 2xy' - 2x^2y = -2e^{-x^2}$ and $y'' + 2xy' + 2y = 0$ have solution $y = e^{-x^2}$ when constrained by either the Dirichlet boundary conditions

$$\begin{cases} y(0) = 1 \\ y(1) = e^{-1} \end{cases}$$

   or the initial conditions

$$\begin{cases} y(0) = 1 \\ y'(0) = 0 \end{cases}$$

   Find the Frobenius approximation up to x^6 for each of these differential equations, constrained by each of these sets of conditions, and compare this approximation to the exact values at $x = \{0.25, 0.50, 0.75, 1.00\}$.

13. Approximate the solution to the Bessel differential equation of zero order $y'' + \frac{1}{x}y' + y = 0$ subject to

$$\begin{cases} y(0) = 1 \\ y'(0) = 0 \end{cases}$$

   by the first four non-zero terms in the Frobenius series.

Chapter 8
PARTIAL DIFFERENTIAL EQUATIONS

Many of the partial differential equations that describe physical systems involve derivatives with respect to space variables (x,y,z) or with respect to space and time variables (x,y,z,t). Such equations include the *diffusion equation* which describes the spread (diffusion) of energy throughout a material medium with diffusion factor K

$$\frac{\partial \psi(x,y,z,t)}{\partial t} - K\nabla^2 \psi(x,y,z,t) = f(x,y,z,t) \qquad (8.1a)$$

the *wave equation* which describes the propagation of a wave traveling at speed c

$$\frac{\partial^2 \psi(x,y,z,t)}{\partial t^2} - c^2\nabla^2 \psi(x,y,z,t) = f(x,y,z,t) \qquad (8.1b)$$

and *Poisson's equation* which describes the electrostatic potential at any point in space due to a distribution of charge, the properties of which are embodied in $\rho(x,y,z)$, the charge density (charge per unit volume).

$$\nabla^2 \psi(x,y,z) = \frac{\rho(x,y,z)}{\varepsilon_0} \qquad (8.1c)$$

For simplicity, we will restrict our discussion to partial differential equations in two variables (x, t) or (x, y).

The generalized form of the diffusion equation in one space dimension

$$\frac{\partial \psi}{\partial t} - P(x,t)\frac{\partial^2 \psi}{\partial x^2} = f(x,t) \qquad (8.2a)$$

is an example of a *parabolic* equation.

H. Cohen, *Numerical Approximation Methods*, DOI 10.1007/978-1-4419-9837-8_8,
© Springer Science+Business Media, LLC 2011

A general form of the wave equation in one space dimension

$$\frac{\partial^2 \psi}{\partial t^2} - P(x,t)\frac{\partial^2 \psi}{\partial x^2} = f(x,t) \tag{8.2b}$$

is an example of a *hyperbolic* equation, and Poisson's equation in two space dimensions is an example of an *elliptic equation*

$$R(x,y)\frac{\partial^2 \psi}{\partial x^2} + S(x,y)\frac{\partial^2 \psi}{\partial y^2} = f(x,y) \tag{8.2c}$$

One can remember the designations of these types of equations by treating the partial derivative operators as coordinates ∂_t, ∂_x, and ∂_y. Then, the "operator coordinate" forms of eqs. 8.2 can be written as

- $\partial_t \approx P\partial_x^{\ 2} + f$
- $\partial_t^{\ 2} \approx P\partial_x^{\ 2} + f$
- $R\partial_x^{\ 2} + S\partial_y^{\ 2} \approx f$

The first of these describes a parabola in a space defined by "coordinates" ∂_t and ∂_x, the second could represent a hyperbola in such "coordinates", and the third could be viewed as the equation of an ellipse in "coordinates" ∂_x and ∂_y.

8.1   Initial and Boundary Conditions

As noted in ch. 7, the time part of a solution is constrained by initial condition(s), and the space part of a solution is constrained by boundary conditions.

For ordinary differential equations, these constraints are given as specific values of the solution. For partial differential equations, what we refer to as initial conditions are generally given as functions of (x, y, z) and what we call boundary conditions are specified as functions of t.

Since a parabolic equation contains a first-order time derivative, its solution is constrained by a single initial condition of the form

$$\psi(x, t_0) \equiv I_0(x) \tag{8.3a}$$

A hyperbolic equation contains a second time derivative. Thus, its solution is subject to two initial conditions: one given in eq. 8.3a and a second expressed as

$$\left(\frac{\partial \psi}{\partial t}\right)_{t=t_0} \equiv I_1(x) \tag{8.3b}$$

Because an elliptic equation contains no time derivatives, its solutions are subject only to boundary conditions.

All three types of equations contain second order derivatives with respect to space variables. Thus, their solutions are always constrained by a two boundary conditions.

For the solutions to the parabolic and hyperbolic equations, the boundary conditions are expressed in the one of three forms. Dirichlet boundary conditions are written as

$$\psi(x,t)\big|_{x=x_0,x_f} \equiv D_{0,f}(t) \tag{8.4a}$$

Neumann boundary conditions are expressed as

$$\left(\frac{\partial \psi(x,t)}{\partial x}\right)_{x=x_0,x_f} \equiv N_{0,f}(t) \tag{8.4b}$$

and Gauss boundary conditions are of the form

$$\left[\psi(x,t) + \gamma \frac{\partial \psi(x,t)}{\partial x}\right]_{x=x_0,x_f} \equiv G_{0,f}(t) \tag{8.4c}$$

The Dirichlet boundary conditions that constrain the solution to an elliptic equation are written as

$$\begin{cases} \psi(x_{0,f}, y) \equiv D_{x_{0,f}}(y) \\ \psi(x, y_{0,f}) \equiv D_{y_{0,f}}(x) \end{cases} \tag{8.5a}$$

Neumann boundary conditions are expressed as

$$\begin{cases} \left(\dfrac{\partial \psi(x,y)}{\partial x}\right)_{x=x_{0,f}} \equiv N_{x_{0,f}}(y) \\ \left(\dfrac{\partial \psi(x,y)}{\partial y}\right)_{y=y_{0,f}} \equiv N_{y_{0,f}}(x) \end{cases} \tag{8.5b}$$

and Gauss boundary conditions are of the form

$$\begin{cases} \psi(x_{0,f}, y) + \gamma_x \left(\dfrac{\partial \psi(x,y)}{\partial x}\right)_{x=x_{0,f}} \equiv G_{x_{0,f}}(y) \\ \psi(x, y_{0,f}) + \gamma_y \left(\dfrac{\partial \psi(x,y)}{\partial y}\right)_{y=y_{0,f}} \equiv G_{y_{0,f}}(x) \end{cases} \tag{8.5c}$$

8.2   Taylor Series Approximations

As noted in ch. 7, a Taylor series is not appropriate for approximating an ordinary differential equation constrained by boundary conditions. Therefore, when the solution to a partial differential equation is subject to initial conditions, so that ψ is analytic at t_0, we can develop a Taylor series expansion in the time variable about t_0 as

$$\psi(x,t) \simeq \psi(x,t_0) + (t - t_0)\left(\frac{\partial \psi}{\partial t}\right)_{t=t_0} + \dots + \frac{(t - t_0)^{\ell}}{\ell!}\left(\frac{\partial^{\ell}\psi}{\partial t^{\ell}}\right)_{t=t_0} \tag{8.6}$$

The boundary conditions only define the values of the spatial variables over which the solution is defined, and the truncated Taylor series is an expression for $\psi(x, t)$ within those boundaries.

The coefficients of various powers of $(t-t_0)$ are found from the initial conditions and from derivatives of the differential equation evaluate at t_0.

Parabolic equations

The solution to the parabolic equation is constrained by a single initial condition

$$\psi(x, t_0) = I_0(x) \tag{8.3a}$$

Writing the parabolic equation for $\psi(x,t)$ equation in the form

$$\frac{\partial \psi}{\partial t} = f(x, t) + P(x, t)\frac{\partial^2 \psi}{\partial x^2} \tag{8.7}$$

we set $t = t_0$ to obtain

$$\left(\frac{\partial \psi(x, t)}{\partial t}\right)_{t_0} = f(x, t_0) + P(x, t_0)\frac{\partial^2 \psi(x, t_0)}{\partial x^2}$$

$$= f(x, t_0) + P(x, t_0)\frac{d^2 I_0(x)}{dx^2} \tag{8.8a}$$

From the time derivative of eq. 8.7, we obtain

$$
\begin{aligned}
\left(\frac{\partial^2 \psi}{\partial t^2}\right)_{t_0} &= \left[\frac{\partial f}{\partial t} + \frac{\partial P}{\partial t}\frac{\partial^2 \psi}{\partial x^2} + P\frac{\partial^2(\partial \psi/\partial t)}{\partial x^2}\right]_{t_0} \\
&= \left[\frac{\partial f}{\partial t} + \frac{\partial P}{\partial t}\frac{\partial^2 \psi}{\partial x^2} + P\frac{\partial^2}{\partial x^2}\left(f + P\frac{\partial^2 \psi}{\partial x^2}\right)\right]_{t_0} \\
&= \left[\left(\frac{\partial f}{\partial t}\right)_{t_0} + \left(\frac{\partial P}{\partial t}\right)_{t_0}\frac{d^2 I_0(x)}{dx^2} + P(x,t_0)\frac{\partial^2 f(x,t_0)}{\partial x^2}\right. \\
&\quad \left. +P(x,t_0)\frac{\partial^2 P(x,t_0)}{\partial x^2}\frac{d^2 I_0(x)}{dx^2} + P^2(x,t_0)\frac{d^4 I_0(x)}{dx^4}\right]
\end{aligned}
\qquad (8.8b)
$$

Coefficients of higher powers of $(t-t_0)$ are found by taking higher derivatives of the differential equation at t_0. Obtaining the coefficient of $(t-t_0)^2$, as in eq. 8.8b, is fairly straightforward but somewhat cumbersome. The calculation of the coefficients of $(t-t_0)^k$ for $k > 2$ can become quite unwieldy.

Example 8.1: Taylor series approximation for the solution to a diffusion equation

In an ideal medium (imperfections in the medium are ignored), the diffusion equation describes how heat energy, for example, distributes throughout the medium. If that medium is a thin heat conducting rod, the equation describes a problem in one space dimension.

The equation that describes the diffusion of thermal energy in a one dimensional medium at all points except at heat source is given by

$$
\frac{\partial \psi(x,t)}{\partial t} = K\frac{\partial^2 \psi(x,t)}{\partial x^2}
\qquad (8.9a)
$$

where K is the thermal diffusion parameter of the medium, which we take to be constant. Taking $K = 1$, we find an approximate solution to

$$
\frac{\partial \psi(x,t)}{\partial t} = \frac{\partial^2 \psi(x,t)}{\partial x^2}
\qquad (8.9b)
$$

Subjecting $\psi(x,t)$ to the initial condition

$$
\psi(x,0) = I_0(x) = e^{-x}
\qquad (8.10)
$$

and the Dirichlet boundary conditions

$$
\begin{cases}
\psi(0,t) = D_0(t) = e^t \\
\psi(1,t) = D_1(t) = e^{t-1}
\end{cases}
\qquad (8.11)
$$

the solution to eq. 8.9b is

$$\psi(x,t) = e^{t-x} \tag{8.12}$$

We point out that the boundary conditions define the range of x values. Thus, the thin rod in this example has a length 1. The initial condition(s) define the instant at which we begin measuring $\psi(x,t)$, which, in this example, is $t_0 = 0$.

Then, the Taylor series approximation up to order t^2, obtained from eq. 8.3a and eqs. 8.8, is

$$\psi(x,t) \simeq \psi(x,0) + t\left(\frac{\partial \psi}{\partial t}\right)_{t=0} + \frac{1}{2!}t^2\left(\frac{\partial^2 \psi}{\partial t^2}\right)_{t=0} \tag{8.13}$$

Setting $t = 0$ in eq. 8.9b, we have

$$\left(\frac{\partial \psi}{\partial t}\right)_{t=0} = \frac{\partial^2 \psi(x,0)}{\partial x^2} = \frac{d^2 I_0(x)}{dx^2} = e^{-x} \tag{8.14a}$$

and from the derivatives of eq. 8.9b, we obtain

$$\left(\frac{\partial^2 \psi}{\partial}\right)_{t=0} = \left[\frac{\partial}{\partial t}\left(\frac{\partial^2 \psi}{\partial x^2}\right)\right]_{t=0} = \left(\frac{\partial^2(\partial \psi/\partial t)}{\partial x^2}\right)_{t=0}$$

$$= \left(\frac{\partial^4 \psi(x,0)}{\partial x^4}\right)_{t=0} = \frac{d^4 I_0(x)}{dx^4} = e^{-x} \tag{8.14b}$$

Therefore, eq. 8.13 becomes

$$\psi(x,t) \simeq \psi(x,0) + t\left(\frac{\partial \psi}{\partial t}\right)_{t=0} + \frac{1}{2!}t^2\left(\frac{\partial^2 \psi}{\partial t^2}\right)_{t=0} = \left[1 + t + \frac{1}{2!}t^2\right]e^{-x} \tag{8.15}$$

The terms in the bracket are recognized as the first three terms in the MacLaurin series for e^t. Therefore, eq. 8.15 is a reasonable approximation to the exact solution

$$\psi(x,t) = e^{t-x} \tag{8.12}$$

Thus, for example, this approximation yields

$$\psi(x,0.50) \simeq 1.62500e^{-x} \tag{8.16a}$$

which is a reasonable approximation to the exact value

$$\psi_{exact}(x,0.50) = e^{(0.50-x)} = 1.64872e^{-x} \tag{8.16b}$$

One can improve the accuracy by using a series with higher order terms in t or by using a multistep Taylor approximation.

For example, to obtain a fourth-order Taylor sum, we require

$$\left(\frac{\partial^3 \psi}{\partial t^3}\right)_{t=0} = \left[\frac{\partial^2}{\partial x^2}\left(\frac{\partial^2 \psi}{\partial t^2}\right)\right]_{t=0} = \left(\frac{\partial^2}{\partial x^2}\frac{\partial}{\partial t}\left(\frac{\partial^2 \psi}{\partial x^2}\right)\right)_{t=0}$$
$$= \left(\frac{\partial^2}{\partial x^2}\frac{\partial^2}{\partial x^2}\frac{\partial \psi}{\partial t}\right)_{t=0} = \frac{\partial^6 \psi(x,0)}{\partial x^6} = \frac{d^6 I_0(x)}{dx^6} = e^{-x} \qquad (8.17a)$$

and

$$\left(\frac{\partial^4 \psi}{\partial t^4}\right)_{t=0} = \left[\frac{\partial^2}{\partial x^2}\left(\frac{\partial^3 \psi}{\partial t^3}\right)\right]_{t=0} = \frac{\partial^8 \psi(x,0)}{\partial x^8} = \frac{d^8 I_0(x)}{dx^8} = e^{-x} \qquad (8.17b)$$

With these and eqs. 8.14, we obtain

$$\psi(x,t) \simeq \psi(x,0) + t\left(\frac{\partial \psi}{\partial t}\right)_{t=0} + \frac{1}{2!}t^2\left(\frac{\partial^2 \psi}{\partial t^2}\right)_{t=0} + \frac{1}{3!}t^3\left(\frac{\partial^3 \psi}{\partial t^3}\right)_{t=0} + \frac{1}{4!}t^4\left(\frac{\partial^4 \psi}{\partial t^4}\right)_{t=0}$$
$$= \left[1 + t + \frac{1}{2!}t^2 + \frac{1}{3!}t^3 + \frac{1}{4!}t^4\right]e^{-x} \qquad (8.18)$$

The sum of terms in the bracket is recognized as the approximation to the MacLaurin series for e^t up to t^4. This yields the approximation

$$\psi(x, 0.50) \simeq 1.64844e^{-x} \qquad (8.19)$$

To develop a two step Taylor series approximation up to order t^2 to find $\psi(x,0.50)$, we begin with $t_0 = 0$ and take $t = 0.25$. With the series approximation up to t^2 given in eq. 8.15, we obtain

$$\psi(x, 0.25) \simeq 1.28125e^{-x} \qquad (8.20)$$

This becomes the initial condition for the second iteration.

With $t_0 = 0.25$ and $t = 0.50$, and using eqs. 8.14 for the time derivatives of $\psi(x,t)$, we obtain

$$\psi(x, 0.50) \simeq \psi(x, 0.25) + (.50 - .25)\left(\frac{\partial \psi}{\partial t}\right)_{t=0.25} + \frac{(.50 - .25)^2}{2}\left(\frac{\partial^2 \psi}{\partial t^2}\right)_{t=0.25}$$
$$= 1.28125e^{-x} + (.25)\frac{d^2 \psi(x,0.25)}{dx^2} + \frac{.25^2}{2}\frac{d^4 \psi(x,0.25)}{dx^4}$$
$$= 1.64160e^{-x} \qquad (8.21)$$

Comparing this result to the exact value

$$\psi(x, 0.50) = e^{0.50-x} = 1.64872e^{-x} \tag{8.17}$$

we see that it is more accurate than the result found using a single step Taylor series up to t^2, but is less accurate than the single step Taylor series up to t^4. $\square$

Hyperbolic equations

Referring to eq. 8.2b, the coefficients of $(t - t_0)^0$ and $(t - t_0)^1$ in the Taylor series for the solution of a hyperbolic equation are given by the two initial conditions of eqs. 8.3. The coefficient of $(t - t_0)^2$ is found by setting $t = t_0$ in the differential equation to obtain

$$\left(\frac{\partial^2 \psi(x, t)}{\partial t^2}\right)_{t=t_0} = f(x, t_0) + P(x, t_0)\frac{d^2 I_0(x)}{dx^2} \tag{8.22}$$

Then,

$$\psi(x, t) \simeq I_0(x) + (t - t_0)I_1(x) + \frac{(t - t_0)^2}{2!}\left[f(x, t_0) + P(x, t_0)\frac{d^2 I_0}{dx^2}\right] + \dots \tag{8.23}$$

The coefficients of higher powers of $(t - t_0)$ are found by taking derivatives of the hyperbolic equation. For example, the coefficients of $(t - t_0)^3$ and $(t - t_0)^4$ are found from

$$
\begin{aligned}
\left(\frac{\partial^3 \psi(x, t)}{\partial t^3}\right)_{t_0} &= \left[\left(\frac{\partial f}{\partial t}\right)_{t_0} + \left(\frac{\partial P}{\partial t}\right)_{t_0}\frac{d^2 I_0(x)}{dx^2} + P(x, t_0)\frac{\partial^2 (\partial \psi / \partial t)_{t_0}}{\partial x^2}\right] \\
&= \left[\left(\frac{\partial f}{\partial t}\right)_{t_0} + \left(\frac{\partial P}{\partial t}\right)_{t_0}\frac{d^2 I_0(x)}{dx^2} + P(x, t_0)\frac{d^2 I_1(x)}{dx^2}\right]
\end{aligned} \tag{8.24a}
$$

and

$$
\begin{aligned}
\left(\frac{\partial^4 \psi(x, t)}{\partial t^4}\right)_{t_0} = &\left[\left(\frac{\partial^2 f}{\partial t^2}\right)_{t_0} + \left(\frac{\partial^2 P}{\partial t^2}\right)_{t_0}\frac{d^2 I_0(x)}{dx^2} + 2\left(\frac{\partial P}{\partial t}\right)_{t_0}\frac{d^2 I_1(x)}{dx^2}\right. \\
&+ P(x, t_0)\left(\frac{d^2 P(x, t_0)}{dx^2}\frac{d^2 I_0(x)}{dx^2} + 2\frac{dP(x, t_0)}{dx}\frac{d^3 I_0(x)}{dx^3}\right. \\
&\left.\left. + P(x, t_0)\frac{d^4 I_0(x)}{dx^4}\right)\right]
\end{aligned} \tag{8.24b}
$$

Example 8.2: Taylor series approximation for the solution to a wave equation

We consider the wave equation for a wave that is propagated in a source-free vacuum in one dimension (e.g., radiation propagated down a narrow evacuated wave guide). Then, eq. 8.1b can be written as

$$\frac{\partial^2 \psi(x,t)}{\partial t^2} - c^2 \frac{\partial^2 \psi(x,t)}{\partial x^2} = 0 \qquad (8.25)$$

where c is the propagation speed of the wave.

With $c = 1$ we subject $\psi(x,t)$ to the initial conditions

$$\begin{cases} \psi(x,0) = I_0(x) = e^{-x} \\ \left(\dfrac{\partial \psi}{\partial t}\right)_{t=0} = I_1(x) = e^{-x} \end{cases} \qquad (8.26)$$

and the Dirichlet boundary conditions

$$\begin{cases} \psi(0,t) = D_0(t) = e^t \\ \psi(1,t) = D_1(t) = e^{t-1} \end{cases} \qquad (8.11)$$

The solution to the wave equation with such constraints is the same as the solution to the diffusion equation of example 8.1:

$$\psi(x,t) = e^{t-x} \qquad (8.12)$$

From the initial conditions and

$$\left(\frac{\partial^2 \psi}{\partial t^2}\right)_{t=0} = \frac{\partial^2 \psi(x,0)}{\partial x^2} = \frac{d^2 I_0(x)}{dx^2} = e^{-x} \qquad (8.27)$$

we obtain the second order Taylor series approximation given by

$$\psi(x,t) \simeq \psi(x,0) + t\left(\frac{\partial \psi}{\partial t}\right)_{t=0} + \frac{1}{2!} t^2 \left(\frac{\partial^2 \psi}{\partial t^2}\right)_{t=0} = \left[1 + t + \frac{1}{2!} t^2\right] e^{-x} \qquad (8.15)$$

To obtain a fourth-order series approximation, we differentiate the wave equation to obtain

$$\left(\frac{\partial^3 \psi}{\partial t^3}\right)_{t=0} = \frac{\partial^2}{\partial x^2}\left(\frac{\partial \psi}{\partial t}\right)_{t=0} = \frac{d^2 I_1(x)}{dx^2} = e^{-x} \qquad (8.28a)$$

and

$$\left(\frac{\partial^4 \psi}{\partial t^4}\right)_{t=0} = \frac{\partial^2}{\partial x^2}\left(\frac{\partial^2 \psi}{\partial t^2}\right)_{t=0} = \frac{\partial^4 \psi(x,0)}{\partial x^4} = \frac{d^4 I_0(x)}{dx^4} = e^{-x} \qquad (8.28b)$$

Then,

$$\psi(x,t) \simeq \left[\psi(x,0) + t\left(\frac{\partial\psi}{\partial t}\right)_{t=0} + \frac{1}{2!}t^2\left(\frac{\partial^2\psi}{\partial t^2}\right)_{t=0}\right.$$
$$\left. + \frac{1}{3!}t^3\left(\frac{\partial^3\psi}{\partial t^3}\right)_{t=0} + \frac{1}{4!}t^4\left(\frac{\partial^4\psi}{\partial t^4}\right)_{t=0}\right] = \left[1 + t + \frac{1}{2!}t^2 + \frac{1}{3!}t^3 + \frac{1}{4!}t^4\right]e^{-x}$$

$$(8.18)$$

As with the analysis given in example 8.1, the terms in the bracket are the first five terms in the MacLaurin series for e^t. $\Box$

We can also develop a multistep Taylor approximation to the hyperbolic equation. To develop such a two step Taylor approximation, we use the initial conditions to obtain $\psi(x, t_0 + h_t)$ [for example, eqs. 8.15 or 8.18]. Then, the second iteration will be of the form

$$\psi(x,t) = \psi(x, t_0 + 2h_t)$$
$$\simeq \psi(x, t_0 + h_t) + [t - (t_0 + h_t)]\left(\frac{\partial\psi(x,t)}{\partial t}\right)_{t=t_0+h_t} + \dots \quad (8.29)$$

An approximate value of the first term in this series is obtained from the truncated Taylor series expanded about t_0 [eqs. 8.15 or 8.18, for example]. Since the second term involves the derivative of ψ at a time other than t_0, it cannot be obtained from the differential equation or from one of the initial conditions. Therefore, it must be approximated by the derivative of the Taylor series approximation developed in the first step. This introduces an addition inaccuracy in the approximation. This process is left as an exercise for the reader (see Problem 2 at the end of this chapter).

Elliptic equations

Since an elliptic equation depends only on space variables x and y, we try to develop a Taylor series approximation to the solution is a sum of terms in powers of

$$\begin{cases} h_x \equiv x - x_0 \\ h_y \equiv y - y_0 \end{cases} \quad (8.30)$$

For example, to expand $\psi(x,y)$ about (x_0, y_0) in a Taylor series quadratic in h, (that is, we keep terms up to h_x^2, h_y^2, and $h_x h_y$), we begin with

$$\psi(x,y) \simeq \psi(x, y_0) + h_y\left(\frac{\partial\psi(x,y)}{\partial y}\right)_{y_0} + \frac{1}{2!}h_y^2\left(\frac{\partial^2\psi(x,y)}{\partial y^2}\right)_{y_0} \quad (8.31a)$$

and

$$\psi(x,y) \simeq \psi(x_0, y) + h_x \left(\frac{\partial \psi(x,y)}{\partial x} \right)_{x_0} + \frac{1}{2!} h_x^2 \left(\frac{\partial^2 \psi(x,y)}{\partial x^2} \right)_{x_0} \qquad (8.31b)$$

Setting $x = x_0$ in eq. 8.31a, we have

$$\psi(x_0, y) \simeq \psi(x_0, y_0) + h_y \left(\frac{\partial \psi(x_0, y)}{\partial y} \right)_{y_0} + \frac{1}{2!} h_y^2 \left(\frac{\partial^2 \psi(x_0, y)}{\partial y^2} \right)_{y_0} \qquad (8.32a)$$

From derivatives of eq. 8.31a, we also obtain

$$\left(\frac{\partial \psi(x,y)}{\partial x} \right)_{x_0} \simeq \left(\frac{\partial \psi(x, y_0)}{\partial x} \right)_{x_0} + h_y \left(\frac{\partial^2 \psi(x,y)}{\partial x \partial y} \right)_{x_0, y_0} + \frac{1}{2!} h_y^2 \left(\frac{\partial^3 \psi(x,y)}{\partial x \partial y^2} \right)_{x_0, y_0}$$

$$(8.32b)$$

and

$$\left(\frac{\partial^2 \psi(x,y)}{\partial x^2} \right)_{x_0, y_0} \simeq \left(\frac{\partial^2 \psi(x, y_0)}{\partial x^2} \right)_{x_0} + h_y \left(\frac{\partial^3 \psi(x,y)}{\partial x^2 \partial y} \right)_{x_0, y_0}$$

$$+ \frac{1}{2!} h_y^2 \left(\frac{\partial^4 \psi(x,y)}{\partial x^2 \partial y^2} \right)_{x_0, y_0} \qquad (8.32c)$$

We substitute these into eq. 8.31b and keep terms up to second order in h to obtain

$$\psi(x,y) \simeq \psi(x_0, y_0) + h_x \left(\frac{\partial \psi(x, y_0)}{\partial x} \right)_{x_0} + h_y \left(\frac{\partial \psi(x_0, y)}{\partial y} \right)_{y_0} + \frac{1}{2!} h_x^2 \left(\frac{\partial^2 \psi(x, y_0)}{\partial x^2} \right)_{x_0}$$

$$+ \frac{1}{2!} h_y^2 \left(\frac{\partial^2 \psi(x_0, y)}{\partial y^2} \right)_{y_0} + h_x h_y \left(\frac{\partial^2 \psi(x,y)}{\partial x \partial y} \right)_{x_0, y_0} \qquad (8.33)$$

If the solution is constrained by the Dirichlet boundary conditions

$$\begin{cases} \psi(x, y_0) \equiv D_{y_0}(x) \\ \psi(x, y_f) \equiv D_{y_f}(x) \\ \psi(x_0, y) \equiv D_{x_0}(y) \\ \psi(x_f, y) \equiv D_{x_f}(y) \end{cases} \qquad (8.34)$$

eq. 8.33 can be written

$$\psi(x,y) \simeq \psi(x_0, y_0) + h_x \left[\frac{dD_{y_0}(x)}{dx} \right]_{x_0} + h_y \left[\frac{dD_{x_0}(y)}{dy} \right]_{y_0}$$

$$+ \frac{1}{2!} h_x^2 \left[\frac{d^2 D_{y_0}(x)}{dx^2} \right]_{x_0} + h_y^2 \left[\frac{d^2 D_{x_0}(y)}{dy^2} \right]_{y_0} + h_x h_y \left(\frac{\partial^2 \psi(x,y)}{\partial x \partial y} \right)_{x_0, y_0} \qquad (8.35)$$

That is, exact expressions can be obtained from these boundary conditions for all the terms on the right hand side of eq. 8.33 except for the last term

$$h_x h_y \left(\frac{\partial^2 \psi(x,y)}{\partial x \partial y} \right)_{x_0, y_0}$$

Therefore, with Dirichlet conditions, $\psi(x,y)$ can only be approximated by the generally inaccurate Taylor series approximation

$$\psi(x,y) \simeq \psi(x_0 + h_x, y_0 + h_y) + h_x \left(\frac{\partial \psi(x, y_0)}{\partial x} \right)_{x_0} + h_y \left(\frac{\partial \psi(x_0, y)}{\partial y} \right)_{y_0} \qquad (8.36)$$

If the solution is constrained by Neumann conditions

$$\begin{cases} \dfrac{\partial \psi(x, y_0)}{\partial x} \equiv N_{y_0}(x) \\[2mm] \dfrac{\partial \psi(x, y_f)}{\partial x} \equiv N_{y_f}(x) \\[2mm] \dfrac{\partial \psi(x_0, y)}{\partial y} \equiv N_{x_0}(y) \\[2mm] \dfrac{\partial \psi(x_f, y)}{\partial y} \equiv N_{x_f}(y) \end{cases} \qquad (8.37)$$

eq. 8.33 can be expressed as

$$\psi(x,y) \simeq \psi(x_0, y_0) + h_x N_{x_0}(y_0) + h_y N_{y_0}(x_0)$$
$$+ \frac{1}{2!} h_x^2 \left(\frac{\partial N_{y_0}(x)}{\partial x} \right)_{x_0} + \frac{1}{2!} h_y^2 \left(\frac{\partial N_{x_0}(y)}{\partial y} \right)_{y_0} + h_x h_y \left(\frac{\partial N_{x_0}(y)}{\partial y} \right)_{y_0} \qquad (8.38a)$$

or equivalently,

$$\psi(x,y) \simeq \psi(x_0, y_0) + h_x N_{x_0}(y_0) + h_y N_{y_0}(x_0)$$
$$+ \frac{1}{2!} h_x^2 \left(\frac{\partial N_{y_0}(x)}{\partial x} \right)_{x_0} + \frac{1}{2!} h_y^2 \left(\frac{\partial N_{x_0}(y)}{\partial y} \right)_{y_0} + h_x h_y \left(\frac{\partial N_{y_0}(x)}{\partial x} \right)_{x_0} \qquad (8.38b)$$

Since $\psi(x_0, y_0)$ cannot be determined from the Neumann boundary conditions, we cannot determine this Taylor sum.

It is straightforward to show that if $\psi(x, y)$ is subject to Gauss boundary conditions, then the derivatives of $\psi(x, y)$ at the boundaries can be expressed in terms of $\psi(x_0, y_0)$, but $\psi(x_0, y_0)$ cannot be determined. Thus, $\psi(x,y)$ cannot be estimated by eq. 8.33 for Gauss boundary conditions. Therefore, some approach other than a truncated Taylor series is used to approximate the solution to an elliptic equation.

8.3 Runge–Kutta Approximations

As the truncated Taylor series approach, the approximate solution to a partial differential equation is achieved by the Runge–Kutta method when the solution is subject to initial conditions. And as described above for the Taylor series approximation, because the Runge–Kutta method yields an approximate solution that is a function of the space variables, the boundary conditions only define the domain of that solution.

Parabolic equations

We again consider the parabolic equation of the form

$$\frac{\partial \psi}{\partial t} = f(x, t) + P(x, t)\frac{\partial^2 \psi}{\partial x^2} \equiv F(x, t, \psi) \tag{8.39}$$

with the initial condition given by

$$\psi(x, t_0) \equiv I_0(x) \tag{8.3a}$$

Referring to eqs. 7.17 and eq. 7.18, the Runge–Kutta parameters, which, for a partial differential equation, are functions of x, are

$$R_1(x) = F(x, t_0, \psi(x, t_0)) = \left[f(x, t_0) + P(x, t_0)\frac{\partial^2 \psi(x, t_0)}{\partial x^2} \right] h_t$$

$$= \left[f(x, t_0) + P(x, t_0)\frac{d^2 I_0(x)}{dx^2} \right] h_t \tag{8.40a}$$

$$R_2(x) = F\left(x, t_0 + \frac{1}{2}h_t, \psi(x, t_0) + \frac{1}{2}R_1(x) \right)$$

$$= \left[f\left(x, t_0 + \frac{1}{2}h_t \right) + P\left(x, t_0 + \frac{1}{2}h_t \right)\frac{d^2}{dx^2}\left(I_0(x) + \frac{1}{2}R_1(x) \right) \right] h_t \tag{8.40b}$$

$$R_3(x) = F\left(x, t_0 + \frac{1}{2}h_t, \psi(x, t_0) + \frac{1}{2}R_2(x) \right)$$

$$= \left[f\left(x, t_0 + \frac{1}{2}h_t \right) + P\left(x, t_0 + \frac{1}{2}h_t \right)\frac{d^2}{dx^2}\left(I_0(x) + \frac{1}{2}R_2(x) \right) \right] h_t \tag{8.40c}$$

and

$$R_4(x) = F(x, t_0 + h_t, \psi(x, t_0) + R_3(x))$$

$$= \left[f(x, t_0 + h_t) + P(x, t_0 + h_t)\frac{d^2}{dx^2}(I_0(x) + R_3(x)) \right] h_t \tag{8.40d}$$

Then,

$$\psi(x, t_0 + h_t) = I_0(x) + \frac{1}{6}[R_1(x) + 2R_2(x) + 2R_3(x) + R_4(x)] \qquad (8.41)$$

Example 8.3: Single step Runge–Kutta approximation for the solution to a diffusion equation

As in example 8.1, we consider

$$\frac{\partial \psi(x, t)}{\partial t} = \frac{\partial^2 \psi(x, t)}{\partial x^2} \qquad (8.9b)$$

With $\psi(x, t)$ subject to the initial condition

$$\psi(x, 0) = I_0(x) = e^{-x} \qquad (8.10)$$

and the Dirichlet boundary conditions

$$\begin{cases} \psi(0, t) = D_0(t) = e^t \\ \psi(1, t) = D_1(t) = e^{t-1} \end{cases} \qquad (8.11)$$

the solution is given by

$$\psi(x, t) = e^{t-x} \qquad (8.12)$$

To find $\psi(x, 0.50)$ by the Runge–Kutta method, with $h_t = 0.50$, we refer to eqs. 8.40 to obtain

$$R_1(x) = h_t \frac{\partial^2 \psi(x, t_0)}{\partial x^2} = 0.5 \frac{d^2 I_0(x)}{dx^2} = 0.5 e^{-x} \qquad (8.42a)$$

$$R_2(x) = h_t \frac{d^2 \left(I_0(x) + \frac{1}{2} R_1(x) \right)}{dx^2} = 0.5 \frac{d^2}{dx^2} \left(e^{-x} + \frac{1}{2} 0.5 e^{-x} \right) = 0.625 e^{-x} \quad (8.42b)$$

$$R_3(x) = h_t \frac{d^2 \left(I_0(x) + \frac{1}{2} R_2(x) \right)}{dx^2} = 0.5 \frac{d^2 (e^{-x} + .3125 e^{-x})}{dx^2} = 0.65625 e^{-x} \quad (8.42c)$$

and

$$R_4(x) = h_t \frac{d^2 (I_0(x) + R_3(x))}{dx^2} = 0.5 \frac{d^2 (e^{-x} + 0.65625 e^{-x})}{dx^2} = 0.82813 e^{-x} \quad (8.42d)$$

Then, from eq. 8.41,

$$\psi(x, 0.50) \simeq e^{-x} + \frac{1}{6} (0.5 + 2 \times 0.625 + 2 \times 0.65625 + 0.82813) e^{-x}$$

$$= 1.64844 e^{-x} \qquad (8.43a)$$

which is a reasonable approximation to the exact result

$$\psi_{exact}(x, 0.50) = e^{-0.50}e^{-x} = 1.64872e^{-x} \tag{8.43b}$$

The accuracy of this result is improved by using a multiple step Runge–Kutta scheme. Taking $t_0 = 0$ and $h_t = 0.25$, the Runge–Kutta parameters for finding $\psi(x, 0.25)$ are

$$R_1^{(0.25)}(x) = 0.25e^{-x} \tag{8.44a}$$

$$R_2^{(0.25)}(x) = 0.28125e^{-x} \tag{8.44b}$$

$$R_3^{(0.25)}(x) = 0.28516e^{-x} \tag{8.44c}$$

and

$$R_4^{(0.25)}(x) = 0.32129e^{-x} \tag{8.44d}$$

Then, from eq. 8.41, we obtain

$$\psi(x, 0.25) \simeq 1.28402e^{-x} \tag{8.45}$$

With $\psi(x, t_0) = \psi(x, 0.25)$, the Runge–Kutta parameters at $t = 0.50$ are then given by

$$R_1^{(0.50)}(x) = 0.32100e^{-x} \tag{8.46a}$$

$$R_2^{(0.50)}(x) = 0.36113e^{-x} \tag{8.46b}$$

$$R_3^{(0.50)}(x) = 0.36615e^{-x} \tag{8.46c}$$

and

$$R_4^{(0.50)}(x) = 0.41254e^{-x} \tag{8.46d}$$

from which

$$\psi(x, 0.50) \simeq 1.64870e^{-x} \tag{8.47}$$

Comparing this result to that obtained by the single step Runge–Kutta method and to the exact result (eqs. 8.43), we see that the two step process yields a noticeable improvement in the accuracy of the results. $\square$

Hyperbolic equation

We again consider the hyperbolic equation in the form

$$\frac{\partial^2 \psi}{\partial t^2} = f(x,t) + P(x,t)\frac{\partial^2 \psi}{\partial x^2} \equiv F(x,t,\psi) \qquad (8.48)$$

with the initial conditions given by

$$\begin{cases} \psi(x,t_0) = I_0(x) \\ \left(\dfrac{\partial \psi}{\partial t}\right)_{t_0} = I_1(x) \end{cases} \qquad (8.3)$$

Using the analysis presented in ch. 7, sec. 7.3 as a guide, we define

$$w(x,t) \equiv \frac{\partial \psi(x,t)}{\partial t} \qquad (8.49)$$

Then the hyperbolic equation can be expressed as

$$\frac{\partial w}{\partial t} = f(x,t) + P(x,t)\frac{\partial^2 \psi}{\partial x^2} = F(x,t,\psi) \qquad (8.50)$$

and the initial conditions become

$$\begin{cases} \psi(x,t_0) = I_0(x) \\ w(x,t_0) = I_1(x) \end{cases} \qquad (8.51)$$

The Runge–Kutta R and S parameters are given by

$$R_1(x) = w(x,t_0)h_t = I_1(x)h_t \qquad (8.52a)$$

$$S_1(x) = F(x,t_0,\psi(x,t_0)) = \left[f(x,t_0) + P(x,t_0)\frac{d^2 I_0(x)}{dx^2}\right]h_t \qquad (8.52b)$$

$$R_2(x) = \left[w(x,t_0) + \frac{1}{2}S_1(x)\right]h_t = \left[I_1(x) + \frac{1}{2}S_1(x)\right]h_t \qquad (8.52c)$$

$$S_2(x) = F\left(x,t_0 + \frac{1}{2}h_t, \psi(x,t_0) + \frac{1}{2}R_1(x)\right)$$

$$= \left[f\left(x,t_0 + \frac{1}{2}h_t\right) + P\left(x,t_0 + \frac{1}{2}h_t\right)\frac{d^2}{dx^2}\left(I_0(x) + \frac{1}{2}R_1(x)\right)\right]h_t \qquad (8.52d)$$

$$R_3(x) = \left[I_0(x) + \frac{1}{2}S_2(x)\right]h_t \qquad (8.52e)$$

$$S_3(x) = F\left(x, t_0 + \frac{1}{2} h_t, \psi(x, t_0) + \frac{1}{2} R_2(x)\right)$$

$$= \left[f\left(x, t_0 + \frac{1}{2} h_t\right) + P\left(x, t_0 + \frac{1}{2} h_t\right) \frac{d^2}{dx^2}\left(I_0(x) + \frac{1}{2} R_2(x)\right) \right] h_t \qquad (8.52f)$$

$$R_4(x) = [I_1(x) + S_3(x)] h_t \qquad (8.52g)$$

and

$$S_4(x) = F(x, t_0 + h_t, \psi(x, t_0) + R_3(x))$$

$$= \left[f(x, t_0 + h_t) + P(x, t_0 + h_t) \frac{d^2}{dx^2}(I_0(x) + R_3(x)) \right] h_t \qquad (8.52h)$$

From these parameters, the solutions for $\psi(x,t)$ and $w(x,t)$ at $t = t_0 + h_t$ are

$$\psi(x, t_0 + h_t) = I_0(x) + \frac{1}{6} [R_1(x) + 2R_2(x) + 2R_3(x) + R_4(x)] \qquad (8.53a)$$

and

$$w(x, t_0 + h_t) = \left(\frac{\partial \psi(x, t)}{\partial t}\right)_{t_0 + h_t}$$

$$= I_1(x) + \frac{1}{6} [S_1(x) + 2S_2(x) + 2S_3(x) + S_4(x)] \qquad (8.53b)$$

Example 8.4: Single step Runge–Kutta approximation for the solution to a wave equation

As in example 8.2, we consider the wave equation with $c = 1$ and $h_t = 0.5$:

$$\frac{\partial^2 \psi(x, t)}{\partial t^2} - \frac{\partial^2 \psi(x, t)}{\partial x^2} = 0 \qquad (8.54)$$

With $\psi(x,t)$ subject to the initial conditions

$$\begin{cases} \psi(x, 0) = I_0(x) = e^{-x} \\ \left(\dfrac{\partial \psi}{\partial t}\right)_{t=0} = I_1(x) = e^{-x} \end{cases} \qquad (8.26)$$

and the Dirichlet boundary conditions

$$\begin{cases} \psi(0, t) = D_0(t) = e^t \\ \psi(1, t) = D_1(t) = e^{t-1} \end{cases} \qquad (8.11)$$

the solution is given by

$$\psi(x, t) = e^{t-x} \qquad (8.12)$$

From eqs. 8.52, the Runge–Kutta parameters are given by

$$R_1(x) = I_1(x)h_t = 0.5e^{-x} \tag{8.55a}$$

$$S_1(x) = \frac{d^2 I_0(x)}{dx^2}h_t = 0.5\frac{d^2 e^{-x}}{dx^2} = 0.5e^{-x} \tag{8.55b}$$

$$R_2(x) = \left[I_1(x) + \frac{1}{2}S_1(x)\right]h_t = 0.5[e^{-x} + 0.25e^{-x}] = 0.625e^{-x} \tag{8.55c}$$

$$S_2(x) = \left[I_0(x) + \frac{1}{2}R_1(x)\right]h_t = 0.5[e^{-x} + 0.25e^{-x}] = 0.625e^{-x} \tag{8.55d}$$

$$R_3(x) = \left[I_1(x) + \frac{1}{2}S_2(x)\right]h_t = 0.5[e^{-x} + 0.3125e^{-x}] = 0.65625e^{-x} \tag{8.55e}$$

$$S_3(x) = \frac{d^2\left(I_0(x) + \frac{1}{2}R_2(x)\right)}{dx^2}h_t = 0.5\frac{d^2(e^{-x} + .3125e^{-x})}{dx^2} = 0.65625e^{-x} \tag{8.55f}$$

$$R_4(x) = [I_1(x) + S_3(x)]h_t = 0.5[e^{-x} + 0.65625e^{-x}] = 0.82813e^{-x} \tag{8.55g}$$

and

$$S_4(x) = \frac{d^2(I_0(x) + R_3(x))}{dx^2}h_t = 0.5\frac{d^2(e^{-x} + 0.65625e^{-x})}{dx^2} = 0.82813e^{-x} \tag{8.55h}$$

from which

$$\psi(x, 0.5) \simeq e^{-x} + \frac{1}{6}e^{-x}(0.5 + 0.625 + 0.65625 + 0.82813) = 1.64844e^{-x} \tag{8.56}$$

and

$$w(x, 0.5) = \left(\frac{\partial\psi}{\partial t}\right)_{t=0.5} \simeq e^{-x} + \frac{1}{6}e^{-x}(0.5 + 0.625 + 0.65625 + 0.82813)$$

$$= 1.64844e^{-x} \tag{8.57a}$$

Comparing these to

$$\psi_{exact}(x, 0.50) = 1.64872e^{-x} \tag{8.42b}$$

and

$$w_{exact}(x, 0.50) = 1.64872e^{-x} \tag{8.57b}$$

we see that these results are fairly accurate and can be improved by using a multiple step Runge–Kutta scheme. It is straightforward to do so and is left as an exercise for the reader (see Problem 4). □

Elliptic equation

As the Taylor series approximation, the Runge–Kutta method is only applicable to differential equations with solution subject to initial conditions. Thus, since it is only constrained by boundary conditions, the Taylor series and Runge–Kutta methods are not appropriate for estimating the solution to an elliptic equation.

8.4 Finite Difference Methods

When approximating a solution by finite differences, we will be determining $\psi(x,t)$ at $x = \{x_0, x_1, \ldots, x_N\}$ and $t = \{t_0, t_1, \ldots, t_M\}$ for the parabolic and hyperbolic equations, and $\psi(x,y)$ at $x = \{x_0, x_1, \ldots, x_N\}$ and $y = \{y_0, y_1, \ldots, y_M\}$ for the elliptic equation. We will express these values in the form of the *solution tables* shown in Table 8.1.

The entries in these tables will be designated using the obvious notation

$$\psi(x_k, t_m) \equiv \psi_{k,m} \tag{8.58a}$$

for the solution to a parabolic or hyperbolic equation, and

$$\psi(x_k, y_m) \equiv \psi_{k,m} \tag{8.58b}$$

for an elliptic equation.

	$x = x_0$	$x = x_1$	• •	$x = x_N$
$t = t_0$				
$t = t_1$				
$\vdots$				
$t = t_M$				

Table 8.1a Form of the solution table for parabolic and hyperbolic equations

	$x = x_0$	$x = x_1$	• •	$x = x_N$
$y = y_0$				
$y = y_1$				
$\vdots$				
$y = y_M$				

Table 8.1b Form of the solution table for elliptic equations

The Euler finite difference approximations to first and second derivatives are

$$\left(\frac{\partial \psi(x_k, t)}{\partial t}\right)_{t_m} \simeq \frac{\psi_{k,m+1} - \psi_{k,m}}{h_t} \tag{8.59a}$$

$$\left(\frac{\partial \psi(x, t_m)}{\partial x}\right)_{x_k} \simeq \frac{\psi_{k+1,m} - \psi_{k,m}}{h_x} \tag{8.59b}$$

$$\left(\frac{\partial^2 \psi(x_k, t)}{\partial t^2}\right)_{t_m} \simeq \frac{\psi_{k,m+2} - 2\psi_{k,m+1} - \psi_{k,m}}{h_t^2} \tag{8.59c}$$

and

$$\left(\frac{\partial^2 \psi(x, t_m)}{\partial x^2}\right)_{x_k} \simeq \frac{\psi_{k+2,m} - 2\psi_{k+1,m} + \psi_{k,m}}{h_x^2} \tag{8.59d}$$

The corresponding Milne finite difference approximations are

$$\left(\frac{\partial \psi(x_k, t)}{\partial t}\right)_{t_m} \simeq \frac{\psi_{k,m+1} - \psi_{k,m-1}}{2h_t} \tag{8.60a}$$

$$\left(\frac{\partial \psi(x, t_m)}{\partial x}\right)_{x_k} \simeq \frac{\psi_{k+1,m} - \psi_{k-1,m}}{2h_x} \tag{8.60b}$$

$$\left(\frac{\partial^2 \psi(x_k, t)}{\partial t^2}\right)_{t_m} \simeq \frac{\psi_{k,m+1} - 2\psi_{k,m} - \psi_{k,m-1}}{h_t^2} \tag{8.60c}$$

and

$$\left(\frac{\partial^2 \psi(x, t_m)}{\partial x^2}\right)_{x_k} \simeq \frac{\psi_{k+1,m} - 2\psi_{k,m} + \psi_{k-1,m}}{h_x^2} \tag{8.60d}$$

Parabolic equations subject to Dirichlet boundary conditions

The parabolic equation is subject to a single initial condition

$$\psi(x, t_0) = I_0(x) \tag{8.3a}$$

from which the entries in the first row of the solution table are given by

$$\psi_{k,0} = I_0(x_k) \tag{8.61}$$

Referring to eq. 8.4a, the entries in the first and last columns of the solution table are given by the Dirichlet boundary conditions

$$\begin{cases} \psi_{0,m} = D_{x_0}(t_m) \\ \psi_{N,m} = D_{x_N}(t_m) \end{cases} \tag{8.62}$$

Euler finite differences

The Euler finite difference approximation to eq. 8.2b is

$$\frac{\psi_{k,m+1} - \psi_{k,m}}{h_t} = f_{k,m} + P_{k,m} \left(\frac{\partial^2 \psi(x, t_m)}{\partial x^2} \right)_{x=x_k} \tag{8.63}$$

With the initial condition, the elements in the second row of the solution table, defined by $t = t_1$, are

$$\begin{aligned} \psi_{k,1} &= \psi_{k,0} + h_t \left[f_{k,1} + P_{k,1} \left(\frac{\partial^2 \psi(x, t_0)}{\partial x^2} \right)_{x=x_k} \right] \\ &= I_0(x_k) + h_t \left[f_{k,1} + P_{k,1} \left(\frac{d^2 I_0(x)}{dx^2} \right)_{x=x_k} \right] \end{aligned} \tag{8.64}$$

The elements in the third and higher rows are obtained by replacing spatial derivatives by finite difference approximations. Since the elements in the first and last columns are specified by the Dirichlet conditions, we take $1 \le k \le N-1$ in

$$\frac{\psi_{k,m+1} - \psi_{k,m}}{h_t} = f_{k,m} + P_{k,m} \frac{\psi_{k+2,m} - 2\psi_{k+1,m} + \psi_{k,m}}{h_x^2} \tag{8.65a}$$

which we write as

$$\psi_{k,m+1} = \psi_{k,m} + h_t \left[f_{k,m} + \frac{P_{k,m}}{h_x^2} \psi_{k+2,m} - 2\frac{P_{k,m}}{h_x^2} \psi_{k+1,m} + \frac{P_{k,m}}{h_x^2} \psi_{k,m} \right] \tag{8.65b}$$

Starting with $m = 1$, we have

$$\psi_{k,2} = \psi_{k,1} + h_t \left[f_{k,1} + \frac{P_{k,1}}{h_x^2} \psi_{k+2,1} - 2\frac{P_{k,1}}{h_x^2} \psi_{k+1,1} + \frac{P_{k,1}}{h_x^2} \psi_{k,1} \right] \tag{8.66}$$

which gives us elements in the third row ($t = t_2$) in terms of elements in the first and second rows.

We note that for $k = N - 1$, the second term in the bracket involves $\psi_{N+1,1}$ which is outside the boundaries for x. Since the Euler finite differences are as accurate as a Taylor series truncated at the first power of h_x, we might consider approximating

$$\psi_{N+1,1} \simeq \psi_{N,1} + h_x \left(\frac{\partial \psi(x, t_1)}{\partial x} \right)_{x=x_N} \tag{8.67}$$

Because $\psi_{N,1}$ is given by the Dirichlet boundary conditions, but $(\partial \psi / \partial x)_{N,1}$ is not, we cannot obtain a value of $\psi_{N+1,1}$ from eq. 8.67. Thus, we cannot complete the solution table for the parabolic equation using Euler finite difference methods.

Milne finite differences

The Milne finite difference approximation to the parabolic equation is

$$\frac{\psi_{k,m+1} - \psi_{k,m-1}}{2h_t} = f_{k,m} + P_{k,m} \frac{\psi_{k+1,m} - 2\psi_{k,m} + \psi_{k-1,m}}{h_x^2} \tag{8.68}$$

With the first row of the solution table ($m = 0$) defined by the initial condition, we see that with $m = 1$ we obtain

$$\frac{\psi_{k,2} - \psi_{k,0}}{2h_t} = f_{k,1} + P_{k,1} \frac{\psi_{k+1,1} - 2\psi_{k,1} + \psi_{k-1,1}}{h_x^2} \tag{8.69}$$

which involves unknown values of ψ in the second ($m = 1$) and third ($m = 2$) rows. Thus, eq. 8.69 cannot be used to find elements in the second row of the solution table.

But the Milne finite difference approximation is as accurate as a Taylor series up to h_t^2. Therefore, we can determine the unfilled entries in the second row using

$$\psi(x, t_1) \simeq \psi(x, t_0) + h_t \left(\frac{\partial \psi(x, t)}{\partial t} \right)_{t=t_0} + \frac{1}{2} h_t^2 \left(\frac{\partial^2 \psi(x, t)}{\partial t^2} \right)_{t=t_0}$$

$$= I_0(x) + h_t \left(\frac{\partial \psi(x, t)}{\partial t} \right)_{t=t_0} + \frac{1}{2} h_t^2 \left(\frac{\partial^2 \psi(x, t)}{\partial t^2} \right)_{t=t_0} \tag{8.70}$$

To evaluate the time derivatives, we refer to eqs. 8.8 to obtain

$$\left(\frac{\partial \psi(x_k, t)}{\partial t}\right)_{t=t_0} = f_{k,0} + P_{k,0}\left(\frac{d^2 I_0}{dx^2}\right)_{x=x_k} \tag{8.71a}$$

and

$$\left(\frac{\partial^2 \psi(x_k, t)}{\partial t^2}\right)_{t=t_0} = \left[\left(\frac{\partial f(x,t)}{\partial t}\right) + \left(\frac{\partial P(x,t)}{\partial t}\right)\left(\frac{d^2 I_0(x)}{dx^2}\right) + P(x, t_0)\left(\frac{\partial^2 f(x,t)}{\partial x^2}\right)\right.$$
$$\left. + P(x, t)\left(\frac{\partial^2 P(x,t)}{\partial x^2}\frac{d^2 I_0(x)}{dx^2}\right) + P^2(x_k, t_0)\frac{d^4 I_0(x)}{dx^4}\right]_{\substack{x=x_k \\ t=t_0}} \tag{8.71b}$$

Then, eq. 8.70 becomes

$$\psi(x_k, t_1) = \psi_{k,1} = I_0(x_k) + h_t\left[f(x_k, t_0) + P(x_k, t_0)I_0''(x_k)\right]$$
$$+ \frac{h_t^2}{2}\left[\left(\frac{f(x,t)}{\partial t}\right) + \left(\frac{\partial P(x,t)}{\partial t}\right)I_0''(x) + P(x, t)\left(\frac{d^2 f(x,t)}{dx^2}\right)\right.$$
$$\left. + P(x, t)\left(\frac{d^2 P(x,t)}{dx^2}\right)I_0''(x) + P^2(x, t)I_0''''(x)\right]_{\substack{x=x_k \\ t=t_0}} \tag{8.72}$$

Thus, the partially filled solution table for a parabolic equation constrained by the Dirichlet boundary conditions is that shown in Table 8.2.

	$x = x_0$	$x = x_1$	$\bullet \bullet$	$x = x_N$
$t = t_0$	$\psi_{0,0}(8.61)$	$\psi_{1,0}(8.61)$	$\bullet \bullet$	$\psi_{N,0}(8.61)$
$t = t_1$	$\psi_{0,1}(8.62)$	$\psi_{1,1}(8.72)$	$\bullet \bullet$	$\psi_{N,1}(8.62)$
$\vdots$	$\vdots$			$\vdots$
$t = t_M$	$\psi_{0,M}(8.62)$			$\psi_{N,M}(8.62)$

Table 8.2 Partial solution table for a parabolic equation constrained by a single initial condition and Dirichlet boundary conditions

The notations $\psi_{k,0}(8.61)$, $\psi_{0/N,m}(8.62)$ and $\psi_{k,m}(8.72)$ indicate that these values of ψ are obtained from the equation number in each parentheses

A second way to estimate $\psi_{k,1}$ is to set $m = 0$ in eq. 8.68 to obtain

$$\frac{\psi_{k,1} - \psi_{k,-1}}{2h_t} = f_{k,0} + P_{k,0}\frac{\psi_{k+1,0} - 2\psi_{k,0} + \psi_{k-1,0}}{h_x^2} \tag{8.73}$$

From the second order Taylor sum, we have

$$\psi_{k,-1} \simeq \psi_{k,0} - h_t\left(\frac{\partial \psi(x_k, t)}{\partial t}\right)_{t=t_0} + \frac{1}{2}h_t^2\left(\frac{\partial^2 \psi(x_k, t)}{\partial t^2}\right)_{t=t_0} \tag{8.74}$$

Substituting this into eq. 8.73, we obtain

$$\frac{\psi_{k,1} - \psi_{k,0}}{2h_t} + \frac{1}{2}\left(\frac{\partial\psi(x_k,t)}{\partial t}\right)_{t=t_0} - \frac{h_t}{4}\left(\frac{\partial^2\psi(x_k,t)}{\partial t^2}\right)_{t=t_0}$$

$$= f_{k,0} + P_{k,0}\frac{\psi_{k+1,0} - 2\psi_{k,0} + \psi_{k-1,0}}{h_x^2} \qquad (8.75)$$

where the derivatives are given in eqs. 8.71.

The unfilled entries in Table 8.2, defined by $1 \leq k \leq N{-}1$ and $2 \leq m \leq M$, are found from

$$\frac{\psi_{k,m+1} - \psi_{k,m-1}}{2h_t} - P_{k,m}\frac{\psi_{k+1,m} - 2\psi_{k,m} + \psi_{k-1,m}}{h_x^2} = f_{k,m} \qquad (8.76a)$$

which we express in the form

$$\psi_{k,m+1} = \psi_{k,m-1} + 2h_t\left[f_{k,m} + P_{k,m}\frac{\psi_{k+1,m} - 2\psi_{k,m} + \psi_{k-1,m}}{h_x^2}\right]$$

$$= 2h_t f_{k,m} + \psi_{k,m-1} + 2CP_{k,m}\left(\psi_{k+1,m} - 2\psi_{k,m} + \psi_{k-1,m}\right) \qquad (8.76b)$$

where

$$C \equiv \frac{h_t}{h_x^2} \qquad (8.77)$$

We see that for given values of k and m in eq. 8.76b, each entry in the row labeled by $m + 1$ is given in terms of entries in rows defined by m and $m{-}1$.

Example 8.5: Finite difference approximation for the solution to a diffusion equation with Dirichlet boundary conditions

We again consider

$$\frac{\partial\psi(x,t)}{\partial t} - \frac{\partial^2\psi(x,t)}{\partial x^2} = 0 \qquad \textit{(8.54)}$$

constrained by the initial condition

$$\psi(x,0) = I_0(x) = e^{-x} \qquad \textit{(8.10)}$$

and the Dirichlet boundary conditions

$$\begin{cases} \psi(0,t) = D_0(t) = e^t \\ \psi(1,t) = D_1(t) = e^{t-1} \end{cases} \qquad \textit{(8.11)}$$

the solution to which is

$$\psi(x,t) = e^{t-x} \qquad (8.12)$$

from which we generate the table of exact values shown in Table 8.3.

	$x = 0.00$	$x = 0.25$	$x = 0.50$	$x = 0.75$	$x = 1.00$
$t = 0.00$	1.00000	0.77880	0.60653	0.47237	0.36788
$t = 0.20$	1.22140	0.95123	0.74082	0.57695	0.44933
$t = 0.40$	1.49182	1.16183	0.90484	0.70469	0.54881
$t = 0.60$	1.82212	1.41907	1.10517	0.86071	0.67032
$t = 0.80$	2.22554	1.73325	1.34986	1.05127	0.81873
$t = 1.00$	2.71828	2.11700	1.64872	1.28403	1.00000

Table 8.3 Table of exact values for e^{t-x}

The Milne finite difference approximation to eq. 8.54 is

$$\psi_{k,m+1} = \psi_{k,m-1} + 2C\left(\psi_{k+1,m} - 2\psi_{k,m} + \psi_{k-1,m}\right) \qquad (8.78)$$

With

$$\psi_{k,0} = I_0(x_k) = e^{-x_k} \qquad (8.79a)$$

$$\left(\frac{\partial\psi(x_k,t)}{\partial t}\right)_{t=0} = \left(\frac{\partial^2\psi(x,0)}{\partial x^2}\right)_{x_k} = I_0''(x_k) = e^{-x_k} \qquad (8.79b)$$

and

$$\left(\frac{\partial^2\psi(x_k,t)}{\partial t^2}\right)_{t=0} = \left(\frac{\partial^2[\partial\psi(x,t)/\partial t]_{t=0}}{\partial x^2}\right)_{x_k}$$

$$= \left(\frac{d^4\psi(x,0)}{dx^4}\right)_{x_k} = I_0''''(x_k) = e^{-x_k} \qquad (8.79c)$$

we obtain

$$\psi_{k,1} = \left(1 + h_t + \frac{h_t^2}{2}\right)e^{-x_k} \qquad (8.80)$$

Taking $h_x = 0.25$ and $h_t = 0.20$, with the Dirichlet boundary conditions

$$\begin{cases} \psi_{0,m} = e^{t_m} \\ \psi_{N,m} = e^{t_m-1} \end{cases} \qquad (8.81)$$

and the initial condition and eq. 8.80, we produce the partial solution shown in Table 8.4(a).

	$x = 0.00$	$x = 0.25$	$x = 0.50$	$x = 0.75$	$x = 1.00$
$t = 0.00$	1.00000	0.77880	0.60653	0.47237	0.36788
$t = 0.20$	1.22140	0.95014	0.73997	0.57629	0.44933
$t = 0.40$	1.49182				0.54881
$t = 0.60$	1.82212				0.67032
$t = 0.80$	2.22554				0.81873
$t = 1.00$	2.71828				1.00000

Table 8.4a   Partial solution table for the diffusion equation

Setting $m = 1$, eq. 8.77 becomes

$$\psi_{k,2} = \psi_{k,0} + 2C\left(\psi_{k+1,1} - 2\psi_{k,1} + \psi_{k-1,1}\right) \tag{8.82}$$

With $1 \le k \le N-1$, this yields the unfilled entries in the third row of the table from the entries in the first and second rows. We continue this process setting $m = 2, 3, 4,$ and 5 to generate the complete solution table for this differential equation given in Table 8.4(b).

	$x = 0.00$	$x = 0.25$	$x = 0.50$	$x = 0.75$	$x = 1.00$
$t = 0.00$	1.00000	0.77880	0.60653	0.47237	0.36788
$t = 0.20$	1.22140	0.95014	0.73997	0.57629	0.44933
$t = 0.40$	1.49182	1.16982	0.90407	0.70739	0.54881
$t = 0.60$	1.82212	1.31016	1.18207	0.82012	0.67032
$t = 0.80$	2.22554	3.62661	−0.59266	2.06511	0.81873
$t = 1.00$	2.71828	−34.66001	45.19519	−24.16649	1.00000

Table 8.4b   Complete solution table for the diffusion equation

Comparing this to the exact results given in Table 8.3, it is obvious that this solution table is completely incorrect. The entries at times later than $t_0 = 0$ and at distances away from $x_0 = 0$ and $x = 1$ are meaningless and unstable (they are very large, some are negative and they vary wildly). □

Such meaningless and unstable behavior in the solution of a parabolic equation is well known (see, for example, Forsythe, G., and Wasow, W., 1960, pp. 27, 92 or Strikwerda, J., 1989, p. 120). These straightforward finite difference methods yield stable results for the parabolic equation only when $C < 1/2$ (see, for example, Kunz, K., 1957, pp. 328–332). Clearly, the values of h_t and h_x used to generate Table 8.4(b) in the above example do not satisfy this condition.

The Crank–Nicolson modification

A commonly used approach to deal with this instability and inaccuracy was developed by Crank, J., and Nicolson, P., (1947). The method approximates the space derivative by the Milne finite difference averaged over the time interval $[t_{m-1}, t_{m+1}]$ as

$$\left(\frac{\partial^2 \psi(x,t)}{\partial x^2}\right)_{x_k,t_m} \simeq$$

$$\frac{1}{2}\left[\frac{\psi_{k+1,m+1} - 2\psi_{k,m+1} + \psi_{k-1,m+1}}{h_x^2} + \frac{\psi_{k+1,m-1} - 2\psi_{k,m-1} + \psi_{k-1,m-1}}{h_x^2}\right]$$

$$= \frac{1}{2h_x^2}\left[(\psi_{k+1,m+1} + \psi_{k+1,m-1}) - 2(\psi_{k,m+1} + \psi_{k,m-1}) + (\psi_{k-1,m+1} + \psi_{k-1,m-1})\right]$$

$$(8.83)$$

Each term in

$$\left(\frac{\partial^2 \psi(x,t)}{\partial x^2}\right)_{x_k,t_m} \simeq \frac{\psi_{k+1,m} - 2\psi_{k,m} + \psi_{k-1,m}}{h_x^2} \qquad (8.59d)$$

contains $\psi_{k+1,m}$, $\psi_{k,m}$, and $\psi_{k-1,m}$. Each such $\psi_{n,m}$ ($n = k + 1, k, k - 1$) is the first term in a Taylor series in powers of h_t. Therefore, each term is as accurate as a Taylor series truncated at h_t^0. However, with the Crank–Nicolson approximation, the truncated Taylor series for each combination in eq. 8.83 is

$$\frac{1}{2}(\psi_{n,m+1} + \psi_{n,m-1}) = \frac{1}{2}\left[\psi_{n,m} + h_t\left(\frac{\partial \psi(x_n,t)}{\partial t}\right)_{t_m} + \frac{1}{2}h_t^2\left(\frac{\partial^2 \psi(x_n,t)}{\partial t^2}\right)_{t_m}\right.$$

$$\left. + \psi_{n,m} - h_t\left(\frac{\partial \psi(x_n,t)}{\partial t}\right)_{t_m} + \frac{1}{2}h_t^2\left(\frac{\partial^2 \psi(x_n,t)}{\partial t^2}\right)_{t_m}\right]$$

$$= \psi_{n,m} + \frac{1}{2}h_t^2\left(\frac{\partial^2 \psi(x_n,t)}{\partial t^2}\right)_{t_m} \qquad (8.84)$$

which is as accurate as the Taylor series up to h_t^2. This modification makes the solution unconditionally stable.

There are other methods for achieving this stability. See, for example, the Jacobi and the Gauss–Seidel methods by Smith, G.D., 1965, pp. 25–28.

As noted in example 8.5, the first row and the first and last columns of the solution table are obtained from the initial and Dirichlet boundary conditions, and the entries in the second row (columns 1 through $N-1$) are found from the truncated Taylor series. With the Crank–Nicolson approximation, the entries in the higher index rows for $1 \leq k \leq N-1$ are found from

$$\psi_{k,m+1} = \psi_{k,m-1} + 2h_t\left[f_{k,m} + \right.$$

$$\left. \frac{P_{k,m}}{2}\left(\frac{\psi_{k+1,m+1} - 2\psi_{k,m+1} + \psi_{k-1,m+1}}{h_x^2} + \frac{\psi_{k+1,m-1} - 2\psi_{k,m-1} + \psi_{k-1,m-1}}{h_x^2}\right)\right]$$

$$(8.85a)$$

which, with eq. 8.77, we write as

$$
\begin{aligned}
&-CP_{k,m}\psi_{k+1,m+1} + \left(1 + 2CP_{k,m}\right)\psi_{k,m+1} - CP_{k,m}\psi_{k-1,m+1} = \\
&2h_t f_{k,m} + \psi_{k,m-1} + CP_{k,m}\left(\psi_{k+1,m-1} - 2\psi_{k,m-1} + \psi_{k-1,m-1}\right)
\end{aligned}
\tag{8.85b}
$$

With $1 \le k \le N-1$, this yields a set of $N-1$ coupled equations for the $N-1$ values of $\{\psi_{1,m+1}, \ldots, \psi_{N-1,m+1}\}$.

Example 8.6: Finite difference approximation for the solution to a diffusion equation with Dirichlet boundary conditions using the Crank–Nicolson modification

We again consider

$$
\frac{\partial \psi(x,t)}{\partial t} = \frac{\partial^2 \psi(x,t)}{\partial x^2}
\tag{8.9b}
$$

with $\psi(x,t)$ constrained as in example 8.5.

With the Crank–Nicolson approximation to $\partial^2 \psi / \partial x^2$, the finite difference approximation to eq. 8.9b is

$$
\begin{aligned}
\psi_{k,m+1} = \psi_{k,m-1} + \\
\frac{h_t}{h_x^2}\left[\left(\psi_{k+1,m+1} - 2\psi_{k,m+1} + \psi_{k-1,m+1}\right) + \left(\psi_{k+1,m-1} - 2\psi_{k,m-1} + \psi_{k-1,m-1}\right)\right]
\end{aligned}
\tag{8.86a}
$$

which we write in the form

$$
\begin{aligned}
C\psi_{k-1,m+1} - (2C+1)\psi_{k,m+1} + C\psi_{k+1,m+1} = \\
-C\left(\psi_{k-1,m-1} + \psi_{k+1,m-1}\right) + (2C-1)\psi_{k,m-1}
\end{aligned}
\tag{8.86b}
$$

As shown in partial solution Table 8.4(a), the values of ψ in the first two rows are obtained from the initial conditions and a truncated Taylor series. Thus, with $h_t = 0.20$, $h_x = 0.25$, using eq. 8.86b, we generate a set of three coupled equations for $\{\psi_{1,m+1}, \psi_{2,m+1}, \psi_{3,m+1}\}$, starting with $m = 1$.

Starting with $m = 1$, for $1 \le k \le 3$, we have

$$
\begin{aligned}
-(2C+1)\psi_{1,m+1} + \frac{h_t}{h_x^2}\psi_{2,m+1} = \\
-C\left(\psi_{0,m-1} + \psi_{2,m-1} + \psi_{0,m+1}\right) + (2C-1)\psi_{1,m-1} \equiv \alpha_{1,m}
\end{aligned}
\tag{8.87a}
$$

$$
\begin{aligned}
C\psi_{1,m+1} - (2C+1)\psi_{2,m+1} + C\psi_{3,m+1} = \\
-C\left(\psi_{1,m-1} + \psi_{3,m-1}\right) + (2C-1)\psi_{2,m-1} \equiv \alpha_{2,m}
\end{aligned}
\tag{8.87b}
$$

and

$$C\psi_{2,m+1} - (2C+1)\psi_{3,m+1} =$$
$$-C(\psi_{4,m+1} + \psi_{2,m-1} + \psi_{4,m-1}) + (2C-1)\psi_{3,m-1} \equiv \alpha_{3,m} \qquad (8.87c)$$

where all $\alpha_{k,m}$ are determined from known entries in the solution table. These equations yield the complete solution table given in Table 8.5.

	$x = 0.00$	$x = 0.25$	$x = 0.50$	$x = 0.75$	$x = 1.00$
$t = 0.00$	1.00000	0.77880	0.60653	0.47237	0.36788
$t = 0.20$	1.22140	0.95014	0.73997	0.57629	0.44933
$t = 0.40$	1.49182	1.16378	0.90710	0.70625	0.54881
$t = 0.60$	1.82212	1.42176	1.10769	0.86263	0.67032
$t = 0.80$	2.22554	1.73544	1.35260	1.05317	0.81873
$t = 1.00$	2.71828	2.11941	1.65228	1.28631	1.00000

Table 8.5 Complete solution table to the diffusion equation with Dirichlet boundary conditions, using the Crank–Nicolson modification to $\partial^2\psi/\partial x^2$

Comparing these results to the exact values given in Table 8.3, we see that the Crank–Nicolson modification yields stable and accurate results. □

Hyperbolic equations subject to Dirichlet boundary conditions

The Milne finite difference approximation to the hyperbolic equation

$$\frac{\partial^2\psi}{\partial t^2} - P(x,t)\frac{\partial^2\psi}{\partial x^2} = f(x,t) \qquad (8.2b)$$

without the Crank–Nicolson modification is given by

$$\frac{\psi_{k,m+1} - 2\psi_{k,m} + \psi_{k,m-1}}{h_t^2} - P_{n,m}\frac{\psi_{k+1,m} - 2\psi_{k,m} + \psi_{k-1,m}}{h_x^2} = f_{k,m} \qquad (8.88a)$$

With the Crank–Nicolson modification, eq. 8.2b is approximated by

$$\frac{\psi_{k,m+1} - 2\psi_{k,m} + \psi_{k,m-1}}{h_t^2} - \frac{P_{k,m}}{2}\left[\frac{\psi_{k+1,m+1} - 2\psi_{k,m+1} + \psi_{k-1,m+1}}{h_x^2}\right.$$
$$\left. + \frac{\psi_{k+1,m-1} - 2\psi_{k,m-1} + \psi_{k-1,m-1}}{h_x^2}\right] = f_{k,m} \qquad (8.88b)$$

We express these as

$$\psi_{k,m+1} = C'f_{k,m} + C'P_{k,m}\psi_{k+1,m} + (2 - C'P_{k,m})\psi_{k,m} + C'P_{k,m}\psi_{k-1,m} - \psi_{k,m-1} \qquad (8.89a)$$

and

$$-\frac{C'}{2}P_{k,m}\psi_{k+1,m+1} + \left(1 + C'P_{k,m}\right)\psi_{k,m+1} - \frac{C'}{2}P_{k,m}\psi_{k-1,m+1}$$

$$= h_t^2 f_{k,m} + 2\psi_{k,m} - \psi_{k,m-1} + \frac{C'}{2}P_{k,m}\left(\psi_{k+1,m-1} - 2\psi_{k,m-1} + \psi_{k-1,m-1}\right) \qquad (8.89b)$$

where

$$C' \equiv \frac{h_t^2}{h_x^2} \qquad (8.90)$$

As will be illustrated by example, the Milne finite difference approximation to the hyperbolic equation, with the Crank–Nicolson modification, yields accurate results, but without the Crank–Nicolson scheme, the results are highly inaccurate. Thus, it is necessary that we use the Crank–Nicolson scheme for the hyperbolic equation.

As we discussed for the parabolic equation, the first and last columns of the solution table for the hyperbolic equation are obtained from the Dirichlet boundary conditions

$$\begin{cases} \psi_{0,m} = D_{x_0}(t_m) \\ \psi_{N,m} = D_{x_N}(t_m) \end{cases} \qquad (8.62)$$

and the entries in the first row of the table are given by the initial condition

$$\psi_{k,0} = I_0(x_k) \qquad (8.61)$$

Entries in the second row are obtained from the truncated Taylor series, which, from both initial conditions, is given by

$$\psi(x_k, t_1) = \psi_{k,1} \simeq I_0(x_n) + h_t I_1(x_k) + \frac{1}{2} h_t^2 \left[f(x_k, t_0) + P(x_k, t_0) I_0''(x_k) \right] \qquad (8.91)$$

Therefore, the partial solution table for the hyperbolic equation subject to Dirichlet boundary conditions is given by Table 8.6

	$x = x_0$	$x = x_1$	• •	$x = x_N$
$t = t_0$	$\psi_{0,0}$(8.61)	$\psi_{1,0}$(8.61)	• •	$\psi_{N,0}$(8.61)
$t = t_1$	$\psi_{0,1}$(8.62)	$\psi_{1,1}$(8.91)	• •	$\psi_{N,1}$(8.62)
$t = t_M$	$\psi_{0,M}$(8.62)			$\psi_{N,M}$(8.62)

Table 8.6 Partial solution table for a hyperbolic equation constrained by two initial conditions and Dirichlet boundary conditions

The unfilled entries are obtained from either of eqs. 8.88 with $1 \leq k \leq N-1$, $1 \leq m \leq M-1$.

Example 8.7: Finite difference approximation for the solution to a wave equation with Dirichlet boundary conditions

We again consider the source-free, one dimensional wave equation with wave velocity $c = 1$.

$$\frac{\partial^2 \psi(x,t)}{\partial t^2} - \frac{\partial^2 \psi(x,t)}{\partial x^2} = 0 \tag{8.54}$$

Imposing the initial and Dirichlet boundary conditions

$$\begin{cases} \psi(x,0) = I_0(x) = e^{-x} \\ \left(\dfrac{\partial \psi(x,t)}{\partial t} \right)_{t=0} = I_1(x) = e^{-x} \end{cases} \tag{8.26}$$

and

$$\begin{cases} \psi(0,t) = D_0(t) = e^t \\ \psi(1,t) = D_1(t) = e^{t-1} \end{cases} \tag{8.11}$$

the solution is

$$\psi(x,t) = e^{t-x} \tag{8.12}$$

Without the Crank–Nicolson modification, the Milne finite difference approximation to eq. 8.54 is given by

$$\frac{\psi_{k,m+1} - 2\psi_{k,m} + \psi_{k,m-1}}{h_t^2} - \frac{\psi_{k+1,m} - 2\psi_{k,m} + \psi_{k-1,m}}{h_x^2} = 0 \tag{8.92a}$$

which we express in the form

$$\psi_{k,m+1} = C'\psi_{k+1,m} + 2(1 - C')\psi_{k,m} + C'\psi_{k-1,m} - \psi_{k,m-1} \tag{8.92b}$$

where

$$C' \equiv \frac{h_t^2}{h_x^2} \tag{8.90}$$

As with the parabolic equation, the entries in the first and last columns of the solution table are given by

$$\begin{cases} \psi_{0,m} = e^{t_m} \\ \psi_{N,m} = e^{t_m - 1} \end{cases} \tag{8.81}$$

and the entries in the first and second rows are found from

$$
\begin{cases}
\psi_{k,0} = e^{-x_k} \\
\psi_{k,1} = \left(1 + h_t + \dfrac{1}{2}\, h_t^2\right) e^{-x_k}
\end{cases}
\tag{8.93}
$$

Thus, the partial solution table, which is identical to Table 8.4(a), is given by Table 8.7(a).

	$x = 0.00$	$x = 0.25$	$x = 0.50$	$x = 0.75$	$x = 1.00$
$t = 0.00$	1.00000	0.77880	0.60653	0.47237	0.36788
$t = 0.20$	1.22140	0.95014	0.73997	0.57629	0.44933
$t = 0.40$	1.49183				0.54881
$t = 0.60$	1.82212				0.67032
$t = 0.80$	2.22554				0.81873
$t = 1.00$	2.71828				1.00000

Table 8.7a Partial solution table for a wave equation

The unfilled entries are found by taking $1 \le k \le 3$ and $1 \le m \le 4$ in eq. 8.92b. The complete solution table is given in Table 8.7(b).

	$x = 0.00$	$x = 0.25$	$x = 0.50$	$x = 0.75$	$x = 1.00$
$t = 0.00$	1.00000	0.77880	0.60653	0.47237	0.36788
$t = 0.20$	1.22140	0.95014	0.73997	0.57629	0.44933
$t = 0.40$	1.49183	1.60028	1.24521	0.97010	0.54881
$t = 0.60$	1.82212	2.49082	2.37772	1.71863	0.67032
$t = 0.80$	2.22554	3.53697	4.05750	3.07404	0.81873
$t = 1.00$	2.71828	4.87814	6.04804	5.07617	1.00000

Table 8.7b Complete solution table for a wave equation constrained by Dirichlet boundary conditions without the Crank–Nicolson modification

Comparing this to the exact values given in Table 8.3, we see that at times t_m for $m > 1$, and at points $x = \{x_1, x_2, x_3\}$, the solutions are completely incorrect. The reason for this is identical to that presented to explain the inaccuracies in the solution to the diffusion equation without the Crank–Nicolson modification. Thus, the Crank–Nicolson correction must be used to obtain accurate results for the hyperbolic as well as the parabolic equation.

With the Crank–Nicolson modification,

$$
\left(\frac{\partial^2 \psi(x,t)}{\partial x^2}\right)_{x_k, t_m} \simeq
$$

$$
\frac{1}{2}\left[\frac{\psi_{k+1,m+1} - 2\psi_{k,m+1} + \psi_{k-1,m+1}}{h_x^2} + \frac{\psi_{k+1,m-1} - 2\psi_{k,m-1} + \psi_{k-1,m-1}}{h_x^2}\right]
\tag{8.83}
$$

the finite difference approximation to the wave equation above is

$$\frac{\psi_{k,m+1} - 2\psi_{k,m} + \psi_{k,m-1}}{h_t^2} \simeq$$

$$\frac{(\psi_{k+1,m+1} - 2\psi_{k,m+1} + \psi_{k-1,m+1}) + (\psi_{k+1,m-1} - 2\psi_{k,m-1} + \psi_{k-1,m-1})}{2h_x^2} \qquad (8.94a)$$

Separating terms with the time index $m + 1$ from all other terms, we express this as

$$\frac{C'}{2}\psi_{k+1,m+1} - (1 + C')\psi_{k,m+1} + \frac{C'}{2}\psi_{k-1,m+1} =$$

$$- 2\psi_{k,m} - \frac{C'}{2}\psi_{k+1,m-1} + (1 + C')\psi_{k,m-1} - \frac{C'}{2}\psi_{k-1,m-1} \qquad (8.94b)$$

Setting $k = 1, 2, 3$, in eq. 8.94b, we obtain

$$\frac{C'}{2}\psi_{2,m+1} - (1 + C')\psi_{1,m+1} =$$

$$-\frac{C'}{2}\psi_{2,m-1} + (1 + C')\psi_{1,m-1} - \frac{C'}{2}\psi_{0,m-1} - 2\psi_{1,m} - \frac{C'}{2}\psi_{0,m+1} \equiv \alpha_{1,m} \qquad (8.95a)$$

$$\frac{C'}{2}\psi_{3,m+1} - (1 + C')\psi_{2,m+1} + \frac{C'}{2}\psi_{1,m+1} =$$

$$-\frac{C'}{2}\psi_{3,m-1} + (1 + C')\psi_{2,m-1} - \frac{C'}{2}\psi_{1,m-1} - 2\psi_{2,m} \equiv \alpha_{2,m} \qquad (8.95b)$$

and

$$- (1 + C')\psi_{3,m+1} + \frac{C'}{2}\psi_{2,m+1} =$$

$$-\frac{C'}{2}\psi_{4,m-1} + (1 + C')\psi_{3,m-1} - \frac{C'}{2}\psi_{2,m-1} - 2\psi_{3,m} - \frac{C'}{2}\psi_{4,m+1} \equiv \alpha_{3,m} \qquad (8.95c)$$

where, for a given m, the quantities $\alpha_{k,m}$ are given in terms of known entries in the solution table. These equations yield the complete solution given in Table 8.7(c).

	$x = 0.00$	$x = 0.25$	$x = 0.50$	$x = 0.75$	$x = 1.00$
$t = 0.00$	1.00000	0.77880	0.60653	0.47237	0.36788
$t = 0.20$	1.22140	0.95014	0.73997	0.57629	0.44933
$t = 0.40$	1.49183	1.16082	0.90387	0.70400	0.54881
$t = 0.60$	1.82212	1.41936	1.10504	0.86072	0.67032
$t = 0.80$	2.22554	1.73554	1.35154	1.05260	0.81873
$t = 1.00$	2.71828	2.12129	1.65324	1.28705	1.00000

Table 8.7c Complete solution table for a wave equation
with the Crank–Nicolson modification to $\partial^2\psi/\partial x^2$

Comparing these to the exact results given in Table 8.3, we see that with the Crank–Nicolson modification, we obtain an accurate numerical solution to the wave equation. □

Elliptic equations subject to Dirichlet boundary conditions

For the elliptic equation

$$R(x,y)\frac{\partial^2\psi}{\partial x^2} + S(x,y)\frac{\partial^2\psi}{\partial y^2} = f(x,y) \qquad (8.2c)$$

constrained by the Dirichlet boundary conditions

$$\begin{cases} \psi(x_0, y_m) = \psi_{0,m} = D_{x_0}(y_m) \\ \psi(x_N, y_m) = \psi_{N,m} = D_{x_N}(y_m) \\ \psi(x_k, y_0) = \psi_{k,0} = D_{y_0}(x_k) \\ \psi(x_k, y_M) = \psi_{k,M} = D_{y_M}(x_k) \end{cases} \qquad (8.96)$$

the partial solution table is that given in Table 8.8.

	$x = x_0$	$x = x_1$	• •	$x = x_N$
$y = y_0$	$\psi_{0,0}(8.96)$	$\psi_{1,0}(8.96)$	• •	$\psi_{N,0}(8.96)$
$y = y_1$	$\psi_{0,1}(8.96)$			$\psi_{N,1}(8.96)$
⋮	⋮			⋮
$y = y_M$	$\psi_{0,M}(8.96)$	$\psi_{1,M}(8.96)$	• •	$\psi_{N,M}(8.96)$

Table 8.8 Partial solution table for an elliptic equation subjected to Dirichlet boundary conditions

The unknowns $\{\psi_{1,1}, \ldots, \psi_{N-1,1}, \psi_{1,2}, \ldots, \psi_{N-1,2}, \ldots, \psi_{1,M-1}, \ldots, \psi_{N-1,M-1}\}$ are obtained from the finite difference approximation to eq. 8.2c. Without the Crank–Nicolson modification, this finite difference equation is

$$\frac{R_{k,m}}{h_x^2}\left(\psi_{k+1,m} - 2\psi_{k,m} + \psi_{k-1,m}\right) + \frac{S_{k,m}}{h_y^2}\left(\psi_{k,m+1} - 2\psi_{k,m} + \psi_{k,m-1}\right) = f_{k,m} \quad (8.97a)$$

With the Crank–Nicolson modification applied to both derivatives, the finite difference approximation to eq. 8.2c is

$$\frac{R_{k,m}}{2h_x^2}\left[\left(\psi_{k+1,m+1} - 2\psi_{k,m+1} + \psi_{k-1,m+1}\right) + \left(\psi_{k+1,m-1} - 2\psi_{k,m-1} + \psi_{k-1,m-1}\right)\right]$$

$$+\frac{S_{k,m}}{2h_y^2}\left[\left(\psi_{k+1,m+1} - 2\psi_{k+1,m} + \psi_{k+1,m-1}\right) + \left(\psi_{k-1,m+1} - 2\psi_{k-1,m} + \psi_{k-1,m-1}\right)\right]$$

$$= f_{k,m} \qquad (8.97b)$$

The unfilled entries are found from these expressions for $1 \leq k \leq N{-}1$ and $1 \leq m \leq M{-}1$.

Example 8.8: Finite difference approximation for the solution to a Poisson equation with Dirichlet boundary conditions

The Poisson equation in two dimensions

$$\frac{\partial^2 \psi}{\partial x^2} + \frac{\partial^2 \psi}{\partial y^2} = 2e^{y-x} \qquad (8.98)$$

when constrained by the Dirichlet boundary conditions

$$\begin{cases} \psi_{0,m} = \psi(0, y_m) = D_{x_0}(y_m) = e^{y_m} \\ \psi_{N,m} = \psi(1, y_m) = D_{x_N}(y_m) = e^{y_m - 1} \\ \psi_{k,0} = \psi(x_k, 0) = D_{y_0}(x_k) = e^{-x_k} \\ \psi_{k,M} = \psi(x_k, 1) = D_{y_M}(x_k) = e^{1 - x_k} \end{cases} \qquad (8.99)$$

has solution

$$\psi(x, y) = e^{y-x} \qquad (8.100)$$

Thus, with $h_x = 0.25$ and $h_y = 0.20$, the exact values are the same as those in Table 8.3, where, of course, the entries that were labeled by t are here labeled by y. This exact solution table is given in Table 8.9.

	$x_0 = 0.00$	$x_1 = 0.25$	$x_2 = 0.50$	$x_3 = 0.75$	$x_4 = 1.00$
$y_0 = 0.00$	1.00000	0.77880	0.60653	0.47237	0.36788
$y_1 = 0.20$	1.22140	0.95123	0.74082	0.57695	0.44933
$y_2 = 0.40$	1.49182	1.16184	0.90484	0.70469	0.54881
$y_3 = 0.60$	1.82212	1.41907	1.10518	0.86071	0.67032
$y_4 = 0.80$	2.22554	1.73325	1.34986	1.05128	0.81873
$y_5 = 1.00$	2.71828	2.11701	1.64873	1.28403	1.00000

Table 8.9 Table of exact results for a Poisson equation

The partial solution table, obtained from the boundary conditions, is given in Table 8.10.

	$x_0 = 0.00$	$x_1 = 0.25$	$x_2 = 0.50$	$x_3 = 0.75$	$x_4 = 1.00$
$y_0 = 0.00$	1.00000	0.77880	0.60653	0.47237	0.36788
$y_1 = 0.20$	1.22140				0.44933
$y_2 = 0.40$	1.49182				0.54881
$y_3 = 0.60$	1.82212				0.67032
$y_4 = 0.80$	2.22554				0.81873
$y_5 = 1.00$	2.71828	2.11701	1.64873	1.28403	1.00000

Table 8.10 Partial solution table for a Poisson equation subject to Dirichlet boundary conditions

We see from this table that we must solve for the twelve unknown quantities $\psi_{k,m}$. Multiplying by $h_y{}^2$ and defining

$$C'' \equiv \frac{h_y^2}{h_x^2} \tag{8.101}$$

the finite difference approximation to eq. 8.98 without the Crank–Nicolson modification is

$$C'' \left(\psi_{k+1,m} - 2\psi_{k,m} + \psi_{k-1,m} \right) + \left(\psi_{k,m+1} - 2\psi_{k,m} + \psi_{k,m-1} \right) = 2h_y^2 e^{y_m - x_k} \tag{8.102a}$$

With the Crank–Nicolson modification, the approximation to eq. 8.98 can be written as

$$C'' \left[\left(\psi_{k+1,m+1} - 2\psi_{k,m+1} + \psi_{k-1,m+1} \right) + \left(\psi_{k+1,m-1} - 2\psi_{k,m-1} + \psi_{k-1,m-1} \right) \right]$$
$$+ \left[\left(\psi_{k+1,m+1} - 2\psi_{k+1,m} + \psi_{k+1,m-1} \right) + \left(\psi_{k-1,m+1} - 2\psi_{k-1,m} + \psi_{k-1,m-1} \right) \right]$$
$$= 4h_y^2 e^{y_m - x_k} \tag{8.102b}$$

Referring to Table 8.10, we see that for all $0 \leq m \leq 5$, $\psi_{0,m}$ and $\psi_{4,m}$ are known entries in the first and last columns of the solution table. Likewise, for all $0 \leq k \leq 4$, $\psi_{k,0}$ and $\psi_{k,5}$ are known entries in the first and last rows. Therefore, setting $1 \leq k \leq 3$ and $1 \leq m \leq 4$ in each of the finite difference equations given in eqs. 8.102, we obtain sets of twelve coupled equations for the twelve unknown values of $\psi_{k,m}$.

The complete table for the solution to eq. 8.98 without the Crank–Nicolson modification is given in Table 8.11(a), and the complete solution table with the Crank–Nicolson approximation is shown in Table 8.11(b).

	$x_0 = 0.00$	$x_1 = 0.25$	$x_2 = 0.50$	$x_3 = 0.75$	$x_4 = 1.00$
$y_0 = 0.00$	1.00000	0.77880	0.60653	0.47237	0.36788
$y_1 = 0.20$	1.22140	0.94767	0.73681	0.57418	0.44933
$y_2 = 0.40$	1.49182	1.15631	0.89853	0.70034	0.54881
$y_3 = 0.60$	1.82212	1.41290	1.09821	0.85591	0.67032
$y_4 = 0.80$	2.22554	1.72834	1.34451	1.04756	0.81873
$y_5 = 1.00$	2.71828	2.11701	1.64873	1.28403	1.00000

Table 8.11a Complete solution table for a Poisson equation with Dirichlet boundary conditions without the Crank–Nicolson modification

	$x_0 = 0.00$	$x_1 = 0.25$	$x_2 = 0.50$	$x_3 = 0.75$	$x_4 = 1.00$
$y_0 = 0.00$	1.00000	0.77880	0.60653	0.47237	0.36788
$y_1 = 0.20$	1.22140	0.94941	0.73880	0.57454	0.44933
$y_2 = 0.40$	1.49182	1.15823	0.90153	0.70207	0.54881
$y_3 = 0.60$	1.82212	1.41499	1.10095	0.85843	0.67032
$y_4 = 0.80$	2.22554	1.73061	1.34607	1.04912	0.81873
$y_5 = 1.00$	2.71828	2.11701	1.64873	1.28403	1.00000

Table 8.11b Complete solution table for a Poisson equation with Dirichlet boundary conditions with the Crank–Nicolson modification

Comparing these two results to the exact values in Table 8.9, we see that while both solutions are acceptable, those obtained using the Crank–Nicolson modification are somewhat more accurate.

To more easily illustrate this, in Table 8.12, we present the percent difference from the exact value of each calculated solution, using a double vertical line to separate this percent difference for the values obtained without the Crank–Nicolson given in eqs. 8.102, we obtain from those found with the Crank–Nicolson scheme (eqs. 8.103).

	$x_0 = 0.00$	$x_1 = 0.25$	$x_2 = 0.50$	$x_3 = 0.75$	$x_4 = 1.00$
$y_0 = 0.00$					
$y_1 = 0.20$		0.37‖0.19	0.54‖0.27	0.48‖0.41	
$y_2 = 0.40$		0.48‖0.31	0.70‖0.37	0.62‖0.37	
$y_3 = 0.60$		0.43‖0.29	0.63‖0.38	0.56‖0.26	
$y_4 = 0.80$		0.28‖0.15	0.40‖0.28	0.35‖0.20	
$y_5 = 1.00$					

Table 8.12 Percent difference of the approximate solution from the exact value without‖with the Crank–Nicolson modification

Since the entries in the first and last rows and first and last columns of the solution table are given by Dirichlet boundary conditions, these values are exact. Therefore, their percent differences are zero and are not shown in the table. We note that for this example, the results obtained using the Crank–Nicolson modification are roughly twice as accurate as those determined without the modification. □

The analysis for solving the three types of equations when subjected to Neumann boundary conditions is essentially the same as when Gauss boundary conditions are imposed. Therefore, we will present the analysis involving both sets of boundary conditions together.

Parabolic equations subject to Neumann and Gauss boundary conditions

We now consider the case for which the constraints on the solution to

$$\frac{\partial \psi}{\partial t} = f(x,t) + P(x,t)\frac{\partial^2 \psi}{\partial x^2} \tag{8.7}$$

are the single initial condition

$$\psi(x,t_0) = I_0(x) \tag{8.3a}$$

and either the Neumann boundary conditions

$$\begin{cases} \left(\dfrac{\partial \psi(x,t)}{\partial x}\right)_{x_0} = N_{x_0}(t) \\[2ex] \left(\dfrac{\partial \psi(x,t)}{\partial x}\right)_{x_N} = N_{x_N}(t) \end{cases} \tag{8.103a}$$

or the Gauss conditions

$$\begin{cases} \psi(x_0,t) + \gamma\left(\dfrac{\partial \psi(x,t)}{\partial x}\right)_{x_0} = G_{x_0}(t) \\[2ex] \psi(x_N,t) + \gamma\left(\dfrac{\partial \psi(x,t)}{\partial x}\right)_{x_N} = G_{x_N}(t) \end{cases} \tag{8.103b}$$

As before, entries in the first row of the solution table are given by the initial condition of eq. 8.3a:

$$\psi_{k,0} = I_0(x_k) \tag{8.61}$$

and those in the second row are obtained from the Taylor series, which we truncate at h_t^2

$$\psi(x_k,t_1) \simeq I_0(x_k) + h_t\left[f(x_k,t_0) + P(x_k,t_0)I_0''(x_k)\right]$$

$$+ \frac{1}{2}h_t^2\left[\left(\frac{f(x_k,t)}{\partial t}\right)_{t_0} + \left(\frac{\partial P(x_k,t)}{\partial t}\right)_{t_0}I_0''(x_k) + P(x_k,t_0)\left(\frac{d^2f(x,t_0)}{dx^2}\right)_{x_k}\right.$$

$$+ P(x_k,t_0)\left(\frac{d^2P(x,t_0)}{dx^2}\right)_{x_k}I_0''(x_k) + P^2(x_k,t_0)I_0''''(x_k)\right] \tag{8.72}$$

When $\psi(x,t)$ is subject to Neumann or Gauss boundary conditions, the entries in the first and last columns of the solution table are not specified. Thus, the partial solution table is that of Table 8.13.

	$x = x_0$	$x = x_1$	$\cdot\ \cdot$	$x = x_N$
$t = t_0$	$\psi_{0,0}$(8.61)	$\psi_{1,0}$(8.61)	$\cdot\ \cdot$	$\psi_{N,0}$(8.61)
$t = t_1$	$\psi_{0,1}$(8.72)	$\psi_{1,1}$(8.72)	$\cdot\ \cdot$	$\psi_{N,1}$(8.72)
$\vdots$				
$t = t_M$				

Table 8.13 Partial solution table for a parabolic equations with solution constrained by Neumann boundary conditions

Using the Crank–Nicolson modification, the Milne finite difference approximation to the parabolic equation is

$$
\begin{aligned}
&- CP_{k,m}\psi_{k+1,m+1} + \left(1 + 2CP_{k,m}\right)\psi_{k,m+1} - CP_{k,m}\psi_{k-1,m+1} \\
&= 2h_t f_{k,m} + \psi_{k,m-1} + CP_{k,m}\left(\psi_{k+1,m-1} - 2\psi_{k,m-1} + \psi_{k-1,m-1}\right)
\end{aligned}
\qquad (8.85b)
$$

where

$$
C \equiv \frac{h_t}{h_x^2}
\qquad (8.77)
$$

To obtain the remaining entries in the first and last columns, we set $k = 0$ and $k = N$ in eq. 8.85b to obtain

$$
\begin{aligned}
&- CP_{0,m}\psi_{1,m+1} + \left(1 + 2CP_{0,m}\right)\psi_{0,m+1} - CP_{0,m}\psi_{-1,m+1} \\
&= 2h_t f_{0,m} + \psi_{0,m-1} + CP_{0,m}\left(\psi_{1,m-1} - 2\psi_{0,m-1} + \psi_{-1,m-1}\right)
\end{aligned}
\qquad (8.104a)
$$

and

$$
\begin{aligned}
&- CP_{N,m}\psi_{N+1,m+1} + \left(1 + 2CP_{N,m}\right)\psi_{N,m+1} - CP_{N,m}\psi_{N-1,m+1} \\
&= 2h_t f_{N,m} + \psi_{N,m-1} + CP_{N,m}\left(\psi_{N+1,m-1} - 2\psi_{N,m-1} + \psi_{N-1,m-1}\right)
\end{aligned}
\qquad (8.104b)
$$

The values of $\psi_{-1,m\pm1}$ and $\psi_{N+1,m\pm1}$ are values of ψ outside the boundaries of x, which can be related to values of ψ within the boundaries using the boundary conditions

For any μ, the Milne finite difference approximation to the Neumann boundary conditions are

$$
N_{x_0}(t_\mu) \equiv N_{0,\mu} \simeq \frac{\psi_{1,\mu} - \psi_{-1,\mu}}{2h_x}
\qquad (8.105a)
$$

from which

$$
\psi_{-1,\mu} = \psi_{1,\mu} - 2h_x N_{0,\mu}
\qquad (8.105b)
$$

and

$$N_{x_N}(t_\mu) \equiv N_{N,\mu} \simeq \frac{\psi_{N+1,\mu} - \psi_{N-1,\mu}}{2h_x} \tag{8.106a}$$

which yields

$$\psi_{N+1,\mu} = \psi_{N-1,\mu} + 2h_x N_{N,\mu} \tag{8.106b}$$

We set $\mu = m + 1$ and $\mu = m-1$ in eq. 8.106b and substitute the results into eq. 8.104a to obtain

$$\begin{aligned}
- 2CP_{0,m}\psi_{1,m+1} &+ \left(1 + 2CP_{0,m}\right)\psi_{0,m+1} = \left[2h_t f_{0,m} + \psi_{0,m-1}\right. \\
&\left. +2CP_{0,m}\left(\psi_{1,m-1} - \psi_{0,m-1}\right) - 2h_x CP_{0,m}\left(N_{0,m+1} + N_{0,m-1}\right)\right]
\end{aligned} \tag{8.107a}$$

Setting $\mu = m + 1$ and $\mu = m - 1$ in eq. 8.106b, eq. 8.104b becomes

$$\begin{aligned}
\left(1 + 2CP_{N,m}\right)\psi_{N,m+1} &- 2CP_{N,m}\psi_{N-1,m+1} = \left[2h_t f_{N,m} + \psi_{N,m-1}\right. \\
&\left. +2CP_{N,m}\left(\psi_{N-1,m-1} - \psi_{N,m-1}\right) + 2h_x CP_{N,m}\left(N_{N,m+1} + N_{N,m-1}\right)\right]
\end{aligned} \tag{8.107b}$$

To obtain the approximate values of $\psi_{-1,\mu}$ and $\psi_{N+1,\mu}$ from the Gauss boundary conditions, we take

$$G_{x_0}(t_\mu) \equiv G_{0,\mu} = \psi_{0,\mu} + \gamma\frac{\psi_{1,\mu} - \psi_{-1,\mu}}{2h_x} \tag{8.108a}$$

from which

$$\psi_{-1,\mu} = \psi_{1,\mu} + \frac{2h_x}{\gamma}\psi_{0,\mu} - \frac{2h_x}{\gamma}G_{0,\mu} \tag{8.108b}$$

and

$$G_{x_N}(t_\mu) \equiv G_{N,\mu} = \psi_{N,\mu} + \gamma\frac{\psi_{N+1,\mu} - \psi_{N,\mu}}{2h_x} \tag{8.109a}$$

which yields

$$\psi_{N+1,\mu} = \psi_{N-1,\mu} - \frac{2h_x}{\gamma}\psi_{N,\mu} + \frac{2h_x}{\gamma}G_{N,\mu} \tag{8.109b}$$

Substituting eqs. 8.108b and 8.109b into eq. 8.85b with $\mu = m + 1$ and with $\mu = m - 1$, we obtain

$$
\left(1 + 2CP_{0,m} - 2CP_{0,m}\frac{h_x}{\gamma}\right)\psi_{0,m+1} - 2CP_{0,m}\psi_{1,m+1} =
$$

$$
\left[2h_t f_{0,m} + 2CP_{0,m}\psi_{1,m-1} + \left(1 - 2CP_{0,m} + 2CP_{0,m}\frac{h_x}{\gamma}\right)\psi_{0,m-1}\right.
$$

$$
\left. - 2CP_{0,m}\frac{h_x}{\gamma}\left(G_{0,m+1} + G_{0,m-1}\right)\right]
\tag{8.110a}
$$

and

$$
- 2CP_{N,m}\psi_{N-1,m+1} + \left(1 + 2CP_{N,m} + 2CP_{N,m}\frac{h_x}{\gamma}\right)\psi_{N,m+1} =
$$

$$
\left[2h_t f_{N,m} + 2CP_{N,m}\psi_{N-1,m-1} + \left(1 - 2CP_{N,m} - 2CP_{N,m}\frac{h_x}{\gamma}\right)\psi_{N,m-1}\right.
$$

$$
\left. + 2CP_{N,m}\frac{h_x}{\gamma}\left(G_{N,m+1} + G_{N,m-1}\right)\right]
\tag{8.110b}
$$

The entries in each row of the solution table labeled by $m > 1$ are found by generating $N-1$ equations from eq. 8.85b with $1 \le k \le N-1$ and the two equations at the boundaries [eqs. 8.107 for Neumann boundary conditions and eqs. 8.110 for Gauss conditions]. These comprise the $N + 1$ equations needed to find the unknowns $\{\psi_{0,m}, \ldots, \psi_{N,m}\}$.

Example 8.9: Finite difference approximation for the solution to a diffusion equation with Neumann and Gauss boundary conditions

We again consider the differential equation

$$
\frac{\partial \psi(x,t)}{\partial t} = \frac{\partial^2 \psi(x,t)}{\partial x^2}
\tag{8.9b}
$$

To obtain the solution

$$
\psi(x,t) = e^{t-x}
\tag{8.12}
$$

we constrain $\psi(x,t)$ by the initial condition

$$
\psi(x,0) = I_0(x) = e^{-x}
\tag{8.10}
$$

Again, we take $h_t = 0.20$, $h_x = 0.25$ so that $0 \le k \le 4$ and $0 \le m \le 5$. From

$$
\begin{cases}
\psi_{k,0} = e^{-x_k} \\
\psi_{k,1} \simeq \left(1 + h_t + \frac{1}{2}h_t^2\right)e^{-x_k}
\end{cases}
\tag{8.93}
$$

the partial solution table is given in Table 8.14

	$x = 0.00$	$x = 0.25$	$x = 0.50$	$x = 0.75$	$x = 1.00$
$t = 0.00$	1.00000	0.77880	0.60653	0.47237	0.36788
$t = 0.20$	1.22000	0.95014	0.73997	0.57629	0.44881
$t = 0.40$					
$t = 0.60$					
$t = 0.80$					
$t = 1.00$					

Table 8.14 Partial solution table for a parabolic equation using the Crank–Nicolson modification

(a) When the solution is constrained by the Neumann boundary conditions

$$
\begin{cases}
\left(\dfrac{\partial \psi(x,t)}{\partial x}\right)_{x=0} = N_0(t) = -e^t \\[4mm]
\left(\dfrac{\partial \psi(x,t)}{\partial x}\right)_{x=1} = N_N(t) = -e^{t-1}
\end{cases}
\tag{8.111}
$$

the set of equations for $\{\psi_{0,m+1}, \ldots, \psi_{4,m+1}\}$, obtained from eqs. 8.85b and 8.107 are

$$
\begin{aligned}
&(1 + 2C)\psi_{0,m+1} - 2C\psi_{1,m+1} \\
&= \psi_{0,m-1} + 2C(\psi_{1,m-1} - \psi_{0,m-1}) - 2h_x C(e^{t_{m-1}} + e^{t_{m+1}}) \equiv \alpha_{0,m}
\end{aligned}
\tag{8.112a}
$$

$$
\begin{aligned}
&-C\psi_{0,m+1} + (1 + 2C)\psi_{1,m+1} - C\psi_{2,m+1} \\
&= \psi_{1,m-1} + C(\psi_{2,m-1} - 2\psi_{1,m-1} + \psi_{0,m-1}) \equiv \alpha_{1,m}
\end{aligned}
\tag{8.112b}
$$

$$
\begin{aligned}
&-C\psi_{1,m+1} + (1 + 2C)\psi_{2,m+1} - C\psi_{3,m+1} \\
&= \psi_{2,m-1} + C(\psi_{3,m-1} - 2\psi_{2,m-1} + \psi_{1,m-1}) \equiv \alpha_{2,m}
\end{aligned}
\tag{8.112c}
$$

$$
\begin{aligned}
&-C\psi_{2,m+1} + (1 + 2C)\psi_{3,m+1} - C\psi_{4,m+1} \\
&= \psi_{3,m-1} + C(\psi_{4,m-1} - 2\psi_{3,m-1} + \psi_{2,m-1}) \equiv \alpha_{3,m}
\end{aligned}
\tag{8.112d}
$$

and

$$
\begin{aligned}
&-2C\psi_{3,m+1} + (1 + 2C)\psi_{4,m+1} \\
&= \psi_{4,m-1} + 2C(\psi_{3,m-1} - \psi_{4,m-1}) + 2h_x C(e^{t_{m-1}-1} + e^{t_{m+1}-1}) \equiv \alpha_{4,m}
\end{aligned}
\tag{8.112e}
$$

Solving these equations for each $1 \le m \le 5$ we obtain the complete solution given in Table 8.15(a).

	$x = 0.00$	$x = 0.25$	$x = 0.50$	$x = 0.75$	$x = 1.00$
$t = 0.00$	1.00000	0.77880	0.60653	0.47237	0.36788
$t = 0.20$	1.22000	0.95014	0.73997	0.57629	0.44881
$t = 0.40$	1.48800	1.16250	0.90796	0.70952	0.55552
$t = 0.60$	1.81678	1.41901	1.10807	0.86567	0.67750
$t = 0.80$	2.22538	1.73676	1.35662	1.06060	0.82986
$t = 1.00$	2.71707	2.12041	1.65611	1.29453	1.01274

Table 8.15a   Complete solution table for a diffusion equation with Neumann boundary conditions using the Crank–Nicolson modification

A comparison of these results with the exact values given in Table 8.3 shows that the solution is stable and is slightly less accurate than those found for the diffusion equation constrained by Dirichlet boundary conditions. This is to be expected since Neumann boundary conditions are approximated by finite differences to obtain the values of ψ at the boundaries, whereas these exact values of ψ are specified by the Dirichlet boundary conditions.

(b) The solution to the diffusion equation of eq. 8.9b given in eq. 8.12 is obtained with the Gauss boundary conditions

$$\begin{cases} \psi(0,t) - \left(\dfrac{\partial \psi(x,t)}{\partial x}\right)_{x=0} = G_0(t) = 2e^t \\[3mm] \psi(1,t) - \left(\dfrac{\partial \psi(x,t)}{\partial x}\right)_{x=1} = G_1(t) = 2e^{t-1} \end{cases} \tag{8.113}$$

The five equations for $\{\psi_{0,m+1}, \ldots, \psi_{4,m+1}\}$, obtained from eq. 8.84b and eqs. 8.110, are

$$[1 + 2C(1 + h_x)]\psi_{0,m+1} - 2C\psi_{1,m+1}$$
$$= [1 - 2C(1 - h_x)]\psi_{0,m-1} + 2C\psi_{1,m-1} + 2Ch_x[2e^{t_{m-1}} + 2e^{t_{m+1}}] \equiv \alpha_{0,m} \tag{8.114a}$$

$$-C\psi_{0,m+1} + (1 + 2C)\psi_{1,m+1} - C\psi_{2,m+1} =$$
$$\psi_{1,m-1} + C(\psi_{2,m-1} - 2\psi_{1,m-1} + \psi_{0,m-1}) \equiv \alpha_{1,m} \tag{8.114b}$$

$$-C\psi_{1,m+1} + (1 + 2C)\psi_{2,m+1} - C\psi_{3,m+1} =$$
$$\psi_{2,m-1} + C(\psi_{3,m-1} - 2\psi_{2,m-1} + \psi_{1,m-1}) \equiv \alpha_{3,m} \tag{8.114c}$$

$$-C\psi_{2,m+1} + (1 + 2C)\psi_{3,m+1} - C\psi_{4,m+1} =$$
$$C\psi_{3,m-1} + C(\psi_{4,m-1} - 2\psi_{3,m-1} + \psi_{2,m-1}) \equiv \alpha_{3,m} \tag{8.114d}$$

and

$$
\begin{aligned}
& -2C\psi_{3,m+1} + [1 + 2C(1 - h_x)]\psi_{4,m+1} = \\
& 2C\psi_{3,m-1} + [1 - 2C(1 - h_x)]\psi_{4,m-1} - 2Ch_x\left[2e^{t_{m-1}-1} + 2e^{t_{m+1}-1}\right] \equiv \alpha_{4,m}
\end{aligned}
$$

(8.114e)

For each value of $1 \le m \le 4$, the solution to this set of equations yields the complete solution table given in Table 8.15(b).

	$x = 0.00$	$x = 0.25$	$x = 0.50$	$x = 0.75$	$x = 1.00$
$t = 0.00$	1.00000	0.77880	0.60653	0.47237	0.36788
$t = 0.20$	1.22000	0.95014	0.73997	0.57629	0.44881
$t = 0.40$	1.48999	1.16434	0.91023	0.71293	0.56113
$t = 0.60$	1.81968	1.42141	1.11071	0.86937	0.68342
$t = 0.80$	2.23087	1.74382	1.36630	1.07451	0.85020
$t = 1.00$	2.72356	2.12868	1.66725	1.31037	1.03592

Table 8.15b Complete solution table for a diffusion equation with Gauss boundary conditions using the Crank–Nicolson modification

Comparing these results to the exact values given in Table 8.3, we see that the solution is stable, and not quite as accurate as when Dirichlet or Neumann conditions are applied. Again, this is due to the fact that for the Neumann and Gauss conditions, ψ and $\partial\psi/\partial x$ are approximated at the boundaries by finite differences. □

Hyperbolic equations subject to Neumann and Gauss boundary conditions

When the solution to the hyperbolic equation

$$
\frac{\partial^2 \psi(x, t)}{\partial t^2} - P(x, t)\frac{\partial^2 \psi(x, t)}{\partial x^2} = f(x, t)
$$

(8.2c)

is subject to the two initial conditions

$$
\begin{cases}
\psi(x, t_0) = I_0(x) \\
\left(\dfrac{\partial \psi(x, t)}{\partial t}\right)_{t_0} = I_1(x)
\end{cases}
$$

(8.3)

entries in the first two rows of the solution table are given by the initial conditions

$$\psi_{k,0} = I_0(x_k) \qquad (8.61)$$

and

$$\psi(x_k, t_1) = \psi_{k,1} \simeq I_0(x_n) + h_t I_1(x_k) + \frac{1}{2} h_t^2 \left[P(x_k, t_0) I_0''(x_k) - f(x_k, t_0) \right] \qquad (8.91)$$

But, as noted for the parabolic equation, when the solution to the hyperbolic equation is subject to Neumann or Gauss boundary conditions, all other entries in the first and last columns must be approximated by finite differences. Thus, the partial solution table is shown in Table 8.16.

	$x = x_0$	$x = x_1$	$\bullet\bullet$	$x = x_N$
$t = t_0$	$\psi_{0,0}(8.61)$	$\psi_{0,1}(8.61)$	$\bullet\bullet$	$\psi_{0,N}(8.61)$
$t = t_1$	$\psi_{0,1}(8.91)$	$\psi_{1,1}(8.91)$	$\bullet\bullet$	$\psi_{N,1}(8.92)$
$t = t_M$				

Table 8.16 Partial solution table for a hyperbolic equation with solution constrained by Neumann or Gauss boundary conditions

With the Crank–Nicolson modification, the Milne finite difference approximation to the hyperbolic equation is

$$-\frac{C'}{2} P_{k,m} \psi_{k+1,m+1} + \left(1 + C' P_{k,m}\right) \psi_{k,m+1} - \frac{C'}{2} P_{k,m} \psi_{k-1,m+1}$$
$$= h_t^2 f_{k,m} + 2\psi_{k,m} - \psi_{k,m-1} + \frac{C'}{2} P_{k,m} \left(\psi_{k+1,m-1} - 2\psi_{k,m-1} + \psi_{k-1,m-1} \right) \qquad (8.89b)$$

where, as before

$$C' \equiv \frac{h_t^2}{h_x^2} \qquad (8.90)$$

To determine the unfilled entries in the first and last columns, we set $k = 0$ and $k = N$ in eq. 8.89b to obtain

$$-\frac{C'}{2} P_{0,m} \psi_{1,m+1} + \left(1 + C' P_{0,m}\right) \psi_{0,m+1} - \frac{C'}{2} P_{0,m} \psi_{-1,m+1}$$
$$= h_t^2 f_{0,m} + 2\psi_{0,m} - \psi_{0,m-1} + \frac{C'}{2} P_{0,m} \left(\psi_{1,m-1} - 2\psi_{0,m-1} + \psi_{-1,m-1} \right) \qquad (8.115a)$$

and

$$-\frac{C'}{2}P_{N,m}\psi_{N+1,m+1} + \left(1 + C'P_{N,m}\right)\psi_{N,m+1} - \frac{C'}{2}P_{N,m}\psi_{N-1,m+1}$$
$$= h_t^2 f_{N,m} + 2\psi_{N,m} - \psi_{N,m-1} + \frac{C'}{2}P_{N,m}\left(\psi_{N+1,m-1} - 2\psi_{N,m-1} + \psi_{N-1,m-1}\right)$$

$$(8.115b)$$

To obtain expressions for $\psi_{-1,m\pm1}$ and $\psi_{N+1,m\pm1}$, we use the finite difference approximation to the boundary conditions.

When the solution is constrained by the Neumann boundary conditions

$$\begin{cases} \left(\dfrac{\partial\psi(x,t)}{\partial x}\right)_{x_0} = N_{x_0}(t) \\[4mm] \left(\dfrac{\partial\psi(x,t)}{\partial x}\right)_{x_N} = N_{x_N}(t) \end{cases} \qquad (8.103)$$

we have shown that for any μ,

$$\psi_{-1,\mu} = \psi_{1,\mu} - 2h_x N_{0,\mu} \qquad (8.105b)$$

and

$$\psi_{N+1,\mu} = \psi_{N-1,\mu} + 2h_x N_{N,\mu} \qquad (8.106b)$$

With $\mu = m+1$ and $\mu = m-1$, we substitute these into eqs. 8.115 to obtain

$$\left(C'P_{0,m} + 1\right)\psi_{0,m+1} - C'P_{0,m}\psi_{1,m+1} = \big[h_t^2 f_{0,m} + 2\psi_{0,m} - \psi_{0,m-1}$$
$$+C'P_{0,m}\psi_{1,m-1} - C'P_{0,m}\psi_{0,m-1} - C'P_{0,m}h_x\left(N_{0,m-1} + N_{0,m+1}\right)\big] \qquad (8.116a)$$

and

$$- C'P_{N,m}\psi_{N-1,m+1} + \left(C'P_{N,m} + 1\right)\psi_{N,m+1} = \big[h_t^2 f_{N,m} + 2\psi_{N,m} - \psi_{N,m-1}$$
$$+C'P_{N,m}\psi_{N-1,m-1} - C'P_{N,m}\psi_{N,m-1} + C'P_{N,m}h_x\left(N_{N,m-1} + N_{N,m+1}\right)\big]$$

$$. \qquad (8.116b)$$

When the solution is subject to Gauss boundary conditions,

$$\begin{cases} \psi(x_0,t) + \gamma\left(\dfrac{\partial\psi(x,t)}{\partial x}\right)_{x_0} = G_{x_0}(t) \\[4mm] \psi(x_N,t) + \gamma\left(\dfrac{\partial\psi(x,t)}{\partial x}\right)_{x_N} = G_{x_N}(t) \end{cases} \qquad (8.111)$$

For any μ, the finite difference approximation to these yields

$$\psi_{-1,\mu} = \psi_{1,\mu} + \frac{2h_x}{\gamma}\psi_{0,\mu} - \frac{2h_x}{\gamma}G_{0,\mu} \qquad (8.108b)$$

and

$$\psi_{N+1,\mu} = \psi_{N-1,\mu} - \frac{2h_x}{\gamma}\psi_{N,\mu} + \frac{2h_x}{\gamma}G_{N,\mu} \qquad (8.109b)$$

Then, for $\mu = m + 1$ and $\mu = m-1$, eqs. 8.116 become

$$-\left(C'P_{0,m} + 1 - \frac{C'P_{0,m}h_x}{\mu}\right)\psi_{0,m+1} + C'P_{0,m}\psi_{1,m+1} =$$

$$\left[h_t^2 f_{0,m} - 2\psi_{0,m} + \left(C'P_{0,m} + 1 + \frac{C'P_{0,m}h_x}{\mu}\right)\psi_{0,m-1} - C'P_{0,m}\psi_{1,m-1}\right.$$

$$\left. - \frac{C'P_{0,m}h_x}{\mu}\left(G_{0,m+1} + G_{0,m-1}\right)\right] \qquad (8.117a)$$

and

$$C'P_{N,m}\psi_{N-1,m+1} - \left(C'P_{N,m} + 1 - \frac{C'P_{N,m}h_x}{\mu}\right)\psi_{N,m+1} =$$

$$\left[h_t^2 f_{N,m} - C'P_{N,m}\psi_{N-1,m-1} - 2\psi_{N,m}\left(C'P_{N,m} + 1 + \frac{C'P_{N,m}h_x}{\mu}\right) + \psi_{N,m-1}\right.$$

$$\left. - \frac{C'P_{N,m}h_x}{\mu}\left(G_{N,m+1} + G_{N,m-1}\right)\right] \qquad (8.117b)$$

The complete solution table is obtained by taking $1 \leq k \leq N-1$ and $m \geq 1$ in eq. 8.89b and using eqs. 8.116 when Neumann or conditions apply, or eqs. 8.117 when Gauss conditions are imposed.

Example 8.10: Finite difference approximation for the solution to a wave equation with Neumann or Gauss boundary conditions

We again consider the wave equation

$$\frac{\partial^2 \psi(x,t)}{\partial t^2} - \frac{\partial^2 \psi(x,t)}{\partial x^2} = 0 \qquad (8.54)$$

with solution

$$\psi(x,t) = e^{t-x} \qquad (8.12)$$

which is constrained by the initial conditions

$$
\begin{cases}
\psi(x,0) = I_0(x) = e^{-x} \\
\left(\dfrac{\partial \psi(x,t)}{\partial t}\right)_{t=0} = I_1(x) = e^{-x}
\end{cases}
\tag{8.26}
$$

As before, we take $h_t = 0.20$, $h_x = 0.25$ so that $0 \le k \le 4$ and $0 \le m \le 5$. The entries in the first and second rows of the solution table are obtained from

$$
\begin{cases}
\psi_{k,0} = I_0(x_k) = e^{-x_k} \\
\psi_{k,1} \simeq \left(1 + h_t + \dfrac{1}{2}h_t^2\right)e^{-x_k}
\end{cases}
\tag{8.92}
$$

The partial solution table for this wave equation is identical to the partial solution table for the diffusion equation of example 8.9 (Table 8.14) which is reproduced as Table 8.17.

	$x = 0.00$	$x = 0.25$	$x = 0.50$	$x = 0.75$	$x = 1.00$
$t = 0.00$	1.00000	0.77880	0.60653	0.47237	0.36788
$t = 0.20$	1.22000	0.95014	0.73997	0.57629	0.44881
$t = 0.40$					
$t = 0.60$					
$t = 0.80$					
$t = 1.00$					

Table 8.17 Partial solution table for a wave equation with Neumann or Gauss boundary conditions using the Crank–Nicolson modification

(a) When the solution is constrained by the Neumann boundary conditions

$$
\begin{cases}
\left(\dfrac{\partial \psi(x,t)}{\partial x}\right)_{x=0} = N_0(t) = -e^{t} \\
\left(\dfrac{\partial \psi(x,t)}{\partial x}\right)_{x=1} = N_N(t) = -e^{t-1}
\end{cases}
\tag{8.111}
$$

the five equations for $\{\psi_{0,m+1}, \ldots, \psi_{4,m+1}\}$, obtained by setting $1 \le k \le 3$ in eq. 8.89b and using eqs. 8.116, are

$$
\begin{aligned}
-(1+C')\psi_{0,m+1} + C'\psi_{1,m+1} = \\
-2\psi_{0,m} + (C'+1)\psi_{0,m-1} - C'\psi_{1,m-1} - C'h_x(e^{t_{m-1}} + e^{t_{m+1}}) \equiv \alpha_{0,m}
\end{aligned}
\tag{8.118a}
$$

$$\frac{C'}{2}\psi_{0,m+1} - (1 + C')\psi_{1,m+1} + \frac{C'}{2}\psi_{2,m+1} =$$

$$-2\psi_{1,m} - \frac{C'}{2}\psi_{0,m-1} + (C' + 1)\psi_{1,m-1} - \frac{C'}{2}\psi_{2,m-1} \equiv \alpha_{1,m} \qquad (8.118b)$$

$$\frac{C'}{2}\psi_{1,m+1} - (1 + C')\psi_{2,m+1} + \frac{C'}{2}\psi_{3,m+1} =$$

$$-2\psi_{2,m} - \frac{C'}{2}\psi_{1,m-1} + (C' + 1)\psi_{2,m-1} - \frac{C'}{2}\psi_{3,m-1} \equiv \alpha_{2,m} \qquad (8.118c)$$

$$\frac{C'}{2}\psi_{2,m+1} - (1 + C')\psi_{3,m+1} + \frac{C'}{2}\psi_{4,m+1} =$$

$$-2\psi_{3,m} - \frac{C'}{2}\psi_{2,m-1} + (C' + 1)\psi_{3,m-1} - \frac{C'}{2}\psi_{4,m-1} \equiv \alpha_{3,m} \qquad (8.118d)$$

and

$$C'\psi_{3,m+1} - (1 + C')\psi_{4,m+1} =$$

$$-2\psi_{4,m} - C'\psi_{3,m-1} + (C' + 1)\psi_{4,m-1} + C'h_x\left(e^{t_{m-1}-1} + e^{t_{m+1}-1}\right) \equiv \alpha_{4,m} \qquad (8.118e)$$

Solving this set of equations for $1 \le m \le 4$, we obtain the solution table given in Table 8.18(a).

	$x = 0.00$	$x = 0.25$	$x = 0.50$	$x = 0.75$	$x = 1.00$
$t = 0.00$	1.00000	0.77880	0.60653	0.47237	0.36788
$t = 0.20$	1.22000	0.95014	0.73997	0.57629	0.44881
$t = 0.40$	1.54588	1.17164	0.90523	0.70017	0.52779
$t = 0.60$	1.96939	1.46245	1.11242	0.84694	0.61683
$t = 0.80$	2.47459	1.84138	1.37572	1.02402	0.73433
$t = 1.00$	3.06407	2.31627	1.71142	1.24830	0.89732

Table 8.18a Complete solution table for a wave equation with Neumann boundary conditions using the Crank–Nicolson scheme

A comparison of these results with the exact values given in Table 8.3 shows that the solution is reasonable and not as accurate as those obtained for the wave equation constrained by Dirichlet boundary conditions. Again this reduction in accuracy results from the fact that the values of ψ at the boundaries are approximate.

(b) When the solution is subjected to Gauss boundary conditions

$$\begin{cases} \psi(0,t) - \left(\dfrac{\partial\psi(x,t)}{\partial x}\right)_{x=0} = G_0(t) = 2e^t \\[4mm] \psi(1,t) - \left(\dfrac{\partial\psi(x,t)}{\partial x}\right)_{x=1} = G_1(t) = 2e^{t-1} \end{cases} \qquad (8.113)$$

the equations for $\{\psi_{0,m+1}, \ldots, \psi_{4,m+1}\}$ for $m > 1$ are

$$- (C' + 1 + Ch_x)\psi_{0,m+1} + C'\psi_{1,m+1} =$$
$$- 2\psi_{0,m} - C'\psi_{1,m-1} + (C' + 1 + C'h_x)\psi_{0,m-1} - 2C'h_x(e^{t_{m-1}} + e^{t_{m+1}}) \equiv \alpha_{0,m}$$
$$(8.119\text{a})$$

$$\frac{C'}{2}\psi_{0,m+1} - (1 + C')\psi_{1,m+1} + \frac{C'}{2}\psi_{2,m+1} =$$
$$- 2\psi_{1,m} - \frac{C'}{2}\psi_{0,m-1} + (1 + C')\psi_{1,m-1} - \frac{C'}{2}\psi_{2,m-1} \equiv \alpha_{1,m} \qquad (8.119\text{b})$$

$$\frac{C'}{2}\psi_{1,m+1} - (1 + C')\psi_{2,m+1} + \frac{C'}{2}\psi_{3,m+1} =$$
$$- 2\psi_{2,m} - \frac{C'}{2}\psi_{1,m-1} + (1 + C')\psi_{2,m-1} - \frac{C'}{2}\psi_{3,m-1} \equiv \alpha_{2,m} \qquad (8.119\text{c})$$

$$\frac{C'}{2}\psi_{2,m+1} - (1 + C')\psi_{3,m+1} + \frac{C'}{2}\psi_{4,m+1} =$$
$$- 2\psi_{3,m} - \frac{C'}{2}\psi_{2,m-1} + (1 + C')\psi_{3,m-1} - \frac{C'}{2}\psi_{4,m-1} \equiv \alpha_{3,m} \qquad (8.119\text{d})$$

and

$$C'\psi_{3,m+1} - (C'1 - C'h_x)\psi_{4,m+1} =$$
$$- 2\psi_{4,m-1} + (C' + 1 - C'h_x)\psi_{4,m-1} - C'\psi_{3,m-1} + 2Ch_x(e^{t_{m-1}-1} + e^{t_{m+1}-1})$$
$$\equiv \alpha_{4,m}$$
$$(8.119\text{e})$$

The solution to these equations yields the complete solution shown in Table 8.18(b).

	$x = 0.00$	$x = 0.25$	$x = 0.50$	$x = 0.75$	$x = 1.00$
$t = 0.00$	1.00000	0.77880	0.60653	0.47237	0.36788
$t = 0.20$	1.22000	0.95014	0.73997	0.57629	0.44881
$t = 0.40$	1.59348	1.18110	0.90610	0.69519	0.50143
$t = 0.60$	2.08133	1.49656	1.11617	0.82657	0.53975
$t = 0.80$	2.62692	1.91440	1.38444	0.97231	0.58789
$t = 1.00$	3.21808	2.42610	1.72301	1.14878	0.66780

Table 8.18b Complete solution table for a wave equation with Gauss boundary conditions using the Crank–Nicolson approximation

A comparison of these results with the exact values given in Table 8.3 shows that the solution is less accurate than the results obtained when Dirichlet or Neumann conditions are applied for the same reasons discussed earlier. □

Elliptic equations subject to Neumann boundary conditions

When the solution to the elliptic equation

$$R(x,y)\frac{\partial^2 \psi}{\partial x^2} + S(x,y)\frac{\partial^2 \psi}{\partial y^2} = f(x,y) \tag{8.2c}$$

is subject to Neumann boundary conditions

$$\begin{cases} \left(\dfrac{\partial \psi(x,y)}{\partial x}\right)_{x_0} = N_{x_0}(y) \\[2ex] \left(\dfrac{\partial \psi(x,y)}{\partial x}\right)_{x_N} = N_{x_N}(y) \end{cases} \tag{8.120a}$$

and

$$\begin{cases} \left(\dfrac{\partial \psi(x,y)}{\partial y}\right)_{y_0} = N_{y_0}(x) \\[2ex] \left(\dfrac{\partial \psi(x,y)}{\partial y}\right)_{y_M} = N_{y_M}(x) \end{cases} \tag{8.120b}$$

none of the boundary values of ψ are specified. Thus, the partial solution table with Neumann boundary conditions is an empty table, and all ψ values $\{\psi_{0,0}, \ldots, \psi_{N,M}\}$ must be determined by finite difference methods.

With the Crank–Nicolson modification, the finite difference approximation to eq. 8.2c is

$$\frac{R_{k,m}}{2h_x^2}\left[\left(\psi_{k+1,m+1} - 2\psi_{k,m+1} + \psi_{k-1,m+1}\right) + \left(\psi_{k+1,m-1} - 2\psi_{k,m-1} + \psi_{k-1,m-1}\right)\right] +$$

$$\frac{S_{k,m}}{2h_y^2}\left[\left(\psi_{k+1,m+1} - 2\psi_{k+1,m} + \psi_{k+1,m-1}\right) + \left(\psi_{k-1,m+1} - 2\psi_{k-1,m} + \psi_{k-1,m-1}\right)\right]$$

$$= f_{k,m}$$

$$\tag{8.97b}$$

with $0 \leq k \leq N$ and $0 \leq m \leq M$.

For $k = 0$, $m = 0$, eq. 8.97b becomes

$$\frac{R_{0,0}}{2h_x^2}\left[\left(\psi_{1,1} - 2\psi_{0,1} + \psi_{-1,1}\right) + \left(\psi_{1,-1} - 2\psi_{0,-1} + \psi_{-1,-1}\right)\right] +$$

$$\frac{S_{0,0}}{2h_y^2}\left[\left(\psi_{1,1} - 2\psi_{1,0} + \psi_{1,-1}\right) + \left(\psi_{-1,1} - 2\psi_{-1,0} + \psi_{-1,-1}\right)\right]$$

$$= f_{0,0}$$

$$\tag{8.121a}$$

With $k = N$, $m = 0$, we have

$$
\frac{R_{N,0}}{2h_x^2}\left[(\psi_{N+1,1} - 2\psi_{N,1} + \psi_{N-1,1}) + (\psi_{N+1,-1} - 2\psi_{N,-1} + \psi_{N-1,-1})\right] +
$$

$$
\frac{S_{N,0}}{2h_y^2}\left[(\psi_{N+1,1} - 2\psi_{N+1,0} + \psi_{N+1,-1}) + (\psi_{N-1,1} - 2\psi_{N-1,0} + \psi_{N-1,-1})\right]
$$

$$
= f_{N,0} \tag{8.121b}
$$

Setting $k = 0$, $m = M$, eq. 8.97b is

$$
\frac{R_{0,M}}{2h_x^2}\left[(\psi_{1,M+1} - 2\psi_{0,M+1} + \psi_{-1,M+1}) + (\psi_{1,M-1} - 2\psi_{0,M-1} + \psi_{-1,M-1})\right] +
$$

$$
\frac{S_{0,M}}{2h_y^2}\left[(\psi_{1,M+1} - 2\psi_{1,M} + \psi_{1,M-1}) + (\psi_{-1,M+1} - 2\psi_{-1,M} + \psi_{-1,M-1})\right]
$$

$$
= f_{0,M} \tag{8.121c}
$$

and with $k = N$, $m = M$, we obtain

$$
\frac{R_{N,M}}{2h_x^2}\left[(\psi_{N+1,M+1} - 2\psi_{N,M+1} + \psi_{N-1,M+1}) + (\psi_{N+1,M-1} - 2\psi_{N,M-1} + \psi_{N-1,M-1})\right] +
$$

$$
\frac{S_{N,M}}{2h_y^2}\left[(\psi_{N+1,M+1} - 2\psi_{N+1,M} + \psi_{N+1,M-1}) + (\psi_{N-1,M+1} - 2\psi_{N-1,M} + \psi_{N-1,M-1})\right]
$$

$$
= f_{N,M}
$$

$$
\tag{8.121d}
$$

These expressions contain values of ψ at x_{-1}, x_{N+1}, y_{-1}, and y_{M+1} which are outside the boundaries of x and y. They can be related to values of ψ within these boundaries using the Neumann boundary conditions.
With

$$
\left(\frac{\partial \psi(x, y_m)}{\partial x}\right)_{x_0} = N_{x_0}(y_m) \simeq \frac{\psi_{1,m} - \psi_{-1,m}}{2h_x} \tag{8.122a}
$$

we obtain

$$
\psi_{-1,m} \simeq \psi_{1,m} - 2h_x N_{x_0}(y_m) \tag{8.122b}
$$

and with

$$
\left(\frac{\partial \psi(x, y_m)}{\partial x}\right)_{x_N} = N_{x_N}(y_m) \simeq \frac{\psi_{N+1,m} - \psi_{N-1,m}}{2h_x} \tag{8.123a}
$$

we have

$$\psi_{N+1,m} \simeq \psi_{N-1,m} + 2h_x N_{x_N}(y_m) \tag{8.123b}$$

Likewise,

$$\left(\frac{\partial\psi(x_k,y)}{\partial y}\right)_{y_0} = N_{y_0}(x_k) \simeq \frac{\psi_{k,1} - \psi_{k,-1}}{2h_y} \tag{8.124a}$$

yields

$$\psi_{k,-1} = \psi_{k,1} - 2h_y N_{y_0}(x_k) \tag{8.124b}$$

and from

$$\left(\frac{\partial\psi(x_k,y)}{\partial y}\right)_{y_M} = N_{y_M}(x_k) \simeq \frac{\psi_{k,M+1} - \psi_{k,M-1}}{2h_y} \tag{8.125a}$$

we obtain

$$\psi_{k,M+1} = \psi_{k,M-1} + 2h_y N_{y_M}(x_k) \tag{8.125b}$$

Eqs. 8.122b, 8.123b, 8.124b, and 8.125b are unique expressions for values of ψ at points (x, y) such that one of these variables is within its prescribed boundary and the other is outside its boundary.

Eqs. 8.121 also contain terms involving $\psi_{-1,-1}$, $\psi_{N+1,-1}$, $\psi_{-1,M+1}$, and $\psi_{N+1,M+1}$. These are values of ψ at points that are outside the boundaries of both x and y.

Since the Neumann boundary functions given in eq. 8.120 are given in closed form, they can be evaluated at x_{-1}, x_{N+1}, y_{-1}, and y_{M+1}. Therefore, we can use eqs. 8.122b, 8.123b, 8.124b, and 8.125b to relate these values of ψ at points (x, y) for which both x and y are outside their boundaries to values of ψ at points within those boundaries.

For example, we first set $m = -1$ in eq. 8.120b to obtain

$$\psi_{-1,-1} = \psi_{1,-1} - 2h_x N_{x_0}(y_{-1}) \tag{8.126a}$$

Then setting $k = 1$ in eq. 8.124b, we have

$$\psi_{1,-1} = \psi_{1,1} - 2h_y N_{y_0}(x_1) \tag{8.126b}$$

With this, eq. 8.126a becomes

$$\psi_{-1,-1} = \psi_{1,1} - 2h_x N_{x_0}(y_{-1}) - 2h_y N_{y_0}(x_1) \tag{8.127a}$$

If we first set $k = -1$ in eq. 8.124b, then set $m = 1$ in eq. 8.122b, we obtain

$$\psi_{-1,-1} = \psi_{1,1} - 2h_x N_{x_0}(y_1) - 2h_y N_{y_0}(x_{-1}) \qquad (8.127b)$$

which differs from the expression for $\psi_{-1,-1}$ in eq. 8.127a.

With identical analysis, using eqs. 8.123b and 8.124b in different order, we find different expressions

$$\psi_{N+1,-1} = \psi_{N-1,1} + 2h_x N_{x_N}(y_{-1}) - 2h_y N_{y_0}(x_{N-1}) \qquad (8.128a)$$

and

$$\psi_{N+1,-1} = \psi_{N-1,1} + 2h_x N_{x_N}(y_1) - 2h_y N_{y_0}(x_{N+1}) \qquad (8.128b)$$

Applying this analysis to eqs. 8.122b and 8.125b results in

$$\psi_{-1,M+1} = \psi_{1,M-1} - 2h_x N_{x_0}(y_{M+1}) + 2h_y N_{y_M}(x_1) \qquad (8.129a)$$

and

$$\psi_{-1,M+1} = \psi_{1,M-1} - 2h_x N_{x_0}(y_{M-1}) + 2h_y N_{y_M}(x_{-1}) \qquad (8.129b)$$

Using eqs. 8.123b and 8.125b in this way, we obtain

$$\psi_{N+1,M+1} = \psi_{N-1,M-1} + 2h_x N_{x_N}(y_{M+1}) + 2h_y N_{y_M}(x_{N-1}) \qquad (8.130a)$$

and

$$\psi_{N+1,M+1} = \psi_{N-1,M-1} + 2h_x N_{x_N}(y_{M-1}) + 2h_y N_{y_M}(x_{N+1}) \qquad (8.130b)$$

Thus, at points where both x and y are outside their defined boundaries, using the same Neumann boundary conditions in different order results in different expressions for ψ. Thus, these values of ψ are not uniquely determined by the Neumann boundary conditions.

When finding the finite difference solutions to an elliptic equation subject to Dirichlet boundary conditions, we found that approximating the derivatives by Milne finite differences without the Crank–Nicolson modification led to reasonably accurate results (see example 8.8).

Without the Crank–Nicolson modification, the finite difference approximation to eq. 8.2c is

$$\frac{R_{k,m}}{h_x^2}\left(\psi_{k+1,m} - 2\psi_{k,m} + \psi_{k-1,m}\right) + \frac{S_{k,m}}{h_y^2}\left(\psi_{k,m+1} - 2\psi_{k,m} + \psi_{k,m-1}\right) = f_{k,m} \quad (8.131)$$

With the pairs of values $(k,m) = \{(0,0),\ (N,0),\ (0,M),\ (N,M)\}$, we obtain the equations

$$\frac{R_{0,0}}{h_x^2}\left(\psi_{1,0} - 2\psi_{0,0} + \psi_{-1,0}\right) + \frac{S_{0,0}}{h_y^2}\left(\psi_{0,1} - 2\psi_{0,0} + \psi_{0,-1}\right) = f_{0,0} \qquad (8.132\text{a})$$

$$\frac{R_{N,0}}{h_x^2}\left(\psi_{N+1,0} - 2\psi_{N,0} + \psi_{N-1,0}\right) + \frac{S_{N,0}}{h_y^2}\left(\psi_{N,1} - 2\psi_{N,0} + \psi_{N,-1}\right) = f_{N,0} \qquad (8.132\text{b})$$

$$\frac{R_{0,M}}{h_x^2}\left(\psi_{1,M} - 2\psi_{0,M} + \psi_{-1,M}\right) + \frac{S_{0,M}}{h_y^2}\left(\psi_{0,M+1} - 2\psi_{0,M} + \psi_{0,M-1}\right) = f_{0,M} \qquad (8.132\text{c})$$

and

$$\frac{R_{N,M}}{h_x^2}\left(\psi_{N+1,M} - 2\psi_{N,M} + \psi_{N-1,M}\right) + \frac{S_{N,M}}{h_y^2}\left(\psi_{N,M+1} - 2\psi_{N,M} + \psi_{N,M-1}\right) = f_{N,M}$$

$$(8.132\text{d})$$

We note that there are no terms in these equations involving ψ at a point that is outside the boundaries of both x and y. The quantities $\psi_{-1,0}$, $\psi_{0,-1}$, $\psi_{N+1,0}$, $\psi_{N,-1}$, $\psi_{-1,M}$, $\psi_{0,M+1}$, $\psi_{N+1,M}$, and $\psi_{N,M+1}$ are values of ψ at points beyond the boundary of only one of the variables. These are obtained in terms of ψ at points within the boundaries of both variables from eqs. 8.122b, 8.123b, 8.124b, and 8.125b.

Referring to eq. 8.122b, we have

$$\psi_{-1,0} \simeq \psi_{1,0} - 2h_x N_{x_0}(y_0) \qquad (8.133\text{a})$$

and

$$\psi_{-1,M} \simeq \psi_{1,M} - 2h_x N_{x_0}(y_M) \qquad (8.133\text{b})$$

From eq. 8.123b, we obtain

$$\psi_{N+1,0} \simeq \psi_{N-1,0} + 2h_x N_{x_N}(y_0) \qquad (8.134\text{a})$$

and

$$\psi_{N+1,M} \simeq \psi_{N-1,M} + 2h_x N_{x_N}(y_M) \qquad (8.134\text{b})$$

Eq. 8.124b yields

$$\psi_{0,-1} \simeq \psi_{0,1} - 2h_y N_{y_0}(x_0) \qquad (8.135\text{a})$$

and

$$\psi_{N,-1} \simeq \psi_{N,1} - 2h_y N_{y_0}(x_N) \tag{8.135b}$$

With eq. 8.125b, we obtain

$$\psi_{0,M+1} \simeq \psi_{0,M-1} + 2h_y N_{y_M}(x_0) \tag{8.136a}$$

and

$$\psi_{N,M+1} \simeq \psi_{N,M-1} + 2h_y N_{y_M}(x_N) \tag{8.136b}$$

With these, eqs. 8.132 can be written as

$$\frac{R_{0,0}}{h_x^2}\left(2\psi_{1,0} - 2\psi_{0,0}\right) + \frac{S_{0,0}}{h_y^2}\left(2\psi_{0,1} - 2\psi_{0,0}\right) =$$

$$f_{0,0} + \frac{2N_{x_0}(y_0)R_{0,0}}{h_x} + \frac{2N_{y_0}(x_0)S_{0,0}}{h_y} \equiv \beta_{0,0} \tag{8.137a}$$

$$\frac{R_{N,0}}{h_x^2}\left(2\psi_{N-1,0} - 2\psi_{N,0}\right) + \frac{S_{N,0}}{h_y^2}\left(2\psi_{N,1} - 2\psi_{N,0}\right) =$$

$$f_{N,0} - \frac{2N_{x_N}(y_0)R_{N,0}}{h_x} + \frac{2N_{y_0}(x_N)S_{N,0}}{h_y} \equiv \beta_{N,0} \tag{8.137b}$$

$$\frac{R_{0,M}}{h_x^2}\left(2\psi_{1,M} - 2\psi_{0,M}\right) + \frac{S_{0,M}}{h_y^2}\left(2\psi_{0,M-1} - 2\psi_{0,M}\right) =$$

$$f_{0,M} + \frac{2N_{x_0}(y_M)R_{0,M}}{h_x} - \frac{2N_{y_M}(x_0)S_{0,M}}{h_y} \equiv \beta_{0,M} \tag{8.137c}$$

and

$$\frac{R_{N,M}}{h_x^2}\left(2\psi_{N-1,M} - 2\psi_{N,M}\right) + \frac{S_{N,M}}{h_y^2}\left(2\psi_{N,M-1} - 2\psi_{N,M}\right) =$$

$$f_{N,M} - \frac{2N_{x_N}(y_M)R_{N,M}}{h_x} - \frac{2N_{y_M}(x_N)S_{N,M}}{h_y} \equiv \beta_{N,M} \tag{8.137d}$$

These equations, along with

$$\frac{R_{k,m}}{h_x^2}\left(\psi_{k+1,m} - 2\psi_{k,m} + \psi_{k-1,m}\right) + \frac{S_{k,m}}{h_y^2}\left(\psi_{k,m+1} - 2\psi_{k,m} + \psi_{k,m-1}\right) =$$

$$f_{k,m} \equiv \beta_{k,m} \tag{8.133}$$

with $(k,m) \neq \{(0,0),\ (N,0),\ (0,M),\ (N,M)\}$, are the coupled equations needed to solve for $\{\psi_{0,0}, \ldots, \psi_{N,M}\}$.

We express these $(N + 1) \times (M + 1)$ equations in matrix form as

$$A\Psi = B \tag{8.138}$$

where

$$\Psi \equiv \begin{pmatrix} \psi_{0,0} \\ \bullet \\ \bullet \\ \bullet \\ \psi_{0,M} \\ \psi_{1,0} \\ \bullet \\ \bullet \\ \bullet \\ \psi_{1,M} \\ \bullet \\ \bullet \\ \bullet \\ \psi_{N,M} \end{pmatrix} \qquad B \equiv \begin{pmatrix} \beta_{0,0} \\ \bullet \\ \bullet \\ \bullet \\ \beta_{0,M} \\ \beta_{1,0} \\ \bullet \\ \bullet \\ \bullet \\ \beta_{1,M} \\ \bullet \\ \bullet \\ \bullet \\ \beta_{N,M} \end{pmatrix} \tag{8.139}$$

and A is the matrix of coefficients of the unknown ψ values.

Referring to eq. 8.128a, the non-zero elements in the first row of A are

$$\frac{2R_{0,0}}{h_x^2}, \quad \frac{2S_{0,0}}{h_y^2}, \quad -2\left(\frac{R_{0,0}}{h_x^2} + \frac{S_{0,0}}{h_y^2}\right)$$

We see from this that the sum of the elements in the first row of A is zero. Likewise, from eq. 8.137b, the non-zero elements in the $(N + 1)$th row of A are

$$\frac{2R_{N,0}}{h_x^2}, \quad \frac{2S_{N,0}}{h_y^2}, \quad -2\left(\frac{R_{N,0}}{h_x^2} + \frac{S_{N,0}}{h_y^2}\right)$$

Thus the elements in this row also sum to zero.

From eqs. 8.137c and 8.137d, it is straightforward to verify that the elements in the rows of A defined by $k = 0$, $m = M$ and by $k = N$, $m = M$ also sum to zero. And, referring to eq. 8.131, in the kth row of A for $1 \leq m \leq M-1$, with $1 \leq k \leq N-1$, the non-zero elements are

$$\frac{R_{k,m}}{2h_x^2}, \quad \frac{S_{k,m}}{2h_y^2}, \quad \frac{R_{k,m}}{2h_x^2}, \quad \frac{S_{k,m}}{2h_y^2}, \quad -2\left(\frac{R_{k,m}}{2h_x^2} + \frac{S_{k,m}}{2h_y^2}\right)$$

The elements in each of these rows also sum to zero.

Thus, when the set of equations for $\{\psi_{0,0}, \ldots, \psi_{N,M}\}$ is expressed in matrix form, the coefficient matrix has the property that the elements in every row sum to zero. It is shown in Appendix 5 that such a matrix is singular. Therefore, the elliptic equation without the Crank–Nicolson modification cannot be solved by Milne finite difference methods.

The reader will demonstrate in Problem 9 that if we take the Milne finite difference approximation to the elliptic equation with the Crank–Nicolson modification, and use the averages of eqs. 8.127a and 8.127b through eqs. 8.130a and 8.130b to approximate the values of $\psi_{-1,-1}$, $\psi_{N+1,-1}$, $\psi_{-1,M+1}$, and $\psi_{N+1,M+1}$ the coefficient matrix is again one for which the sum of the elements in every row is zero. Thus, we conclude that the elliptic equation, subject to Neumann boundary conditions, cannot be solved by finite difference methods.

Elliptic equations with Gauss boundary conditions

When the elliptic equation of eq. 8.2c is subject to Gauss boundary conditions

$$\begin{cases} \psi(x_0, y_m) + \gamma_x \left(\dfrac{\partial \psi(x, y_m)}{\partial x} \right)_{x_0} = G_{x_0}(y_m) \\[4mm] \psi(x_N, y_m) + \gamma_x \left(\dfrac{\partial \psi(x, y_m)}{\partial x} \right)_{x_N} = G_{x_N}(y_m) \end{cases} \tag{8.140a}$$

and

$$\begin{cases} \psi(x_k, y_0) + \gamma_y \left(\dfrac{\partial \psi(x_k, y)}{\partial y} \right)_{y_0} = G_{y_0}(x_k) \\[4mm] \psi(x_k, y_M) + \gamma_y \left(\dfrac{\partial \psi(x_k, y)}{\partial y} \right)_{y_M} = G_{y_M}(x_k) \end{cases} \tag{8.140b}$$

these constraints do not specify values of ψ at the boundaries. Therefore, as with Neumann boundary conditions, the partial solution table is empty and all ψ values $\{\psi_{0,0}, \ldots, \psi_{N,M}\}$ must be determined from the finite difference equations.

The finite difference approximation to the elliptic equation with the Crank–Nicolson modification is

$$\frac{R_{k,m}}{2h_x^2} \left[\left(\psi_{k+1,m+1} - 2\psi_{k,m+1} + \psi_{k-1,m+1} \right) + \left(\psi_{k+1,m-1} - 2\psi_{k,m-1} + \psi_{k-1,m-1} \right) \right] +$$

$$\frac{S_{k,m}}{2h_y^2} \left[\left(\psi_{k+1,m+1} - 2\psi_{k+1,m} + \psi_{k+1,m-1} \right) + \left(\psi_{k-1,m+1} - 2\psi_{k-1,m} + \psi_{k-1,m-1} \right) \right]$$

$$= f_{k,m}$$

$$\tag{8.97b}$$

Setting $(k, m) = \{(0, 0), (N, 0), (0, M), (N, M)\}$ in eq. 8.97b we obtain eqs. 8.121, which contain $\psi_{-1,-1}$, $\psi_{N+1,-1}$, $\psi_{-1,M+1}$, and $\psi_{N+1,M+1}$.

From the finite difference approximations to the Gauss boundary conditions, we have

$$\psi_{-1,m} \simeq \psi_{1,m} + \frac{2h_x}{\gamma_x}\psi_{0,m} - \frac{2h_x}{\gamma_x}G_{x_0}(y_m) \qquad (8.141a)$$

and

$$\psi_{N+1,m} \simeq \psi_{N-1,m} - \frac{2h_x}{\gamma_x}\psi_{N,m} + \frac{2h_x}{\gamma_x}G_{x_N}(y_m) \qquad (8.141b)$$

From eq. 8.140b, we find

$$\psi_{k,-1} \simeq \psi_{k,1} + \frac{2h_y}{\gamma_y}\psi_{k,0} - \frac{2h_y}{\gamma_y}G_{y_0}(x_k) \qquad (8.141c)$$

and

$$\psi_{k,M+1} \simeq \psi_{k,M-1} - \frac{2h_y}{\gamma_y}\psi_{k,M} + \frac{2h_y}{\gamma_y}G_{y_M}(x_k) \qquad (8.141d)$$

To obtain an approximation to $\psi_{-1,-1}$, we first set $m = -1$ in eq. 8.141a, then use eq. 8.141c to estimate $\psi_{1,-1}$ and $\psi_{0,-1}$. This results in

$$\psi_{-1,-1} = \psi_{1,1} + \frac{2h_y}{\gamma_y}\psi_{1,0} + \frac{2h_x}{\gamma_x}\psi_{0,1} + \frac{4h_xh_y}{\gamma_x\gamma_y}\psi_{0,0}$$
$$- \frac{2h_y}{\gamma_y}G_{y_0}(x_1) - \frac{2h_x}{\gamma_x}G_{x_0}(y_{-1}) - \frac{4h_xh_y}{\gamma_x\gamma_y}G_{y_0}(x_0) \qquad (8.142a)$$

Using eq. 8.141c with $k = -1$ first, then using eq. 8.141a to obtain $\psi_{-1,1}$ and $\psi_{-1,0}$, we have

$$\psi_{-1,-1} = \psi_{1,1} + \frac{2h_y}{\gamma_y}\psi_{1,0} + \frac{2h_x}{\gamma_x}\psi_{0,1} + \frac{4h_xh_y}{\gamma_x\gamma_y}\psi_{0,0}$$
$$- \frac{2h_y}{\gamma_y}G_{y_0}(x_{-1}) - \frac{2h_x}{\gamma_x}G_{x_0}(y_1) - \frac{4h_xh_y}{\gamma_x\gamma_y}G_{x_0}(y_0) \qquad (8.142b)$$

As we found with the Neumann boundary conditions, Milne approximations to the Gauss boundary conditions yield an approximation to $\psi_{-1,-1}$ that is not unique. As such, we consider the Milne approximation without the Crank–Nicolson modification, which is

$$\frac{R_{k,m}}{h_x^2}\left(\psi_{k+1,m} - 2\psi_{k,m} + \psi_{k-1,m}\right) + \frac{S_{k,m}}{h_y^2}\left(\psi_{k,m+1} - 2\psi_{k,m} + \psi_{k,m-1}\right) = f_{k,m} \quad \textbf{(8.97a)}$$

or, with

$$C'' = \frac{h_y^2}{h_x^2} \qquad\qquad (8.101)$$

we write this as

$$\begin{aligned}[C'' R_{k,m}\psi_{k+1,m} + C'' R_{k,m}\psi_{k-1,m} + S_{k,m}\psi_{k,m+1} + S_{k,m}\psi_{k,m-1} \\ -2(C'' R_{k,m} + S_{k,m})\psi_{k,m}] = h_y^2 f_{k,m}\end{aligned} \qquad (8.143)$$

Using eqs. 8.141, we see that for $(k, m) = (0, 0)$, eq. 8.143 becomes

$$\begin{aligned}\Big[2C'' R_{0,0}\psi_{1,0} + 2S_{0,0}\psi_{0,1} \\ -2\Big(C'' R_{0,0} + S_{0,0} - C'' R_{0,0}\frac{h_x}{\gamma_x} - S_{0,0}\frac{h_y}{\gamma_y}\Big)\psi_{0,0}\Big] \\ = h_y^2 f_{0,0} + 2C'' R_{0,0}\frac{h_x}{\gamma_x}G_{x_0}(y_0) + 2S_{0,0}\frac{h_y}{\gamma_y}G_{y_0}(x_0) \equiv \beta_{0,0}\end{aligned} \qquad (8.144a)$$

With $(k, m) = (N, 0)$, we obtain

$$\begin{aligned}\Big[2C'' R_{N,0}\psi_{N-1,0} + 2S_{N,0}\psi_{N,1} \\ -2\Big(C'' R_{N,0} + S_{N,0} + C'' R_{N,0}\frac{h_x}{\gamma_x} - S_{N,0}\frac{h_y}{\gamma_y}\Big)\psi_{N,0}\Big] \\ = h_y^2 f_{N,0} - 2C'' R_{N,0}\frac{h_x}{\gamma_x}G_{x_N}(y_0) + 2S_{N,0}\frac{h_y}{\gamma_y}G_{y_0}(x_N) \equiv \beta_{N,0}\end{aligned} \qquad (8.144b)$$

Setting $(k, m) = (0, M)$, we find

$$\begin{aligned}\Big[2C'' R_{0,M}\psi_{1,M} + 2S_{0,M}\psi_{0,M-1} \\ -2\Big(C'' R_{0,M} + S_{0,M} - C'' R_{0,M}\frac{h_x}{\gamma_x} + S_{0,M}\frac{h_y}{\gamma_y}\Big)\psi_{0,M}\Big] \\ = h_y^2 f_{0,M} + 2C'' R_{0,M}\frac{h_x}{\gamma_x}G_{x_0}(y_M) - 2S_{0,M}\frac{h_y}{\gamma_y}G_{y_M}(x_0) \equiv \beta_{0,M}\end{aligned} \qquad (8.144c)$$

and with $(k, m) = (N, M)$, we have

$$
\begin{aligned}
& \Big[2C''R_{N,M}\psi_{N-1,M} + 2S_{N,M}\psi_{N,M-1} \\
& \quad -2\Big(C''R_{N,M} + S_{N,M} + C''R_{N,M}\frac{h_x}{\gamma_x} + S_{N,M}\frac{h_y}{\gamma_y} \Big)\psi_{N,M} \Big] \\
& = h_y^2 f_{N,M} - 2C''R_{N,M}\frac{h_x}{\gamma_x}G_{x_N}(y_M) - 2S_{N,M}\frac{h_y}{\gamma_y}G_{y_M}(x_N) \equiv \beta_{N,M}
\end{aligned}
\tag{8.144d}
$$

The equations for $\psi_{n,m}$ with $(k, m) \neq \{(0, 0), (N, 0), (0, M), (N, M)\}$ are obtained directly from

$$
\begin{aligned}
& \big[C''R_{k,m}\psi_{k+1,m} + C''R_{k,m}\psi_{k-1,m} + S_{k,m}\psi_{k,m+1} + S_{k,m}\psi_{k,m-1} \\
& \quad -2\big(C''R_{k,m} + S_{k,m} \big)\psi_{k,m} \big] = h_y^2 f_{k,m} \equiv \beta_{k,m}
\end{aligned}
\tag{8.145}
$$

We again write the equations for $\{\psi_{0,0}, \ldots, \psi_{N,M}\}$ in matrix form

$$
A\Psi = B
\tag{8.140}
$$

where Ψ and B are the column vectors given in eq. 8.139. Referring to eq. 8.144a, we see that the non-zero elements in the first row of A are

$$
2C''R_{0,0}, \ 2S_{0,0}, \ -\Big(C''R_{0,0} + S_{0,0} - C''R_{0,0}\frac{h_x}{\gamma_x} - S_{0,0}\frac{h_y}{\gamma_y} \Big)
$$

We see that the sum of these terms, and thus the sum of the elements in the first row of A, is not zero. Identical analysis of eqs. 8.144b, 8.144c, and 8.144d yields elements of A in other rows, the sums of which are also non-zero. Therefore, A is not singular, and it is possible to obtain a solution to the coupled equations for $\{\psi_{0,0}, \ldots, \psi_{N,M}\}$.

Thus, since the Gauss boundary conditions contain a term ψ added to the derivative and the Neumann conditions do not, the elliptic equation with Gauss boundary conditions can be solved by finite difference methods whereas the elliptic equation with Neumann conditions cannot.

Example 8.11: Finite difference approximation for the solution to a Poisson equation with Gauss boundary conditions

We again consider

$$
\frac{\partial^2\psi}{\partial x^2} + \frac{\partial^2\psi}{\partial y^2} = 2e^{y-x}
\tag{8.98}
$$

which, when constrained by the Gauss boundary conditions

$$
\begin{cases}
\psi(0,y) + 2\left(\dfrac{\partial \psi(x,y)}{\partial x}\right)_{x=0} = G_0^x(y) = -e^y \\[4mm]
\psi(1,y) + 2\left(\dfrac{\partial \psi(x,y)}{\partial x}\right)_{x=1} = G_N^x(y) = -e^{y-1}
\end{cases}
\tag{8.146a}
$$

and

$$
\begin{cases}
\psi(x,0) + 2\left(\dfrac{\partial \psi(x,y)}{\partial y}\right)_{y=0} = G_0^y(y) = 3e^{-x} \\[4mm]
\psi(x,1) + 2\left(\dfrac{\partial \psi(x,y)}{\partial y}\right)_{y=1} = G_M^y(y) = 3e^{1-x}
\end{cases}
\tag{8.146b}
$$

has solution

$$
\psi(x,y) = e^{y-x} \tag{8.100}
$$

With $h_x = 0.25$ and $h_y = 0.20$ (and therefore $0 \le k \le 4$ and $0 \le m \le 5$), eq. 8.145b becomes

$$
C''\psi_{n+1,m} + C''\psi_{n-1,m} - 2(C''1)\psi_{n,m} + \psi_{n,m+1} + \psi_{n,m-1} = 2h_y^2 e^{y_m - x_n} \tag{8.147}
$$

We use the finite difference approximation to the boundary conditions

$$
\psi_{-1,m} = \psi_{1,m} + h_x \psi_{0,m} + h_x e^{y_m} \tag{8.148a}
$$

$$
\psi_{5,m} = \psi_{3,m} - h_x \psi_{4,m} - h_x e^{y_m - 1} \tag{8.148b}
$$

$$
\psi_{k,-1} = \psi_{k,1} + h_y \psi_{k,0} - 3h_y e^{-x_k} \tag{8.148c}
$$

and

$$
\psi_{k,6} = \psi_{k,4} - h_y \psi_{k,5} + 3h_y e^{1-x_k} \tag{8.148d}
$$

to relate values of ψ at points outside the x or y boundary to values at points within that boundary.

   With eq. 8.146 and eqs. 8.147, we generate the set of 30 equations for the unknowns $\{\psi_{0,0}, \ldots, \psi_{0,5}, \ldots, \psi_{4,0}, \ldots, \psi_{4,5}\}$, the solution to which is given in Table 8.19.

	$x = 0.00$	$x = 0.25$	$x = 0.50$	$x = 0.75$	$x = 1.00$
$y = 0.00$	1.03427	0.81193	0.63856	0.50291	0.39630
$y = 0.20$	1.25040	0.97986	0.76886	0.60393	0.47466
$y = 0.40$	1.51505	1.18579	0.92892	0.72829	0.57133
$y = 0.60$	1.83939	1.43833	1.12539	0.88108	0.69021
$y = 0.80$	2.23702	1.74799	1.36634	1.06851	0.83611
$y = 1.00$	2.72464	2.12760	1.66164	1.29817	1.01480

Table 8.19 Solution table for Poisson's equation subject to Gauss boundary conditions without the Crank–Nicolson modification

Comparing these to the exact values given in Table 8.3, we see that, unlike to solution constrained by Dirichlet boundary conditions, these results are accurate to only a few decimal places, but the results are reasonable and stable. Again, this inaccuracy stems from the fact that we find approximate values of ψ at the boundaries from approximations of the Gauss boundary conditions. $\square$

Problems

1. The parabolic differential equation $x\dfrac{\partial^2\psi}{\partial x^2} - t^2\dfrac{\partial\psi}{\partial t} = 2xt^2 e^{xt}$ subject to initial condition $\psi(x, 1) - e^x$ and the Neumann boundary conditions

$$\begin{cases} \left(\dfrac{\partial\psi}{\partial x}\right)_{x=1} = te^t \\ \left(\dfrac{\partial\psi}{\partial x}\right)_{x=2} = te^{2t} \end{cases}$$

has solution $\psi(x, t) = e^{xt}$

(I) What Dirichlet boundary conditions constrain this solution?
(II) Find $\psi(x,t)$ at $t = 2$ for $x = \{1.00, 1.25, 1.50, 1.75, 2.00\}$ by

(a) a single step Taylor series in t of 2nd order in $(t-1)$.
(b) a single step Taylor series in t of 4th order in $(t-1)$.
(c) a two step Taylor series of 2nd order in $(t-1)$.

In each case, compare your results to the exact solution.

2. The wave equation $\dfrac{\partial^2\psi}{\partial x^2} - \dfrac{1}{4}\dfrac{\partial^2\psi}{\partial t^2} = 0$ subject to the Dirichlet boundary conditions

$$\begin{cases} \psi(0, t) = \sin(2\pi t) \\ \psi(1, t) = -\sin(2\pi t) \end{cases}$$

and the initial conditions

$$
\begin{cases}
\psi(x,0) = \sin(\pi x) \\
\left(\dfrac{\psi(x,t)}{\partial t} \right)_{t=0} = 2\pi\cos(\pi x)
\end{cases}
$$

has solution $\psi(x,t) = \sin(\pi x + 2\pi t)$

    (I) What Neumann boundary conditions constrain this solution?
    (II) Find $\psi(x,t)$ at $t = 0.5$ for $x = \{0.00, 0.25, 0.50, 0.75, 1.00\}$ by

      (a) single step Taylor series in t of 2nd order in t.
      (b) a single step Taylor series in t of 4th order in t.

    (III) Use the results of part II(b) to develop a two step Taylor series of 2nd
        order in t. Use that result to determine $\psi(x,t)$ at $t = 0.5$ at the values of
        x given above.
        In each case, compare your results to the exact solution.

3. The parabolic differential equation $x\dfrac{\partial^2 \psi}{\partial x^2} - t^2\dfrac{\partial \psi}{\partial t} = 2xt^2 e^{xt}$ subject to initial
condition $\psi(x,1) = e^x$ and the Neumann boundary conditions

$$
\begin{cases}
\left(\dfrac{\partial \psi}{\partial x} \right)_{x=1} = te^t \\[2ex]
\left(\dfrac{\partial \psi}{\partial x} \right)_{x=2} = te^{2t}
\end{cases}
$$

has solution $\psi(x,t) = e^{xt}$
Find $\psi(x,t)$ at $t = 2$ for $x = \{1.00, 1.25, 1.50, 1.75, 2.00\}$ by

(a) a single step Runge–Kutta scheme.
(b) a two step Runge–Kutta scheme.
    In each case, compare your results to the exact solution.

4. The wave equation $\dfrac{\partial^2 \psi}{\partial x^2} - \dfrac{1}{4}\dfrac{\partial^2 \psi}{\partial t^2} = 0$ subject to the Dirichlet boundary
conditions

$$
\begin{cases}
\psi(0,t) = \sin(2\pi t) \\
\psi(1,t) = -\sin(2\pi t)
\end{cases}
$$

and the initial conditions

$$
\begin{cases}
\psi(x,0) = \sin(\pi x) \\
\left(\dfrac{\psi(x,t)}{\partial t} \right)_{t=0} = 2\pi\cos(\pi x)
\end{cases}
$$

has solution $\psi(x,t) = \sin(\pi x + 2\pi t)$

Find $\psi(x,t)$ at $t = 0.5$ for $x = \{0.00, 0.25, 0.50, 0.75, 1.00\}$ by

(a) a single step Runge–Kutta scheme.
(b) a two step Runge–Kutta scheme.
   In each case, compare your results to the exact solution.

5. The parabolic equation $x\dfrac{\partial^2\psi}{\partial x^2} - t^2\dfrac{\partial\psi}{\partial t} = 2xt^2e^{xt}$ subject to initial condition $\psi(x, 1) = e^x$ and the Dirichlet boundary conditions

$$\begin{cases} \psi(1,t) = e^t \\ \psi(2,t) = e^{2t} \end{cases}$$

has solution $\psi(x, t) = e^{xt}$
For $h_x = 0.25$ and $h_t = 0.20$, use Milne finite differences to determine the solution table for this differential equation for $1 \le x \le 2, 1 \le t \le 2$.

(a) Without the Crank–Nicolson modification.
(b) With the Crank–Nicolson modification.
   In each case, compare your results to the exact solution table.

6. The parabolic equation $x\dfrac{\partial^2\psi}{\partial x^2} - t^2\dfrac{\partial\psi}{\partial t} = 2xt^2e^{xt}$ subject to initial condition $\psi(x, 1) = e^x$ and the Neumann boundary conditions

$$\begin{cases} \left(\dfrac{\partial\psi}{\partial x}\right)_{x=1} = te^t \\ \left(\dfrac{\partial\psi}{\partial x}\right)_{x=2} = te^{2t} \end{cases}$$

has solution $\psi(x, t) = e^{xt}$
For $h_x = 0.25$ and $h_t = 0.20$, use Milne finite differences to determine the solution table for this differential equation for $1 \le x \le 2, 1 \le t \le 2$.

(a) Without the Crank–Nicolson modification.
(b) With the Crank–Nicolson modification.
   In each case, compare your results to the exact solution.

7. The wave equation $\dfrac{\partial^2\psi}{\partial x^2} - \dfrac{1}{4}\dfrac{\partial^2\psi}{\partial t^2} = 0$ subject to the Dirichlet boundary conditions

$$\begin{cases} \psi(0,t) = \sin(2\pi t) \\ \psi(1,t) = -\sin(2\pi t) \end{cases}$$

and the initial conditions

$$\begin{cases} \psi(x,0) = \sin(\pi x) \\ \left(\dfrac{\psi(x,t)}{\partial t}\right)_{t=0} = 2\pi\cos(\pi x) \end{cases}$$

has solution $\psi(x, t) = \sin(\pi x + 2\pi t)$

For $h_x = 0.25$ and $h_t = 0.20$, use Milne finite differences to determine the solution table for this differential equation for $0 \leq x \leq 1, 0 \leq t \leq 1$.

(a) Without the Crank–Nicolson modification.
(b) With the Crank–Nicolson modification.
   In each case, compare your results to the exact solution.

8. The wave equation $\dfrac{\partial^2 \psi}{\partial x^2} - \dfrac{1}{4}\dfrac{\partial^2 \psi}{\partial t^2} = 0$ subject to the Gauss boundary conditions

$$
\begin{cases}
\psi(0,t) + \dfrac{1}{\pi}\left(\dfrac{\partial \psi}{\partial x}\right)_{x=0} = \sin(2\pi t) + \cos(2\pi t) \\[2mm]
\psi(1,t) + \dfrac{1}{\pi}\left(\dfrac{\partial \psi}{\partial x}\right)_{x=1} = -\sin(2\pi t) - \cos(2\pi t)
\end{cases}
$$

and the initial conditions

$$
\begin{cases}
\psi(x,0) = \sin(\pi x) \\[2mm]
\left(\dfrac{\partial \psi(x,t)}{\partial t}\right)_{t=0} = 2\pi\cos(\pi x)
\end{cases}
$$

has solution $\psi(x,t) = \sin(\pi x + 2\pi t)$

For $h_x = 0.25$ and $h_t = 0.20$, use Milne finite differences to determine the solution table for this differential equation for $0 \leq x \leq 1, 0 \leq t \leq 1$.

(a) Without the Crank–Nicolson modification.
(b) With the Crank–Nicolson modification.
   In each case, compare your results to the exact solution.

9. Show that a solution to the elliptic equation $R(x,y)\dfrac{\partial^2 \psi}{\partial x^2} + S(x,y)\dfrac{\partial^2 \psi}{\partial y^2} = f(x,y)$ subject to

$$
\begin{cases}
\left(\dfrac{\partial \psi}{\partial x}\right)_{x_0} = N_{x_0}(y) \\[2mm]
\left(\dfrac{\partial \psi}{\partial x}\right)_{x_N} = N_{x_N}(y)
\end{cases}
$$

and

$$
\begin{cases}
\left(\dfrac{\partial \psi}{\partial y}\right)_{y_0} = N_{y_0}(x) \\[2mm]
\left(\dfrac{\partial \psi}{\partial y}\right)_{y_M} = N_{y_M}(x)
\end{cases}
$$

cannot be obtained using Milne finite difference approximations with the Crank–Nicolson modification.

10. Show that a solution to the elliptic equation $R(x,y)\dfrac{\partial^2 \psi}{\partial x^2} + S(x,y)\dfrac{\partial^2 \psi}{\partial y^2} +$
    $T(x,y)\dfrac{\partial \psi}{\partial x} + U(x,y)\dfrac{\partial \psi}{\partial y} = f(x,y)$ subject to

$$\begin{cases} \left(\dfrac{\partial \psi}{\partial x}\right)_{x_0} = N_{x_0}(y) \\[2mm] \left(\dfrac{\partial \psi}{\partial x}\right)_{x_N} = N_{x_N}(y) \end{cases}$$

and

$$\begin{cases} \left(\dfrac{\partial \psi}{\partial y}\right)_{y_0} = N_{y_0}(x) \\[2mm] \left(\dfrac{\partial \psi}{\partial y}\right)_{y_M} = N_{y_M}(x) \end{cases}$$

cannot be obtained using Milne finite difference methods, either with or without the Crank–Nicolson modification.

11. The elliptic equation $\dfrac{\partial^2 \psi}{\partial x^2} + \dfrac{\partial^2 \psi}{\partial y^2} = 2$ subject to the Gauss boundary conditions

$$\begin{cases} \psi(0,y) + \left(\dfrac{\partial \psi(x,y)}{\partial x}\right)_{x=0} = y^2 \\[2mm] \psi(1,y) + \left(\dfrac{\partial \psi(x,y)}{\partial x}\right)_{x=1} = y^2 + 3 \end{cases}$$

and

$$\begin{cases} \psi(x,0) - \left(\dfrac{\partial \psi(x,y)}{\partial y}\right)_{y=0} = x^2 \\[2mm] \psi(x,1) - \left(\dfrac{\partial \psi(x,y)}{\partial y}\right)_{y=1} = x^2 - 1 \end{cases}$$

has solution $\psi(x,y) = x^2 + y^2$

For $h_x = 0.25$ and $h_y = 0.20$, use Milne finite differences to determine the solution table for this differential equation for $0 \le x \le 1, 0 \le y \le 1$.

(a) Without the Crank–Nicolson modification.
(b) With the Crank–Nicolson modification.
   In each case, compare your results to the exact solution.

12. The time dependent Schroedinger equation in one dimension $i\hbar \dfrac{\partial \psi(x,t)}{\partial t} = -\dfrac{\hbar^2}{2m}\dfrac{\partial^2 \psi(x,t)}{\partial x^2} + V(x)\psi(x,t)$ describes the quantum mechanical behavior of a particle of mass m under the influence of a potential energy function $V(x)$. Write down the Milne finite difference approximation to this equation with the Crank–Nicolson modification.

Chapter 9
LINEAR INTEGRAL EQUATIONS IN ONE VARIABLE

The general form of an integral equation for the unknown function of one variable $\psi(x)$ is

$$A(x)\psi(x) = \psi_0(x) + \lambda \int_{a(x)}^{b(x)} J[x, y, \psi(y)]dy \qquad (9.1a)$$

When $J[x, y, \psi(y)]$ is of the form $K(x, y)\psi(y)$, eq. 9.1a is the linear integral equation

$$A(x)\psi(x) = \psi_0(x) + \lambda \int_{a(x)}^{b(x)} K(x, y)\psi(y)dy \qquad (9.1b)$$

where $A(x)$, $\psi_0(x)$, $a(x)$, $b(x)$, and $K(x, y)$ are known functions and λ is a specified constant. $\psi_0(x)$ is referred to as the *inhomogeneous function* and $K(x,y)$ is the *kernel of the equation*.

If both limits a and b are constants, the integral equation is called a *Fredholm integral equation*. If a is constant and $b(x) = x$, the integral equation is the *Volterra equation*.

Except when $K(x, y)$ is one of a few special cases, there are no techniques for solving eqs. 9.1 in a closed form. As such, for many integral equations, numerical and other approximation methods are essential for estimating $\psi(x)$.

9.1 Fredholm Equations of the First Kind with Non-singular Kernel

If $A(x) = 0$ for all $x \in [a, b]$, the *Fredholm equation of the first kind* is of the form

$$\psi_0(x) = \lambda \int_a^b K(x, y)\psi(y)dy \qquad (9.2)$$

where $K(x, y)$ is analytic at all x and y in $[a, b]$.

H. Cohen, *Numerical Approximation Methods*, DOI 10.1007/978-1-4419-9837-8_9,
© Springer Science+Business Media, LLC 2011

Solution by Quadrature Sum

One commonly used approach for solving Fredholm equations is to approximate the integral by a quadrature sum

$$\int_a^b K(x,y)\psi(y)dy \simeq \sum_{m=1}^{N} w_m K(x,y_m)\psi(y_m) \tag{9.3}$$

With this, eq. 9.2 becomes

$$\psi_0(x) = \lambda \sum_{m=1}^{N} w_m K(x,y_m)\psi(y_m) \tag{9.4}$$

Setting x to each abscissa point in the set $\{y_k\}$, and using the notation

$$K(y_k, y_m) \equiv K_{km} \tag{9.5}$$

we obtain a set of coupled equations, the matrix form of which is

$$\begin{pmatrix} \psi_0(y_1) \\ \bullet \\ \bullet \\ \bullet \\ \psi_0(y_N) \end{pmatrix} = \lambda \begin{pmatrix} w_1 K_{11} & \bullet\bullet & w_N K_{1N} \\ \bullet & & \bullet \\ \bullet & \bullet & \bullet \\ \bullet & & \bullet \\ w_1 K_{N1} & \bullet\bullet & w_N K_{NN} \end{pmatrix} \begin{pmatrix} \psi(y_1) \\ \bullet \\ \bullet \\ \bullet \\ \psi(y_N) \end{pmatrix} \tag{9.6a}$$

It is straightforward to obtain the solution as

$$\begin{pmatrix} \psi(y_1) \\ \bullet \\ \bullet \\ \bullet \\ \psi(y_N) \end{pmatrix} = \frac{1}{\lambda} \begin{pmatrix} w_1 K_{11} & \bullet\bullet & w_N K_{1N} \\ \bullet & & \bullet \\ \bullet & \bullet & \bullet \\ \bullet & & \bullet \\ w_1 K_{N1} & \bullet\bullet & w_N K_{NN} \end{pmatrix}^{-1} \begin{pmatrix} \psi_0(y_1) \\ \bullet \\ \bullet \\ \bullet \\ \psi_0(y_N) \end{pmatrix} \tag{9.6b}$$

If one of the abscissa points (e.g., y_n) is such that $K_{nm} = 0$, then all the elements in the nth row of the coefficient matrix are zero. Thus, the determinant of the matrix of eq. 9.6a is

$$\begin{vmatrix} w_1 K_{11} & & \bullet\bullet & & w_N K_{1N} \\ \bullet & & & & \bullet \\ \bullet & & & & \bullet \\ w_1 K_{(n-1)1} & & \bullet\bullet & & w_N K_{(n-1)N} \\ 0 & 0 \ 0 & \bullet\bullet & 0 \ 0 & 0 \\ w_1 K_{(n+1)1} & & \bullet\bullet & & w_N K_{(n+1)N} \\ \bullet & & & & \bullet \\ \bullet & & & & \bullet \\ w_1 K_{N1} & & \bullet\bullet & & w_N K_{NN} \end{vmatrix} = 0 \tag{9.7}$$

Therefore, the kernel matrix is singular. Thus, one must choose a quadrature rule such that K_{nm} is non-zero for all abscissae of the quadrature rule.

Example 9.1: Solution to a Fredholm equation of the first kind by quadratures

Since the solution to the Fredholm equation of the first kind

$$\int_{-1}^{1} xe^{y(1-x)}\psi(y)dy = e^x - e^{-x} \tag{9.8a}$$

is

$$\psi(x) = e^{-x} \tag{9.8b}$$

Using a quadrature rule to approximate the integral

$$\int_{0}^{1} xe^{y(1-x)}\psi(y)dy \simeq \sum_{k=1}^{N} w_k xe^{x_k(1-x)}\psi(x_k) \tag{9.9}$$

we see that for $x = 0$, the kernel $xe^{y(1-x)}$ is zero. Thus, if we use a Gauss–Legendre quadrature rule (the most appropriate for an integral from -1 to 1), we must choose an even order rule, so that $x = 0$ is not one of the abscissa.

To obtain a sense of the accuracy obtained using an even order Gauss–Legendre rule, we approximate the integral as in eq. 9.9 by a six-point quadrature. We obtain

$$\begin{pmatrix} \psi(0.93247) \\ \psi(0.66121) \\ \psi(0.23862) \\ \psi(-0.23862) \\ \psi(-0.66121) \\ \psi(-0.93247) \end{pmatrix} = \begin{pmatrix} -4.00895 \\ 2.49535 \\ -7.24554 \\ -11.67704 \\ 9.36377 \\ -25.87992 \end{pmatrix} \tag{9.10a}$$

The exact values at these points are given by

$$\begin{pmatrix} \psi_{exact}(0.93247) \\ \psi_{exact}(0.66121) \\ \psi_{exact}(0.23862) \\ \psi_{exact}(-0.23862) \\ \psi_{exact}(-0.66121) \\ \psi_{exact}(-0.93247) \end{pmatrix} = \begin{pmatrix} 0.39358 \\ 0.51623 \\ 0.78771 \\ 1.26950 \\ 1.93713 \\ 2.54078 \end{pmatrix} \tag{9.10b}$$

Clearly, the results presented in eq. 9.10a do not represent e^{-x}. They are highly inaccurate and oscillate wildly from point to point. In addition, the results are not improved by using a larger quadrature set to approximate to the integral. $\square$

This example illustrates the well known fact that the Fredholm integral of the first kind is an ill-conditioned problem (see, for example, Baker, C., et. al., 1964).

There are several methods presented in the literature for smoothing these results. For example, the *Galerkin* approach involves choosing the value of x at which the solution is to be obtained and approximating $\psi(y)$ by a sum over known *basis functions* $\phi_k(y)$, which are chosen by the user. Then, with

$$\psi(y) \simeq \sum_{k=1}^{N} a_k \phi_k(y) \tag{9.11a}$$

the *error parameter*

$$\varepsilon\left(\vec{a}\right) \equiv \left| \psi_0(x) - \sum_{k=1}^{N} a_k \int_a^b K(x,y)\phi_k(y)dy \right|^2 \tag{9.11b}$$

is minimized (for example, by the method of least squares, minimizing with respect to the coefficients a_k). Excellent treatments of such approaches are given by Twomey, S., 1963, Baker, C., et. al., 1964, Hanson, R., 1971, and Hansen, P., 1992.

Series approximation

If $K(x, y)$ and $\psi_0(x)$ are analytic at all points in $[a, b]$, then $\psi(x)$ is also analytic everywhere in $[a, b]$. If $x_0 \, \varepsilon \, [a, b]$, we can $\psi(x)$ as

$$\psi(x) = \sum_{k=0}^{\infty} c_k(x - x_0)^k \tag{9.12a}$$

that is valid at all x in $[a, b]$. A truncated series

$$\psi(x) = \sum_{k=0}^{N} c_k(x - x_0)^k \tag{9.12b}$$

yields an approximate solution to eq. 9.2.

Substituting eq. 9.12b into eq. 9.2, we obtain

$$\psi_0(x) \simeq \lambda \sum_{k=0}^{N} c_k \int_a^b K(x,y)(y - x_0)^k dy \equiv \lambda \sum_{k=0}^{N} c_k I_k(x) \tag{9.13}$$

Because the integrand is a known function, the integral $I_k(x)$ can be evaluated in a closed form or can be accurately approximated by some quadrature rule. Therefore, $I_k(x)$ is known at every x.

To determine the coefficients c_k, $\psi_0(x)$ and each $I_k(x)$ are expanded in their Taylor series and we equate the coefficients of corresponding powers of $(x-x_0)$ up to $(x-x_0)^N$.

Using an approach that is suggested by the Galerkin method, a more convenient way to develop the series method is to write $K(x, y)$, $\psi_0(x)$, and $\psi(x)$ as sums over orthogonal polynomials, where the polynomials used depend on the limits of the integral.

Referring to ch. 4, if x and y vary over $[0, \infty]$ and if $\psi_0(x)$ and $K(x, y)$ can be written as

$$K(x, y) \equiv e^{-y} L(x, y) \tag{9.14a}$$

and

$$\psi_0(x) = e^{-x} \Omega_0(x) \tag{9.14b}$$

such that

$$\lim_{y \to \infty} e^{-y} L(x, y) = 0 \tag{9.15a}$$

and

$$\lim_{x \to \infty} e^{-x} \Omega_0(x) = 0 \tag{9.15b}$$

then one might expand these functions as a series involving Laguerre polynomials. Likewise, if $-\infty \le x$ and $y \le \infty$, and if $K(x,y)$ and $\psi_0(x)$ can be written as

$$K(x, y) \equiv e^{-y^2} L(x, y) \tag{9.16a}$$

and

$$\psi_0(x) = e^{-x^2} \Omega_0(x) \tag{9.16b}$$

with

$$\lim_{y \to \infty} e^{-y^2} L(x, y) = 0 \tag{9.17a}$$

and

$$\lim_{x \to \infty} e^{-x^2} \Omega(x) = 0 \tag{9.17b}$$

one might write the series for these functions in terms of the Hermite polynomials. If x and y vary over finite limits a and b, one can transform this interval to $[-1, 1]$ and write the series for $\psi_0(x)$ and $K(x, y)$ in terms of Legendre polynomials.

It was shown in ch. 4, eqs. 4.9 through 4.16, which for any limits of integration (finite or infinite) any integral can be converted to one integrated over $[-1, 1]$. It was also shown that if the exponential factors for $x \ \varepsilon \ [0, \ \infty]$ or $x \ \varepsilon \ [-\infty,\infty]$ are not explicitly part of the integrand, one usually achieves accurate results by transforming integrals over an infinite domain to $[-1, 1]$ and using a Gauss–Legendre quadrature. Thus, for a kernel and/or inhomogeneous term that does not contain the exponential explicitly, one should consider transforming the integral to $[-1, 1]$ and expanding the functions over Legendre polynomials. This is an example of what Baker calls *expansion methods* (Baker, C. T. H., 1977, pp. 205–214).

Legendre polynomials form an infinite set of mutually orthogonal polynomial functions over the interval $[-1, 1]$ such that

$$\int_{-1}^{1} P_n(x)P_m(x)dx = \frac{2}{(2n + 1)} \delta_{nm} \qquad (4.100b)$$

As such, we transform the domains of ψ, ψ_0, and K to $x, y \ \varepsilon \ [-1, 1]$ then expand the unknown function, the inhomogeneous function and the kernel as

$$\psi(x) = \sum_{k=0}^{\infty} c_k P_k(x) \qquad (9.18a)$$

$$\psi_0(x) = \sum_{k=0}^{\infty} \phi_k P_k(x) \qquad (9.18b)$$

and

$$K(x, y) = \sum_{\substack{k=0 \\ m=0}}^{\infty} \mu_{km} P_k(x) P_m(y) \qquad (9.18c)$$

from which we determine the coefficients

$$\phi_k = \frac{(2k + 1)}{2} \int_{-1}^{1} \psi_0(x) P_k(x) dx \qquad (9.19a)$$

and

$$\mu_{km} = \frac{(2k + 1)(2m + 1)}{4} \int_{-1}^{1} \int_{-1}^{1} K(x, y) P_k(x) P_m(y) dx dy \qquad (9.19b)$$

Then, the Fredholm equation of the first kind becomes

$$\sum_{k=0}^{\infty}\phi_k P_k(x) = \sum_{\substack{k=0\\m=0\\n=0}}^{\infty}\mu_{km}c_n P_k(x)\int_{-1}^{1}P_m(y)P_n(y)dy = \sum_{\substack{k=0\\m=0\\n=0}}^{\infty}\mu_{km}c_n P_k(x)\frac{2}{(2m+1)}\delta_{mn}$$

$$= \sum_{\substack{k=0\\m=0}}^{\infty}\mu_{km}c_m P_k(x)\frac{2}{(2m+1)} \tag{9.20}$$

Equating the coefficients of the various Legendre polynomials, we obtain

$$\phi_k = \sum_{m=0}^{\infty}\frac{2}{(2m+1)}\mu_{km}c_m \tag{9.21a}$$

Approximating the infinite series by a finite sum, we obtain a finite set of linear equations

$$\phi_k = \sum_{m=0}^{N}\frac{2}{(2m+1)}\mu_{km}c_m \tag{9.21b}$$

In this way, the only integrals we may have to approximate by quadratures are those of eqs. 9.19, which are integrals of known functions. It has been shown in ch. 4 that when the integrand is analytic over the range of integration, quadrature rules usually yield accurate results for such integrals. And approximating an infinite series by a finite sum is equivalent to approximating a function by a truncated Taylor sum. Such an approximation of an analytic function has been shown to be quite accurate. Thus, as will be seen, this method yields stable and accurate results.

We define the $N \times N$ matrix A with elements

$$a_{km} = \frac{2}{(2m+1)}\mu_{km} \tag{9.22}$$

and the column vectors

$$F \equiv \begin{pmatrix} \phi_1 \\ \bullet \\ \bullet \\ \phi_N \end{pmatrix} \tag{9.23a}$$

and

$$C \equiv \begin{pmatrix} c_1 \\ \vdots \\ c_N \end{pmatrix} \tag{9.23b}$$

to write eq. 9.21b as a matrix equation which is solved straightforwardly by matrix inversion.

Once one has obtained values of the coefficients c_k, one can determine the approximation to $\psi(x)$ at any x in the domain of ψ.

Example 9.2: Solution to a Fredholm equation of the first kind by series expansion

We again consider

$$\int_{-1}^{1} x e^{y(1-x)} \psi(y) dy = e^x - e^{-x} \tag{9.8a}$$

the solution to which is

$$\psi(x) = e^{-x} \tag{9.8b}$$

from which

$$\begin{pmatrix} \psi_{exact}(1.0) \\ \psi_{exact}(0.5) \\ \psi_{exact}(0.0) \\ \psi_{exact}(-0.5) \\ \psi_{exact}(-1.0) \end{pmatrix} = \begin{pmatrix} 0.36788 \\ 0.60653 \\ 1.00000 \\ 1.64872 \\ 2.71828 \end{pmatrix} \tag{9.24}$$

Taking five terms in the sums of eqs. 9.18, the approximations to $\psi(x)$, $\psi_0(x)$, and $K(x, y)$ are

$$\psi(x) \simeq \sum_{k=0}^{4} c_k P_k(x) \tag{9.25a}$$

$$\psi_0(x) \simeq \sum_{k=0}^{4} \phi_k P_k(x) \tag{9.25b}$$

and

$$K(x, y) = \sum_{\substack{k=0 \\ m=0}}^{4} \mu_{km} P_k(x) P_m(y) \tag{9.25c}$$

We note that $\psi_0(x)$ given in eq. 9.8a is an odd function. Referring to eq. 9.19a, since the integral is evaluated over symmetric limits, ϕ_0, ϕ_2, and ϕ_4 are zero since their integrands are odd functions. The non-zero coefficients are

$$\phi_1 = \frac{1}{2}\int_{-1}^{1}(e^x - e^{-x})P_1(x)dx = \frac{1}{2}\int_{-1}^{1}x(e^x - e^{-x})dx = 6e^{-1} \tag{9.26a}$$

and

$$\phi_3 = \frac{7}{2}\int_{-1}^{1}(e^x - e^{-x})P_3(x)dx = \frac{7}{4}\int_{-1}^{1}(e^x - e^{-x})(5x^3 - 3x)dx$$
$$= 7(37e^{-1} - 5e^1) \tag{9.26b}$$

Rather than write down all 25 coefficients μ_{mk}, we present two as examples:

$$\mu_{21} = \frac{15}{4}\int_{-1}^{1}xP_2(x)\int_{-1}^{1}e^{y(1-x)}P_1(y)dydx\frac{15}{8}\int_{-1}^{1}x(3x^2 - 1)\int_{-1}^{1}e^{y(1-x)}ydydx$$
$$= \frac{15}{4}\int_{-1}^{1}x(3x^2 - 1)\frac{[(2-x)e^{-(1-x)} - xe^{(1-x)}]}{(1-x)^2}dx \tag{9.27a}$$

and

$$\mu_{43} = \frac{63}{4}\int_{-1}^{1}xP_4(x)\int_{-1}^{1}e^{y(1-x)}P_3(y)dydx = \frac{63}{64}\int_{-1}^{1}x(35x^4 - 30x^2 + 3)\times$$
$$\left[e^{(1-x)}\left(\frac{1}{(1-x)} - \frac{6}{(1-x)^2} + \frac{15}{(1-x)^3} - \frac{15}{(1-x)^4}\right) + \right.$$
$$\left. e^{-(1-z)}\left(\frac{1}{(1-x)} - \frac{6}{(1-x)^2} + \frac{15}{(1-x)^3} - \frac{15}{(1-x)^4}\right)\right]dydx \tag{9.27b}$$

Such integrals cannot be evaluated in closed form.

It is straightforward to show that these integrands are not singular at $x = 1$. Therefore, they are well approximated by quadrature sums. In this example, we approximate them using a 20-point Gauss–Legendre quadrature rule.

For this five polynomial series approximation, we obtain

$$\Psi = \begin{pmatrix} \psi(1.0) \\ \psi(0.5) \\ \psi(0.0) \\ \psi(-0.5) \\ \psi(-1.0) \end{pmatrix} = \begin{pmatrix} 0.36251 \\ 0.60773 \\ 0.99665 \\ 1.65340 \\ 2.70208 \end{pmatrix} \tag{9.28}$$

The accuracy of these results is indicated by their percent differences from the exact values given in eq. 9.24. These differences are

$$
\begin{pmatrix} \Delta(1.0) \\ \Delta(0.5) \\ \Delta(0.0) \\ \Delta(-0.5) \\ \Delta(-1.0) \end{pmatrix} = \begin{pmatrix} 1.5\% \\ 0.2\% \\ 0.3\% \\ 0.3\% \\ 0.6\% \end{pmatrix}
\tag{9.29}
$$

which indicates that this five term series approximation yields a fairly accurate approximation to the solution to a Fredholm equation of the first kind. □

9.2 Fredholm Equations of the Second Kind with Non-singular Kernel

Referring to eq. 9.1, when $A(x)$ is non-zero for all $x \, \varepsilon \, [a, \, b]$, we can divide the Fredholm equation by $A(x)$, rename the inhomogeneous function and kernel as

$$
\frac{\psi_0(x)}{A(x)} \rightarrow \psi_0(x)
\tag{9.30}
$$

and

$$
\frac{K(x,y)}{A(x)} \rightarrow K(x,y)
\tag{9.31}
$$

to obtain the *Fredholm equation of the second kind*

$$
\psi(x) = \psi_0(x) + \lambda \int_a^b K(x,y)\psi(y)dy
\tag{9.32}
$$

Solution by quadrature sum

We approximate the integral of eq. 9.32 by a quadrature sum to obtain

$$
\psi(x) \simeq \psi_0(x) + \lambda \sum_{m=1}^N w_m K(x, y_m)\psi(y_m)
\tag{9.33}
$$

With at least one value of $\psi_0(y_k)$ non-zero, we set x to each of the quadrature points in the set $\{y_m\}$, to obtain

$$
\begin{pmatrix} \psi(y_1) \\ \vdots \\ \psi(y_N) \end{pmatrix} = \begin{pmatrix} \psi_0(y_1) \\ \vdots \\ \psi_0(y_N) \end{pmatrix} + \lambda \begin{pmatrix} w_1 K_{11} & \bullet\bullet & w_N K_{1N} \\ & & \\ \vdots & & \vdots \\ & & \\ w_1 K_{N1} & \bullet\bullet & w_N K_{NN} \end{pmatrix} \begin{pmatrix} \psi(y_1) \\ \vdots \\ \psi(y_N) \end{pmatrix}
\tag{9.34}
$$

which is solved straightforwardly by matrix inversion.

An approximation to $\psi(x)$ at any x can then be obtained by substituting these values of $\psi(y_k)$ into the sum in eq. 9.33.

Example 9.3: Solution to a Fredholm equation of the second kind by quadratures

The solution to

$$
\psi(x) = e^x - \int_{-1}^{1} xe^{y(1-x)}\psi(y)dy
\tag{9.35}
$$

is

$$
\psi(x) = e^{-x}
\tag{9.8b}
$$

Using a four-point Gauss–Legendre quadrature rule, the integral equation is approximated by

$$
\psi(x) \simeq e^x - \sum_{m=1}^{4} w_m xe^{y_m(1-x)}\psi(y_m)
\tag{9.36a}
$$

Setting x to each y_m, this becomes

$$
\psi(y_k) \simeq e^{y_k} - \sum_{m=1}^{4} w_m y_k e^{y_m(1-y_k)}\psi(y_m)
\tag{9.36b}
$$

We define the 4×4 kernel matrix with elements

$$
a_{km} = w_m y_k e^{y_m(1-y_k)}
\tag{9.37}
$$

Then, eq. 9.36b can be expressed as

$$
\begin{pmatrix} \psi(y_1) \\ \vdots \\ \psi(y_4) \end{pmatrix} = \begin{pmatrix} e^{y_1} \\ \vdots \\ e^{y_4} \end{pmatrix} - \begin{pmatrix} w_1 y_1 e^{y_1(1-y_1)} & \bullet\bullet & w_4 y_1 e^{y_4(1-y_1)} \\ & & \\ \vdots & & \vdots \\ & & \\ w_1 y_4 e^{y_1(1-y_4)} & \bullet\bullet & w_4 y_4 e^{y_4(1-y_4)} \end{pmatrix} \begin{pmatrix} \psi(y_1) \\ \vdots \\ \psi(y_4) \end{pmatrix}
\tag{9.38}
$$

the solution to which is given by

$$
\begin{pmatrix} \psi(y_1) \\ \vdots \\ \vdots \\ \psi(y_4) \end{pmatrix} = \left[1 + \begin{pmatrix} w_1 y_1 e^{y_1(1-y_1)} & \cdots & w_4 y_1 e^{y_4(1-y_1)} \\ & \vdots & \\ & \vdots & \\ w_1 y_4 e^{y_1(1-y_4)} & \cdots & w_4 y_4 e^{y_4(1-y_4)} \end{pmatrix} \right]^{-1} \begin{pmatrix} e^{y_1} \\ \vdots \\ \vdots \\ e^{y_4} \end{pmatrix} \tag{9.39}
$$

From this, we obtain

$$
\Psi = \begin{pmatrix} \psi(0.86114) \\ \psi(0.33998) \\ \psi(-0.33998) \\ \psi(-0.86114) \end{pmatrix} = \begin{pmatrix} 0.42268 \\ 0.71178 \\ 1.40492 \\ 2.36585 \end{pmatrix} \tag{9.40a}
$$

Comparing this to

$$
\Psi_{exact} = \begin{pmatrix} \psi_{exact}(0.86114) \\ \psi_{exact}(0.33998) \\ \psi_{exact}(-0.33998) \\ \psi_{exact}(-0.86114) \end{pmatrix} = \begin{pmatrix} 0.42268 \\ 0.71178 \\ 1.40492 \\ 2.36586 \end{pmatrix} \tag{9.40b}
$$

we see that the method of quadratures yields an extremely accurate result. The largest of the 4% differences from the exact values is 1.3×10^{-5}.

To obtain values of $\psi(x)$ at any value of x, we substitute the values given in eq. 9.40a into the sum of eq. 9.36a. From this we find

$$
\begin{pmatrix} \psi(1.0) \\ \psi(0.5) \\ \psi(0.0) \\ \psi(-0.5) \\ \psi(-1.0) \end{pmatrix} = \begin{pmatrix} 0.36788 \\ 0.60653 \\ 0.00000 \\ 1.64872 \\ 2.71828 \end{pmatrix} \tag{9.41}
$$

which is identical to the exact values of $\psi(x)$ at $x = \{1.0, 0.5, 0.0, -0.5, -1.0\}$ to five decimals. The largest percent difference from the exact values is 7.4×10^{-5}. □

Approximating ψ(y) by spline interpolation

Another approach is to approximate $\psi(y)$ by a constant in the Fredholm integral. To do so, we divide $[a, b]$ into small segments, writing the integral as

$$\int_a^b K(x,y)\psi(y)dy = \sum_{m=0}^{N-1} \int_{x_m}^{x_{m+1}} K(x,y)\psi(y)dy \qquad (9.42)$$

where

$$x_0 = a \qquad (9.43a)$$

and

$$x_N = b \qquad (9.43b)$$

We note that the points x_m are chosen by the user and are not necessarily abscissae of a quadrature rule.

Although it is not necessary to do so, for the sake of simplicity, we take these small intervals to be evenly spaced by defining

$$x_{m+1} - x_m \equiv \Delta x \qquad (9.44)$$

with Δx independent of m. By taking Δx to be small enough, we can approximate $\psi(y)$ over each segment $x_m \le y \le x_{m+1}$ by the constant

$$\psi(y) \simeq \alpha\psi(x_m) + \beta\psi(x_{m+1}) \qquad (9.45)$$

where α and β are chosen by the user. This is a cardinal spline interpolation over $[x_m, x_{m+1}]$. Unless there is some reason to choose otherwise, a reasonable choice would be to take

$$\alpha = \beta = \frac{1}{2} \qquad (9.46)$$

With these values for α and β, the Fredholm equation of the second kind becomes

$$\psi(x) \simeq \psi_0(x) + \frac{\lambda}{2}\sum_{m=0}^{N-1}[\psi(x_m) + \psi(x_{m+1})]\int_{x_m}^{x_{m+1}} K(x,y)dy$$

$$\equiv \psi_0(x) + \frac{\lambda}{2}\sum_{m=0}^{N-1}[\psi(x_m) + \psi(x_{m+1})]I_m(x) \qquad (9.47)$$

where

$$I_m(x) \equiv \int_{x_m}^{x_{m+1}} K(x,y)dy \qquad (9.48)$$

Then, with $0 \le k \le N-1$, we set $x = \{x_0, \ldots, x_{N-1}\}$ to obtain

$$\psi(x_k) \simeq \psi_0(x_k) + \frac{\lambda}{2} \sum_{m=0}^{N-1} [\psi(x_m) + \psi(x_{m+1})] I_m(x_k)$$

$$= \psi_0(x_k) + \frac{\lambda}{2} \{ I_0(x_k)\psi(x_0) + [I_0(x_k) + I_1(x_k)]\psi(x_1) +$$

$$\cdots + [I_{N-2}(x_k) + I_{N-1}(x_k)]\psi(x_{N-1}) + I_{N-1}(x_k)\psi(x_N) \} \qquad (9.49)$$

which yields a set of simultaneous linear equations for the set $\{\psi(x_k)\}$.

Example 9.4: Solution to a Fredholm equation of the second kind by a cardinal spline interpolation

The integral equation

$$\psi(x) = 1 + \int_0^1 xe^{-y(1-x)}\psi(y)dy \qquad (9.50a)$$

has solution

$$\psi(x) = e^x \qquad (9.50b)$$

For simplicity of illustration, we obtain the solution at $x = \{0, 1/3, 2/3, 1\}$ taking $\Delta x = 1/3$. Then, with $\lambda = \psi_0 = 1$ and with $x_0 = 0$, $x_N = x_3 = 1$, and

$$I_0(x) = \int_0^{1/3} xe^{-y(1-x)}dy = \begin{cases} \dfrac{x\left[1 - e^{-(1-x)/3}\right]}{(1-x)} & x \ne 1 \\ \dfrac{x}{3} & x = 1 \end{cases} \qquad (9.51a)$$

$$I_1(x) = \int_{1/3}^{2/3} xe^{-y(1-x)}dy = \begin{cases} \dfrac{x\left[e^{-(1-x)/3} - e^{-2(1-x)/3}\right]}{(1-x)} & x \ne 1 \\ \dfrac{x}{3} & x = 1 \end{cases} \qquad (9.51b)$$

and

$$I_2(x) = \int_{2/3}^1 xe^{-y(1-x)}dy = \begin{cases} \dfrac{x\left[e^{-2(1-x)/3} - e^{-(1-x)}\right]}{(1-x)} & x \ne 1 \\ \dfrac{x}{3} & x = 1 \end{cases} \qquad (9.51c)$$

eq. 9.49, for this example, is

$$\left[1 - \frac{1}{2}I_0(0)\right]\psi(0) - \frac{1}{2}[I_0(0) + I_1(0)]\psi\left(\frac{1}{3}\right) - \frac{1}{2}[I_1(0) + I_2(0)]\psi\left(\frac{2}{3}\right) - \frac{1}{2}I_2(0)\psi(1) = 1$$

(9.52a)

$$-\frac{1}{2}I_0\left(\frac{1}{3}\right)\psi(0) + \left[1 - \frac{1}{2}I_0\left(\frac{1}{3}\right) - \frac{1}{2}I_1\left(\frac{1}{3}\right)\right]\psi\left(\frac{1}{3}\right) - \frac{1}{2}\left[I_1\left(\frac{1}{3}\right) + I_2\left(\frac{1}{3}\right)\right]\psi\left(\frac{2}{3}\right) - \frac{1}{2}I_2\left(\frac{1}{3}\right)\psi(1) = 1$$

(9.52b)

$$-\frac{1}{2}I_0\left(\frac{2}{3}\right)\psi(0) + \frac{1}{2}\left[I_0\left(\frac{2}{3}\right) + I_1\left(\frac{2}{3}\right)\right]\psi\left(\frac{1}{3}\right) + \left[1 - \frac{1}{2}I_1\left(\frac{2}{3}\right) - \frac{1}{2}I_2\left(\frac{2}{3}\right)\right]\psi\left(\frac{2}{3}\right) - \frac{1}{2}I_2\left(\frac{2}{3}\right)\psi(1) = 1$$

(9.52c)

$$-\frac{1}{2}I_0(1)\psi(0) - \frac{1}{2}[I_0(1) + I_1(1)]\psi\left(\frac{1}{3}\right) - \frac{1}{2}[I_1(1) + I_2(1)]\psi\left(\frac{2}{3}\right) + \left[1 - \frac{1}{2}I_2(1)\right]\psi(1) = 1$$

(9.52d)

We obtain

$$\begin{pmatrix} \psi(0) \\ \psi\left(\frac{1}{3}\right) \\ \psi\left(\frac{2}{3}\right) \\ \psi(1) \end{pmatrix} = \begin{pmatrix} 1.00000 \\ 1.40497 \\ 1.96740 \\ 2.74895 \end{pmatrix}$$

(9.53a)

Comparing these results to

$$\begin{pmatrix} \psi_{exact}(0) \\ \psi_{exact}\left(\frac{1}{3}\right) \\ \psi_{exact}\left(\frac{2}{3}\right) \\ \psi_{exact}(1) \end{pmatrix} = \begin{pmatrix} 1.00000 \\ 1.39561 \\ 1.94773 \\ 2.71828 \end{pmatrix}$$

(9.53b)

we see that the method yields reasonably accurate results. □

We note that this approach is not applicable to Fredholm equations of the first kind. Writing the Fredholm integral as

$$\int_a^b K(x,y)\psi(y)dy = \sum_{k=0}^{N-1} \int_{x_k}^{x_{k+1}} K(x,y)\psi(y)dy$$

(9.42)

and approximating $\psi(y)$ as

$$\psi(y) \simeq \frac{1}{2}[\psi(x_k) + \psi(x_{k+1})] \quad x_k \leq y \leq x_{k+1}$$

(9.54)

the Fredholm integral of the first kind becomes

$$\psi_0(x) \simeq \frac{\lambda}{2} \sum_{m=0}^{N-1} [\psi(x_k) + \psi(x_{k+1})] I_k(x) \qquad (9.55a)$$

Setting x to each of the x_k, we obtain the set of equations

$$\psi_0(x_m) \simeq \frac{\lambda}{2} \sum_{m=0}^{N-1} [\psi(x_k) + \psi(x_{k+1})] I_k(x_m) \qquad (9.55b)$$

Expressing these equations in matrix form, we obtain the solution

$$\Psi = M^{-1}\Psi_0 \qquad (9.56)$$

where the matrix M is given by

$$M =$$

$$\frac{\lambda}{2} \begin{pmatrix} I_0(x_1) & [I_0(x_1)+I_1(x_1)] & \bullet\bullet & [I_{N-2}(x_1)+I_{N-1}(x_1)] & I_{N-1}(x_1) \\ \bullet & & & & \\ \bullet & & & & \\ I_0(x_k) & [I_0(x_k)+I_1(x_k)] & \bullet\bullet & [I_{N-2}(x_k)+I_{N-1}(x_k)] & I_{N-1}(x_k) \\ \bullet & & & & \\ \bullet & & & & \\ I_0(x_N) & [I_0(x_N)+I_1(x_N)] & \bullet\bullet & [I_{N-2}(x_N)+I_{N-1}(x_N)] & I_{N-1}(x_N) \end{pmatrix} \qquad (9.57)$$

Using the Gauss–Jordan elimination method for finding M^{-1} (see ch. 5), we can cast M into a form such that two columns are identical. Thus, M is singular. We demonstrate this by example.

Example 9.5: Matrix for a Fredholm equation of the first kind is singular

The 4×4 matrix M is given by

$$M = \frac{\lambda}{2} \begin{pmatrix} I_0(x_1) & [I_0(x_1)+I_1(x_1)] & [I_1(x_1)+I_2(x_1)] & I_2(x_1) \\ I_0(x_2) & [I_0(x_2)+I_1(x_2)] & [I_1(x_2)+I_2(x_2)] & I_2(x_2) \\ I_0(x_3) & [I_0(x_3)+I_1(x_3)] & [I_1(x_3)+I_2(x_3)] & I_2(x_3) \\ I_0(x_4) & [I_0(x_4)+I_1(x_4)] & [I_1(x_4)+I_2(x_4)] & I_2(x_4) \end{pmatrix} \qquad (9.58)$$

Performing the Gauss–Jordan operations on the columns of M,

$$col_2 \to col_2 - col_1 \qquad (9.59a)$$

and

$$col_3 \to col_3 - col_4 \qquad (9.59b)$$

we obtain

$$M \to \frac{\lambda}{2} \begin{pmatrix} I_0(x_1) & I_1(x_1) & I_1(x_1) & I_2(x_1) \\ I_0(x_2) & I_1(x_2) & I_1(x_2) & I_2(x_2) \\ I_0(x_3) & I_1(x_3) & I_1(x_3) & I_2(x_3) \\ I_0(x_4) & I_1(x_4) & I_1(x_4) & I_2(x_4) \end{pmatrix} \tag{9.60}$$

Since the second and third columns of this matrix are identical, M is singular. $\square$

Interpolation of the kernel

Another approach to solving the Fredholm equation of the second kind is to interpolate the kernel over the interval of integration.

We approximate

$$K(x,y) \simeq \sum_{m=1}^{N} K(x,y_m)v_m(y) \tag{9.61}$$

with

$$v_k(x) = \frac{[q(x)-q(x_1)]...[q(x)-q(x_{k-1})][q(x)-q(x_{k+1})]...[q(x)-q(x_N)]}{[q(x_k)-q(x_1)]...[q(x_k)-q(x_{k-1})][q(x_k)-q(x_{k+1})]...[q(x_k)-q(x_N)]} \tag{1.18}$$

where $q(x)$ is some function appropriate to the kernel being represented. Then, the Fredholm equation of the second kind becomes

$$\psi(x) \simeq \psi_0(x) + \lambda \sum_{m=1}^{N} K(x,y_m) \int_a^b v_m(y)\psi(y)dy \tag{9.62}$$

We define

$$\beta_m \equiv \int_a^b v_m(y)\psi(y)dy \tag{9.63}$$

so that eq. 9.62 is written as

$$\psi(x) = \psi_0(x) + \lambda \sum_{m=1}^{N} K(x,y_m)\beta_m \tag{9.64}$$

Substituting this into eq. 9.63, we have

$$\beta_k = \int_a^b v_k(y) \left[\psi_0(y) + \lambda \sum_{m=1}^N K(y, y_m) \beta_m \right] dy \qquad (9.65)$$

With

$$\alpha_k \equiv \int_a^b v_k(y) \psi_0(y) dy \qquad (9.66a)$$

and

$$\Gamma_{km} \equiv \int_a^b v_k(y) K(y, y_m) dy \qquad (9.66b)$$

eq. 9.65 can be written as the set of linear equations

$$\beta_k = \alpha_k + \lambda \sum_{m=1}^N \Gamma_{km} \beta_m \qquad (9.67)$$

The solution for the set $\{\beta_k\}$ is found by standard methods. Then, $\psi(x)$ at any x is given by eq. 9.64.

Example 9.6: Solution to a Fredholm equation of the second kind by interpolation of the kernel

We again consider

$$\psi(x) = e^x - \int_{-1}^1 x e^{y(1-x)} \psi(y) dy \qquad (9.35)$$

which has solution

$$\psi(x) = e^{-x} \qquad (9.8)$$

Since the kernel is an exponential function with a non-negative exponent for $x \; \varepsilon$ $[-1, 1]$, we choose

$$q(y) = e^y \qquad (9.68)$$

in the interpolating function $v_k(y)$ in eq. 1.18. We then interpolate the kernel over the nine points $\{-1, -0.75, \ldots, 0.75, 1\}$ to obtain

$$\alpha_k = \int_{-1}^1 e^y v_k(y) dy = \int_{-1}^1 e^y \frac{\ldots [e^y - e^{x_{k-1}}][e^y - e^{x_{k+1}}] \ldots}{\ldots [e^{x_k} - e^{x_{k-1}}][e^{x_k} - e^{x_{k+1}}] \ldots} dy \qquad (9.69a)$$

and

$$\Gamma_{km} = \int_{-1}^{1} ye^{x_m(1-y)}v_k(y)dy = \int_{-1}^{1} ye^{x_m(1-y)}\frac{\cdots[e^y - e^{x_{k-1}}][e^y - e^{x_{k+1}}]\cdots}{\cdots[e^{x_k} - e^{x_{k-1}}][e^{x_k} - e^{x_{k+1}}]\cdots}dy \quad (9.69b)$$

Although these integrals for α_k and Γ_{km} can be evaluated in closed form, to save ourselves effort, we evaluate them numerically using a 20-point Gauss–Legendre rule.

Solving for the set $\{\beta_k\}$ and substituting these values into eq. 9.64, we obtain

$$\begin{pmatrix} \psi(-1.00) \\ \psi(-0.75) \\ \psi(-0.25) \\ \psi(0.25) \\ \psi(0.75) \\ \psi(1.00) \end{pmatrix} = \begin{pmatrix} 2.71800 \\ 2.11678 \\ 1.28396 \\ 0.77881 \\ 0.47217 \\ 0.36809 \end{pmatrix} \quad (9.70a)$$

which is an accurate approximation to

$$\begin{pmatrix} \psi_{exact}(-1.00) \\ \psi_{exact}(-0.75) \\ \psi_{exact}(-0.25) \\ \psi_{exact}(0.25) \\ \psi_{exact}(0.75) \\ \psi_{exact}(1.00) \end{pmatrix} = \begin{pmatrix} 2.71828 \\ 2.11700 \\ 1.28402 \\ 0.77880 \\ 0.47237 \\ 0.36788 \end{pmatrix} \quad (9.70b)\square$$

Expansion in orthogonal polynomials

As discussed for Fredholm equations of the first kind, we can solve equations of the second kind by expanding $\psi(x)$, $\psi_0(x)$, and $K(x,y)$ in series involving orthogonal polynomials. If, for example, we transform the domain of these functions to $[-1, 1]$, then expand these functions in terms of Legendre polynomials, we have

$$\psi(x) = \sum_{k=0}^{\infty} c_k P_k(x) \qquad (9.18a)$$

$$\psi_0(x) = \sum_{k=0}^{\infty} \phi_k P_k(x) \qquad (9.18b)$$

and

$$K(x,y) = \sum_{\substack{k=0 \\ m=0}}^{\infty} \mu_{km} P_k(x) P_m(y) \qquad (9.18c)$$

Using the orthogonality condition for Legendre polynomials eq. 4.100b, the Fredholm equation of the second kind can be expressed as

$$\sum_{k=0}^{\infty} c_k P_k(x) = \sum_{k=0}^{\infty} \phi_k P_k(x) + \sum_{\substack{k=0 \\ m=0}}^{\infty} \mu_{km} c_m P_k(x) \frac{2}{(2m+1)} \qquad (9.71)$$

Thus, the set of linear equations for the coefficients set $\{c_k\}$ are

$$c_k = \phi_k + \sum_{m=0}^{\infty} \frac{2}{(2m+1)} \mu_{km} c_m \qquad (9.72a)$$

which we approximate by the finite sum

$$c_k = \phi_k + \sum_{m=0}^{N} \frac{2}{(2m+1)} \mu_{km} c_m \qquad (9.72b)$$

Solving for the set $\{c_k\}$, we obtain the approximate solution

$$\psi(x) \simeq \sum_{k=0}^{N} c_k P_k(x) \qquad (9.73)$$

Example 9.7: Solution to a Fredholm equation of the second kind by series expansion

We again consider

$$\psi(x) = e^x - \int_{-1}^{1} x e^{y(1-x)} \psi(y) dy \qquad (9.35)$$

which has solution

$$\psi(x) = e^{-x} \tag{9.8b}$$

With

$$\psi(x) \simeq \sum_{k=0}^{4} c_k P_k(x) \tag{9.74a}$$

$$e^x \simeq \sum_{k=0}^{4} \phi_k P_k(x) \tag{9.74b}$$

and

$$xe^{y(1-x)} \simeq \sum_{\substack{k=0 \\ m=0}}^{4} \mu_{km} P_k(x) P_m(y) \tag{9.74c}$$

we solve the matrix equation

$$\begin{pmatrix} c_0 \\ \bullet \\ \bullet \\ \bullet \\ c_4 \end{pmatrix} = \begin{pmatrix} e^{x_0} \\ \bullet \\ \bullet \\ \bullet \\ e^{x_4} \end{pmatrix} - \begin{pmatrix} 2\mu_{00} & \bullet\bullet & \frac{2}{9}\mu_{04} \\ & \bullet & \\ & \bullet & \\ 2\mu_{40} & \bullet\bullet & \frac{2}{9}\mu_{44} \end{pmatrix} \begin{pmatrix} c_0 \\ \bullet \\ \bullet \\ \bullet \\ c_4 \end{pmatrix} \tag{9.75}$$

for the coefficients c_k. Substituting these into eq. 9.74a, we can approximate $\psi(x)$ at any $x \; \varepsilon \; [-1,1]$. We find

$$\begin{pmatrix} \psi(1.00) \\ \psi(0.50) \\ \psi(0.00) \\ \psi(-0.50) \\ \psi(-1.00) \end{pmatrix} = \begin{pmatrix} 0.34486 \\ 0.61254 \\ 0.99804 \\ 1.64621 \\ 2.70190 \end{pmatrix} \tag{9.76a}$$

which compares reasonably well with

$$\begin{pmatrix} \psi_{exact}(1.00) \\ \psi_{exact}(0.50) \\ \psi_{exact}(0.00) \\ \psi_{exact}(-0.50) \\ \psi_{exact}(-1.00) \end{pmatrix} = \begin{pmatrix} 0.36788 \\ 0.60653 \\ 1.00000 \\ 1.64872 \\ 2.71828 \end{pmatrix} \tag{9.76b}$$

except at $x = 1.00$, where the largest error is 6.3%. □

Neumann series

The *Neumann series in λ* for the inhomogeneous Fredholm equation is obtained by replacing $\psi(y)$ in the integral of eq. 9.32 by

$$\psi_0(y) + \lambda \int_a^b K(y, z)\psi(z)dz$$

to obtain

$$\psi(x) = \psi_0(x) + \lambda \int_a^b K(x, y)\left[\psi_0(y) + \lambda \int_a^b K(y, z)\psi(z)dz\right]dy =$$

$$\psi_0(x) + \lambda \int_a^b K(x, y)\psi_0(y)dy + \lambda^2 \int_a^b \int_a^b K(x, y)K(y, z)\psi(z)dydz \quad (9.77)$$

Repeating this process ad infinitum, we obtain the Neumann series in λ,

$$\psi(x) =$$

$$\psi_0(x) + \lambda \int_a^b K(x, y)\psi_0(y)dy + \lambda^2 \int_a^b \int_a^b K(x, y)K(y, z)\psi_0(z)dydz$$

$$+ \lambda^3 \int_a^b \int_a^b \int_a^b K(x, y)K(y, z)K(z, w)\psi_0(w)dydzdw + \dots$$

$$\equiv \psi_0(x) + \sum_{m=1}^{\infty} \lambda^m I_m(x) \quad (9.78)$$

where

$$I_m(x) \equiv \underbrace{\int_a^b \int_a^b \bullet \bullet \int_a^b K(x, y)K(y, z) \bullet \bullet K(t, u)\psi_0(u)dydz \bullet \bullet dtdu}_{m \text{ integrals}} \quad (9.79)$$

If the value of λ is such that the infinite series of eq. 9.78 converges (see, for example, the *Cauchy ratio test* for convergence, Cohen, H., 1992, pp. 128–129), we can approximate the Neumann series by truncating it to obtain

$$\psi(x) \simeq \psi_0(x) + \sum_{m=1}^{N} \lambda^m I_m(x) \quad (9.80)$$

As discussed in Appendix 1, such a series in λ also allows us to create a Pade Approximant, a diagonal form of which, should be more accurate than the truncated Neumann series.

Example 9.8: Neumann series for an inhomogeneous Fredholm equation of the second kind

It is straightforward to show that the solution to

$$\psi(x) = e^{x/5} - \frac{1}{5}\int_{-1}^{1} xe^{y(1-x)/5}\psi(y)dy \tag{9.81}$$

is

$$\psi(x) = e^{-x/5} \tag{9.82}$$

Writing eq. 9.81 as

$$\psi(x) = e^{x/5} - \lambda\int_{-1}^{1} xe^{y(1-x)/5}\psi(y)dy \tag{9.83}$$

the three term Neumann sum is

$$\psi(x) \simeq e^{x/5} - \lambda\int_{-1}^{1} xe^{y(1-x)/5}e^{y/5}dy + \lambda^2\int_{-1}^{1}\int_{-1}^{1} xe^{y(1-x)/5}ye^{z(1-y)/5}e^{z/5}dydz \tag{9.84}$$

With

$$I_1(x) \equiv \int_{-1}^{1} xe^{y(1-x)/5}e^{y}dy = \frac{5x}{(2-x)}\left[e^{(2-x)/5} - e^{-(2-x)/5}\right] \tag{9.85a}$$

and

$$I_2(x) \equiv \int_{-1}^{1}\int_{-1}^{1} xe^{y(1-x)/5}ye^{z(1-y)/5}e^{z/5}dzdy$$

$$= \int_{-1}^{1}\frac{xy}{(2-y)}e^{y(1-x)/5}\left[e^{(2-y)/5} - e^{-(2-y)/5}\right]dy \tag{9.85b}$$

(which, at a given x, we approximate by a quadrature sum), we obtain an approximation to $\psi(x)$ at any $x \in [-1, 1]$. With $\lambda = 1/5$, we obtain

$$\psi(x) \simeq e^{x/5} - \frac{x}{(2-x)}\left[e^{(2-x)/5} - e^{-(2-x)/5}\right]$$

$$+ \frac{1}{25}\int_{-1}^{1}\frac{xy}{(2-y)}e^{y(1-x)/5}\left[e^{(2-y)/5} - e^{-(2-y)/5}\right]dy \tag{9.86}$$

To approximate the integrals $I_m(x)$ of eq. 9.79, we use a 20-point Gauss–Legendre quadrature rule. Setting $x = \{-0.75, -0.25, 0.25, 0.75\}$, we find

$$\begin{pmatrix} \psi(-0.75) \\ \psi(-0.25) \\ \psi(0.25) \\ \psi(0.75) \end{pmatrix} = \begin{pmatrix} 1.13651 \\ 1.04559 \\ 0.95420 \\ 0.86155 \end{pmatrix} \tag{9.87a}$$

which is a reasonable approximation to

$$\begin{pmatrix} \psi_{exact}(-0.75) \\ \psi_{exact}(-0.25) \\ \psi_{exact}(0.25) \\ \psi_{exact}(0.75) \end{pmatrix} = \begin{pmatrix} 1.16183 \\ 1.05127 \\ 0.95123 \\ 0.86071 \end{pmatrix} \tag{9.87b}$$

The [2, 2] Pade Approximant is found by requiring that

$$\psi^{[2,2]}(x) = \frac{p_0(x) + \lambda p_1(x)}{1 + \lambda q_1(x)} \tag{9.88}$$

be identical to

$$\psi_2(x) = \psi_0(x) + \lambda I_1(x) + \lambda^2 I_2(x) \tag{9.89}$$

This results in

$$\psi^{[2,2]}(x) = \frac{\psi_0(x) + \lambda \left[I_1(x) - \psi_0(x) \frac{I_2(x)}{I_1(x)} \right]}{1 - \lambda \frac{I_2(x)}{I_1(x)}} \tag{9.90}$$

From this, we obtain

$$\begin{pmatrix} \psi(-0.75) \\ \psi(-0.25) \\ \psi(0.25) \\ \psi(0.75) \end{pmatrix} = \begin{pmatrix} 1.14092 \\ 1.04632 \\ 0.95397 \\ 0.86152 \end{pmatrix} \tag{9.91}$$

which, as expected, is slightly more accurate than the Neumann series. $\square$

9.3   Eigensolutions of Fredholm Equations of the Second Kind with Non-singular Kernel

When $\psi_0(x) = 0$ for all $x \; \varepsilon \; [a, b]$, the Fredholm equation of the second kind becomes the *homogeneous Fredholm equation*

$$\psi(x) = \lambda \int_a^b K(x, y)\psi(y)dy \qquad (9.92)$$

which only has solutions for specific values of λ. The *eigenvalue of the kernel* is $1/\lambda$ and $\psi_\lambda(x)$, the solution to eq. 9.92 for that specific value of λ, is the corresponding *eigenfunction of the kernel*.

Solution by quadratures

Using a N-point quadrature rule to approximate the integral, eq. 9.92 becomes

$$\psi(x) \simeq \lambda \sum_{m=1}^{N} w_m K(x, y_m)\psi(y_m) \qquad (9.93)$$

Setting x in this expression to each quadrature point in the set $\{y_k\}$, eq. 9.93 can be written in matrix form as

$$\left[1 - \lambda \begin{pmatrix} w_1 K_{11} & \bullet\bullet & w_N K_{1N} \\ \bullet & & \bullet \\ \bullet & & \bullet \\ w_1 K_{N1} & \bullet\bullet & w_N K_{NN} \end{pmatrix} \right] \begin{pmatrix} \psi(y_1) \\ \bullet \\ \bullet \\ \psi(y_N) \end{pmatrix} = 0 \qquad (9.94)$$

The eigenvalues of the kernel are obtained from

$$\begin{vmatrix} (1 - \lambda w_1 K_{11}) & \bullet\bullet & -w_N K_{1N} \\ \bullet & & \bullet \\ \bullet & & \bullet \\ -w_1 K_{N1} & \bullet\bullet & (1 - \lambda w_N K_{NN}) \end{vmatrix} = 0 \qquad (9.95)$$

An approximation to the corresponding eigenfunctions are obtained by solving $(N - 1)$ of the equations of eq. 9.94 for $(N - 1)$ values $\psi(y_k)$ in terms of one of the ψ quantities. For example, we can solve for each $\psi(y_k)$ for $k \geq 2$ in terms of $\psi(y_1)$. Thus, the value of each ratio $\psi(y_k)/\psi(y_1)$ is known and eq. 9.93 can be expressed as

$$\psi_\lambda(x) =$$
$$\lambda[w_1 K(x,y_1)\psi_\lambda(y_1) + w_2 K(x,y_2)\psi_\lambda(y_2) + \dots + w_N K(x,y_N)\psi_\lambda(y_N)]$$
$$= \lambda \psi_\lambda(y_1)\left[w_1 K(x,y_1) + \frac{\psi_\lambda(y_2)}{\psi_\lambda(y_1)} w_2 K(x,y_2) + \dots + \frac{\psi_\lambda(y_N)}{\psi_\lambda(y_1)} w_N K(x,y_N)\right] \quad (9.96)$$

The undetermined coefficient $\psi_\lambda(y_1)$ is obtained from a normalization condition defined by the user.

Example 9.9: Eigensolution by quadratures

Let us consider

$$\psi(x) = \lambda \int_0^1 e^{xy}\psi(y)dy \quad (9.97)$$

We transform the Gauss–Legendre data to points over the interval [0, 1] (as described in ch. 4). These are

$$y_k^{[0,1]} = \frac{1 + y_k^{[-1,1]}}{2} \quad (9.98a)$$

and

$$w_k^{[0,1]} = \frac{w_k^{[-1,1]}}{2} \quad (9.98b)$$

Approximating the integral in eq. 9.97 by a three-point quadrature rule we obtain

$$\psi(x) \simeq \lambda \sum_{m=1}^{3} w_k e^{xy_k}\psi(y_k) \quad (9.99)$$

Setting x to each of the points $\{y_1, y_2, y_3\}$, we have

$$\begin{pmatrix} 1 - \lambda w_1 e^{y_1 y_1} & -\lambda w_2 e^{y_1 y_2} & -\lambda w_3 e^{y_1 y_3} \\ -\lambda w_1 e^{y_2 y_1} & 1 - \lambda w_2 e^{y_2 y_2} & -\lambda w_3 e^{y_2 y_3} \\ -\lambda w_1 e^{y_3 y_1} & -\lambda w_2 e^{y_3 y_2} & 1 - \lambda w_3 e^{y_3 y_3} \end{pmatrix} \begin{pmatrix} \psi_1 \\ \psi_2 \\ \psi_3 \end{pmatrix} = 0 \quad (9.100a)$$

The values of λ, found from

$$\begin{vmatrix} 1 - \lambda w_1 e^{y_1 y_1} & -\lambda w_2 e^{y_1 y_2} & -\lambda w_3 e^{y_1 y_3} \\ -\lambda w_1 e^{y_2 y_1} & 1 - \lambda w_2 e^{y_2 y_2} & -\lambda w_3 e^{y_2 y_3} \\ -\lambda w_1 e^{y_3 y_1} & -\lambda w_2 e^{y_3 y_2} & 1 - \lambda w_3 e^{y_3 y_3} \end{vmatrix} = 0 \quad (9.100b)$$

are given by

$$\{\lambda_1, \lambda_2, \lambda_3\} = \{0.73908, 9.43870, 290.88072\} \tag{9.101}$$

Referring to eq. 9.96, the corresponding eigenfunctions are

$$\psi_{\lambda_1}(x) = \psi_1^{\lambda_1}\left[0.20530e^{0.88730x} + 0.25994e^{0.50000x} + 0.13009e^{0.11270x}\right] \tag{9.102a}$$

$$\psi_{\lambda_2}(x) = \psi_1^{\lambda_2}\left[2.62186e^{0.88730x} - 1.05829e^{0.50000x} - 2.81603e^{0.11270x}\right] \tag{9.102b}$$

$$\psi_{\lambda_3}(x) = \psi_1^{\lambda_3}\left[80.80020e^{0.88730x} - 199.95011e^{0.50000x} + 122.19328e^{0.11270x}\right] \tag{9.102c}$$

As noted in example 3.4 of Baker, C. T. H., 1977, p. 177, it has been shown that λ_1, the smallest value of λ for this kernel, is bounded by

$$0.73888 < \lambda < 0.73926 \tag{9.103}$$

Our result is consistent with this. $\square$

Eigensolution of a degenerate kernel

A kernel that can be written as

$$K(x, y) = \sum_{k=1}^{N} A_k(x)B_k(y) \tag{9.104}$$

is called a *degenerate kernel*. The eigensolutions for such a kernel can be determined exactly.

With eq. 9.104, the homogeneous Fredholm equation becomes

$$\psi(x) = \lambda \sum_{k=1}^{N} A_k(x) \int_a^b B_k(y)\psi(y)dy \equiv \lambda \sum_{k=1}^{N} A_k(x)\beta_k \tag{9.105}$$

Substituting this expression for $\psi(y)$ into

$$\beta_k \equiv \int_a^b B_k(y)\psi(y)dy \tag{9.106}$$

eq. 9.105 becomes

$$\beta_k = \lambda \sum_{m=1}^{N} \beta_m \int_a^b B_k(y) A_m(y) dy \equiv \lambda \sum_{m=1}^{N} \mu_{km} \beta_m \qquad (9.107)$$

This set of equations only has solution for specific values λ (the inverses of which are the eigenvalues of the kernel). Referring to eq. 9.105, the elements of the set $\{\beta_k{}^\lambda\}$ yield the corresponding eigenfunction given by

$$\psi_\lambda(x) = \lambda \sum_{k=1}^{N} \beta_k^\lambda A_k(x) \qquad (9.108)$$

Example 9.10: Eigensolution to a Fredholm equation of the second kind with a degenerate kernel

Writing

$$\psi(x) = \lambda \int_0^1 (x+y)\psi(y) dy = \lambda \left[x \int_0^1 \psi(y) dy + \int_0^1 y\psi(y) dy \right] \equiv \lambda[\beta_1 x + \beta_2]$$

$$(9.109)$$

we define

$$\beta_1 \equiv \int_0^1 \psi(y) dy \qquad (9.110a)$$

and

$$\beta_2 \equiv \int_0^1 y\psi(y) dy \qquad (9.110b)$$

Substituting $\psi(y)$ given in eq. 9.109 into eqs. 9.110, we obtain

$$\beta_1 = \lambda \int_0^1 [\beta_1 y + \beta_2] dy = \lambda \left[\frac{\beta_1}{2} + \beta_2 \right] \qquad (9.111a)$$

and

$$\beta_2 = \lambda \int_0^1 y[\beta_1 y + \beta_2] dy = \lambda \left[\frac{\beta_1}{3} + \frac{\beta_2}{2} \right] \qquad (9.111b)$$

from which

$$\lambda_\pm = 2\left(1 \pm \frac{1}{\sqrt{3}}\right) \tag{9.112}$$

With

$$\beta_2^\pm = \left(\frac{1}{\lambda_\pm} - \frac{1}{2}\right)\beta_1^\pm \tag{9.113}$$

the eigenfunctions are given by eq. 9.109 to be

$$\psi_\pm = \beta_1^\pm \lambda_\pm \left[x + \frac{1}{\lambda_\pm} - \frac{1}{2}\right] \tag{9.114}\square$$

Approximate eigensolutions by interpolating the kernel

When a nondegenerate kernel has an infinite number of eigensolutions to eq. 9.92, any finite approximation method will yield a finite subset of such solutions.

Referring to the discussion given in example 9.6, we interpolate the kernel over a set of points $\{y_k\}$ selected by the user. We write

$$K(x,y) \simeq \sum_{k=1}^N K(x, y_k) v_k(y) \tag{9.61}$$

where

$$v_k(y) = \frac{[q(y) - q(y_1)]...[q(y) - q(y_{k-1})][q(y) - q(y_{k+1})]...[q(y) - q(y_N)]}{[q(y_k) - q(y_1)]...[q(y_k) - q(y_{k-1})][q(y_k) - q(y_{k+1})]...[q(y_k) - q(y_N)]} \tag{1.18}$$

In this way, the kernel is approximated by a degenerate kernel, and the homogeneous Fredholm equation becomes

$$\psi(x) = \lambda \sum_{m=1}^N K(x, x_m) \int_a^b v_m(y)\psi(y)dy \tag{9.115}$$

As before, we define

$$\beta_k \equiv \int_a^b v_k(y)\psi(y)dy \tag{9.63}$$

with which eq. 9.115 can be written as

$$\psi(x) = \lambda \sum_{k=1}^{N} K(x, x_k)\beta_k \tag{9.116}$$

This is an approximation to the eigenfunction corresponding to λ.

Using the expression of eq. 9.116 for $\psi(y)$, eq. 9.63 becomes

$$\beta_k = \lambda \sum_{m=1}^{N} \left[\int_a^b v_k(y)K(y, x_m)dy \right] \beta_m \tag{9.117}$$

With

$$\Gamma_{km} \equiv \int_a^b v_k(y)K(y, x_m)dy \tag{9.66b}$$

eq. 9.117 is the eigenvalue equation

$$\beta_k = \lambda \sum_{m=1}^{N} \Gamma_{km}\beta_m \tag{9.118}$$

The eigenvalues and the values of the β_k are determined from eq. 9.118 by standard methods (see, for example, matrix methods of Jacobi, Givens and House-holder introduced in ch. 5). The corresponding eigenfunctions are then given by eq. 9.116.

The accuracy of the interpolation of eq. 9.61, which can be determined independently of the method used to find the eigenpairs, will be a measure of the accuracy of the results.

Example 9.11: Eigensolution by interpolation of the kernel

We again consider

$$\psi(x) = \lambda \int_0^1 e^{xy}\psi(y)dy \tag{9.97}$$

Since the kernel is an exponential function of y with a non-negative exponent for $x \, \varepsilon \, [0, 1]$, we choose

$$q(y) = e^y \tag{9.68}$$

as the interpolation function in $v_k(y)$. For the purpose of illustration, we interpolate the kernel over the three points $\{y_1, y_2, y_3\} = \{0, .5, 1\}$. As such, we will determine three eigensolutions of the kernel.

The level of accuracy of the interpolation of the kernel is represented in Table 9.1.

y	Interpolated	Exact
0.1	1.03884	1.04081
0.2	1.08043	1.08329
0.3	1.12479	1.12750
0.4	1.17183	1.17351
0.5	1.22140	1.22140
0.6	1.27324	1.27125
0.7	1.32693	1.32313
0.8	1.38187	1.37713
0.9	1.43722	1.43333

Table 9.1 Interpolated values of e^{xy} at $x = 0.4$

In generating this table for $x = 0.4$, the largest percent difference between the interpolated and exact values of the kernel is 0.34%. Thus, the accuracy of the eigensolutions is expected to be at this level.

With k, r, and s taking on the values 1, 2, and 3, and with $r \neq s \neq k$, we have

$$
\Gamma_{km} = \int_0^1 e^{yx_m} v_k(y)\,dy = \int_0^1 e^{yx_m} \frac{[e^y - e^{x_r}][e^y - e^{x_s}]}{[e^{x_k} - e^{x_r}][e^{x_k} - e^{x_s}]}\,dy
$$

$$
= \frac{\left[\frac{(e^{(2+x_m)}-1)}{(2+x_m)} - (e^{x_r} + e^{x_s})\frac{(e^{(1+x_m)}-1)}{(1+x_m)} + e^{x_r + x_s}\frac{(e^{x_m}-1)}{x_m} \right]}{[e^{x_k} - e^{x_r}][e^{x_k} - e^{x_s}]}
\tag{9.119}
$$

where

$$
\frac{(e^{x_m} - 1)}{x_m} = 1
\tag{9.120}
$$

for $x_m = 0$.

From

$$
|1 - \lambda\Gamma| = \left| 1 - \lambda \begin{pmatrix} 0.15473 & 0.13579 & 0.10060 \\ 0.68639 & 0.90912 & 1.21862 \\ 0.15888 & 0.25253 & 0.39906 \end{pmatrix} \right| = 0
\tag{9.121}
$$

we obtain

$$
(\lambda_1, \lambda_2, \lambda_3) = (0.73874, 9.46937, 273.84781)
\tag{9.122}
$$

Referring to eq. 9.116, the corresponding eigenfunctions are

$$
\psi_1(x) = \beta_1 \lambda_1 \left[K(x,0) + \frac{\beta_2}{\beta_1} K(x, 0.5) + \frac{\beta_3}{\beta_1} K(x,1) \right]
$$
$$
= \beta_1 \left[0.73874 + 5.37741 e^{0.5x} + 1.54551 e^x \right]
\tag{9.123a}
$$

$$\psi_2(x) = \beta_1 \lambda_2 \left[K(x,0) + \frac{\beta_2}{\beta_1} K(x,0.5) + \frac{\beta_3}{\beta_1} K(x,1) \right]$$
$$= \beta_1 \left[9.46937 + 1.02738 e^{0.5x} - 6.01102 e^x \right] \qquad (9.123b)$$

and

$$\psi_3(x) = \beta_1 \lambda_3 \left[K(x,0) + \frac{\beta_2}{\beta_1} K(x,0.5) + \frac{\beta_3}{\beta_1} K(x,1) \right]$$
$$= \beta_1 \left[273.84781 - 423.54533 e^{0.5x} + 160.46100 e^x \right] \qquad (9.123c)$$

where β_1 is determined by a user-defined normalization condition.

Again, we see that λ_1, the smallest value of λ, is consistent with

$$0.73888 < \lambda < 0.73926 \qquad \qquad \textbf{\textit{(9.103)}}$$

as noted in example 3.4 of Baker , C. T. H., 1977, p. 177. □

9.4   Volterra Equations with Non-singular Kernel

The general form of a linear Volterra integral equation of the first kind is

$$\phi_0(x) = \lambda \int_a^x L(x,y)\psi(y)dy \qquad (9.124)$$

and the Volterra equation of the second kind is

$$\psi(x) = \psi_0(x) + \lambda \int_a^x K(x,y)\psi(y)dy \qquad (9.125)$$

where $K(x, y)$ is analytic at all x and at all $y \leq x$.

Since the integral is zero when $x = a$, we note from eq. 9.124 that in order for the equation of the first kind to have a solution, $\phi_0(x)$ must satisfy

$$\phi_0(a) = 0 \qquad (9.126a)$$

The solution to the Volterra equation of the second kind satisfies

$$\psi(a) = \psi_0(a) \qquad (9.126b)$$

Converting the Volterra equation of the first kind
into the Volterra equation of the second kind

Differentiating the Volterra of the first kind, we have

$$\phi_0'(x) = \lambda \int_a^x \frac{\partial L(x,y)}{\partial x} \psi(y) dy + L(x,x)\psi(x) \tag{9.127a}$$

If $L(x,x) = 0$ for every x, this becomes a Volterra equation of the first kind. If $L(x, x) \neq 0$ for all x, this can be written as

$$\psi(x) = \frac{\phi_0'(x)}{L(x,x)} - \lambda \int_a^x \frac{1}{L(x,x)} \frac{\partial L(x,y)}{\partial x} \psi(y) dy \tag{9.127b}$$

Defining

$$\frac{\phi_0'(x)}{L(x,x)} \equiv \psi_0(x) \tag{9.128a}$$

and

$$-\frac{1}{L(x,x)} \frac{\partial L(x,y)}{\partial x} = K(x,y) \tag{9.128b}$$

eq. 9.127b becomes

$$\psi(x) = \psi_0(x) + \lambda \int_a^x K(x,y)\psi(y) dy \tag{9.125}$$

which is a Volterra equation of the second kind.

   Thus, we can solve a Volterra equation of the first kind by solving the equivalent Volterra equation of the second kind.

Example 9.12: Converting a Volterra equation of the first kind to a Volterra equation of the second kind

The solution to

$$xe^x = \int_0^x e^{(x-y)} \psi(y) dy \tag{9.129}$$

is

$$\psi(x) = e^x \tag{9.50b}$$

With

$$\phi_0'(x) = (1+x)e^x \qquad (9.130)$$

and

$$L(x,x) = e^0 = 1 \qquad (9.131)$$

eq. 9.129 becomes

$$\psi(x) = (1+x)e^x - \int_0^x e^{(x-y)}\psi(y)dy \qquad (9.132)$$

It is a trivial exercise to show that

$$\psi(x) = e^x \qquad (\textbf{9.50b})$$

is the solution to eq. 9.132. □

Taylor series approximation for the Volterra equation of the second kind

The Taylor series for $\psi(x)$ expanded around a is

$$\psi(x) = \psi(a) + \psi'(a)(x-a) + \frac{1}{2!}\psi''(a)(x-a)^2 + \frac{1}{3!}\psi'''(a)(x-a)^3 + \dots \qquad (9.133)$$

The first term in the series is

$$\psi(a) = \psi_0(a) \qquad (\textbf{9.126b})$$

and from the derivative of $\psi(x)$ given in eq. 9.125, we have

$$\psi'(x) = \psi_0'(x) + \lambda z \int_a^x \frac{\partial K(x,y)}{\partial x}\psi(y)dy + \lambda K(x,x)\psi(x) \qquad (9.134a)$$

Thus

$$\psi'(a) = \psi_0'(a) + \lambda K(a,a)\psi(a) = \psi_0'(a) + \lambda K(a,a)\psi_0'(a) \qquad (9.134b)$$

The derivative of $\psi'(x)$ given in eq. 9.134a yields

$$\psi''(x) = \psi_0''(x) + \lambda \int_a^x \frac{\partial^2 K(x,y)}{\partial x^2} \psi(y)dy +$$

$$\lambda \frac{\partial K(x,y)}{\partial x}\bigg|_{y=x} \psi(x) + \lambda \frac{\partial K(x,y)}{\partial y}\bigg|_{y=x} \psi(x) + \lambda K(x,x)\psi'(x) \qquad (9.135a)$$

from which

$$\psi''(a) = \psi_0''(a) + \lambda \frac{\partial K(x,y)}{\partial x}\bigg|_{\substack{y=a \\ x=a}} \psi_0(a) + \lambda \frac{\partial K(x,y)}{\partial y}\bigg|_{\substack{y=a \\ x=a}} \psi_0(a)$$

$$+ \lambda K(a,a)\psi'(a)$$

where $\psi'(a)$ is given in eq. 9.134b.

Example 9.13: Solution to a Volterra equation by Taylor series

The solution to

$$\psi(x) = 1 + x + \int_0^x e^{(x-y)}\psi(y)dy \qquad (9.136)$$

is

$$\psi(x) = \frac{1}{4} + \frac{x}{2} + \frac{3}{4}e^{2x} \qquad (9.137)$$

With

$$\psi_0(0) = 1 \qquad (9.138a)$$

and

$$K(0,0) = 1 \qquad (9.138b)$$

eq. 9.134b yields

$$\psi'(0) = 2 \qquad (9.139a)$$

and from eq. 9.135b

$$\psi''(a) = 3 \qquad (9.139b)$$

Therefore,

$$\psi(x) \simeq \psi_0(0) + \psi'(0)x + \frac{1}{2!}\psi''(0)x^2 = 1 + 2x + \frac{3}{2}x^2 \qquad (9.140)$$

We compare this to the MacLaurin expansion of the solution given in eq. 9.137,

$$\psi_{exact}(x) = \frac{1}{4} + \frac{1}{2}x + \frac{3}{4}\left(1 + 2x + \frac{4x^2}{2!} + \ldots\right) = 1 + 2x + \frac{3}{2}x^2 + \ldots \qquad (9.141)$$

Thus, the Taylor series approximation is identical to the series representation of the known solution up to x^2. □

Approximating $\psi(y)$ by a spline interpolation

As we did for the Fredholm equation, we develop an approach in which we approximate $\psi(y)$ by a constant in the Volterra integral. We begin by denoting the limits of the integral $[a, x]$ as $[x_0, x_N]$. To approximate the solution to the Volterra equation at x_N, we divide $[x_0, x_N]$ into small segments, writing the integral in the Volterra equation as

$$\int_{x_0}^{x_N} K(x_N, y)\psi(y)dy = \sum_{m=0}^{N-1} \int_{x_m}^{x_{m+1}} K(x_N, y)\psi(y)dy \qquad (9.142)$$

where, as before, for convenience, we take these intervals to be evenly spaced of width Δx.

By taking Δx to be small enough, we can approximate $\psi(y)$ over each segment $x_k \leq y \leq x_{k+1}$ by the constant

$$\psi(y) \simeq \alpha\psi(x_m) + \beta\psi(x_{m+1}) \qquad (\textbf{9.45})$$

where α and β, chosen by the user, are here taken to be 1/2 as note in eq. 9.46. Then eq. 9.142 becomes

$$\int_{x_0}^{x_N} K(x_N, y)\psi(y)dy \simeq \frac{1}{2}\sum_{m=0}^{N-1} [\psi(x_m) + \psi(x_{m+1})] \int_{x_m}^{x_{m+1}} K(x_N, y)dy \qquad (9.143)$$

Defining

$$I_m(x_N) \equiv \int_{x_m}^{x_{m+1}} K(x_N, y)dy \qquad (9.144)$$

the Volterra equation of the second kind becomes

$$\psi(x_N) \simeq \psi_0(x_N) + \frac{\lambda}{2}\sum_{m=0}^{N-1} [\psi(x_m) + \psi(x_{m+1})]I_m(x_N) \qquad (9.145)$$

Referring to eq. 9.126b,

$$\psi(x_0) = \psi(a) = \psi_0(a) \tag{9.146}$$

Then, eq. 9.145 becomes

$$\left(1 - \frac{\lambda}{2}I_{N-1}(x_N)\right)\psi(x_N) \simeq \psi_0(x_N) + \frac{\lambda}{2}\psi_0(a)I_0(x_N) +$$

$$\frac{\lambda}{2}\{\psi(x_1)I_0(x_N) + [\psi(x_1) + \psi(x_2)]I_1(x_N) + \dots$$

$$+ [\psi(x_{N-2}) + \psi(x_{N-1})]I_{N-2}(x_N) + \psi(x_{N-1})I_{N-1}(x_N)$$

To determine the set $\{\psi(x_k)\}$, we begin with $x = x_1$ in the Volterra equation to obtain

$$\psi(x_1) = \psi_0(x_1) + \lambda \int_{x_0}^{x_1} K(x_1, y)\psi(y)dy \simeq \psi_0(x_1) + \frac{\lambda}{2}[\psi_0(x_0) + \psi(x_1)]I_0(x_1)$$

$$\tag{9.148}$$

from which

$$\left(1 - \frac{\lambda}{2}I_0(x_1)\right)\psi(x_1) \simeq \psi_0(x_1)\left(1 + \frac{\lambda}{2}I_0(x_1)\right) \tag{9.149}$$

Then, with $\psi(x_1)$ given by eq. 9.149, we set $x = x_2$ in the Volterra equation to obtain

$$\psi(x_2) = \psi_0(x_2) + \lambda \int_{x_0}^{x_1} K(x_2, y)\psi(y)dy + \lambda \int_{x_1}^{x_2} K(x_2, y)\psi(y)dy$$

$$\simeq \psi_0(x_2) + \frac{\lambda}{2}[\psi(x_0) + \psi(x_1)]I_0(x_2) + \frac{\lambda}{2}[\psi(x_1) + \psi(x_2)]I_1(x_2) \tag{9.150}$$

Then, $\psi(x_2)$ is found from

$$\left(1 - \frac{\lambda}{2}I_1(x_2)\right)\psi(x_2) \simeq \psi_0(x_2) + \frac{\lambda}{2}[\psi(x_0) + \psi(x_1)]I_0(x_2) + \frac{\lambda}{2}I_1(x_2)\psi(x_1) \tag{9.151}$$

and so on.

Example 9.14: Solution to a Volterra equation of the second kind

As noted earlier,

$$\psi(x) = 1 + x + \int_0^x e^{(x-y)}\psi(y)dy \tag{9.136}$$

has solution

$$\psi(x) = \frac{1}{4} + \frac{x}{2} + \frac{3}{4}e^{2x} \qquad (9.137)$$

For simplicity of illustration, we obtain the solution at $x = 0.9$ taking $\Delta x = 0.3$. Then, with $x_0 = 0$ and $x_N = x_3 = 0.9$, and

$$\psi(x_0) = \psi_0(0) = 1 \qquad (9.152)$$

we obtain $\psi(x_3)$ from eq. 9.147 in terms of $\psi(x_1)$ and $\psi(x_2)$. These quantities are determined from eqs. 9.149 and 9.151.

Approximating $\psi(y)$ by

$$\psi(y) \simeq \frac{1}{2}[\psi(x_m) + \psi(x_{m+1})] \quad x_m \le y \le x_{m+1} \qquad (9.153)$$

eq. 9.145 becomes

$$\psi(x_3) \simeq \psi_0(x_3) + \frac{1}{2}[\psi(x_0) + \psi(x_1)]I_0(x_3)$$
$$+ \frac{1}{2}[\psi(x_1) + \psi(x_2)]I_1(x_3) + \frac{1}{2}[\psi(x_2) + \psi(x_3)]I_2(x_3) \qquad (9.154a)$$

from which

$$\left[1 - \frac{1}{2}I_2(x_3)\right]\psi(x_3) = \psi_0(x_3) +$$
$$\frac{1}{2}\psi_0(x_0)I_0(x_3) + \frac{1}{2}[I_0(x_3) + I_1(x_3)]\psi(x_1) + \frac{1}{2}[I_1(x_3) + I_2(x_3)]\psi(x_2) \qquad (9.154b)$$

From eq. 9.149, $\psi(x_1)$ is given by

$$\left[1 - \frac{1}{2}I_0(x_1)\right]\psi(x_1) \simeq \psi_0(x_1) + \frac{1}{2}\psi_0(x_0)I_0(x_1) \qquad (9.155)$$

and with this value of $\psi(x_1)$, eq. 9.151 yields

$$\left[1 - \frac{1}{2}I_1(x_2)\right]\psi(x_2) \simeq \psi_0(x_2) + \frac{1}{2}\psi_0(x_0)I_0(x_2) + \frac{1}{2}[I_0(x_2) + I_1(x_2)]\psi(x_1)$$

$$(9.156)$$

With

$$I_k(x_m) = \int_{x_k}^{x_{K+1}} K(x_m, y)dy = \int_{x_k}^{x_{k+1}} e^{(x_m - y)}dy = e^{x_m}(e^{-x_k} - e^{-x_{k+1}}) \qquad (9.157)$$

we obtain

$$\begin{pmatrix} \psi(0.3) \\ \psi(0.6) \\ \psi(0.9) \end{pmatrix} = \begin{pmatrix} 1.78764 \\ 3.11604 \\ 5.44382 \end{pmatrix} \qquad (9.158a)$$

Comparing these to

$$\begin{pmatrix} \psi_{exact}(0.3) \\ \psi_{exact}(0.6) \\ \psi_{exact}(0.9) \end{pmatrix} = \begin{pmatrix} 1.76659 \\ 3.04009 \\ 5.23724 \end{pmatrix} \qquad (9.158b)$$

we see that this method yields a reasonably accurate approximate solution. $\square$

9.5 Fredholm Equations with Weakly Singular Kernel

A weakly singular kernel satisfies

$$\lim_{y \to x} K(x, y) = \infty \qquad (9.159a)$$

and

$$\lim_{y \to x} (x - y)K(x, y) = 0 \qquad (9.159b)$$

Equations of this type, which arises in applied problems, are of the form

$$\psi(x) = \psi_0(x) + \lambda \int_a^b \frac{L(x, y)}{|x - y|^p} \psi(y)dy \qquad 0 < p < 1 \qquad (9.160a)$$

and

$$\psi(x) = \psi_0(x) + \lambda \int_a^b L(x, y)\ell n|x - y|\psi(y)dy \qquad (9.160b)$$

where $L(x, y)$ is analytic at all x and y in $[a, b]$.

An example of such an equation is the Kirkwood–Riseman formula, from which one determines the viscosity and the translational diffusion constants of macromolecules:

$$\psi(x) = \psi_0(x) - \lambda \int_{-1}^{1} \frac{1}{\sqrt{|x-y|}} \psi(y)dy \qquad (9.161)$$

(see Kirkwood, J. G., and Riseman, J., 1948).

Lagrange interpolation methods

To approximate the solution to either of eq. 9.160 by Lagrange-like interpolation, we write

$$\psi(y) \simeq \sum_{m=1}^{M} \psi(y_m) v_m(y) \qquad (9.162)$$

where

$$v_m(y) = \frac{(q(y)-q(y_1))...(q(y)-q(y_{m-1}))(q(y)-q(y_{m+1}))...(q(y)-q(y_M))}{(q(y_m)-q(y_1))...(q(y_m)-q(y_{m-1}))(q(y_m)-q(y_{m+1}))...(q(y_m)-q(y_M))} \qquad (1.18)$$

With this approximation of $\psi(y)$, eqs. 9.160 become

$$\psi(x) = \psi_0(x) + \lambda \sum_{m=1}^{M} \psi(x_m) \int_{a}^{b} \frac{L(x,y)}{|x-y|^p} v_m(y)dy \quad 0<p<1 \qquad (9.163a)$$

and

$$\psi(x) = \psi_0(x) + \lambda \sum_{m=1}^{M} \psi(x_m) \int_{a}^{b} L(x,y) \ell n|x-y| v_m(y)dy \qquad (9.163b)$$

Setting x to each y_k, we obtain

$$\psi(y_k) = \psi_0(y_k) + \lambda \sum_{m=1}^{M} \psi(y_m) \int_{a}^{b} \frac{L(y_k,y)}{|y_k-y|^p} v_m(y)dy \qquad (9.163c)$$

and

$$\psi(y_k) = \psi_0(y_k) + \lambda \sum_{m=1}^{M} \psi(y_m) \int_a^b L(y_k, y)\ell n |y_k - y| v_m(y) dy \qquad (9.163d)$$

Depending on $L(x, y)$, the integrals in eqs. 9.163c and 9.163d are then evaluated in closed form or are approximated numerically by quadrature sums.

Referring to ch. 4, eqs. 4.152–4.155, when approximating these integrals by quadrature sums, the singularity structure of the integrand should be "smoothed out" by writing

$$\int_a^b \frac{L(y_k, y)v_m(y)}{|y_k - y|^p} dy = \int_a^b \frac{[L(y_k, y)v_m(y) - L(y_k, y_k)v_m(y_k)]}{|y_k - y|^p} dy$$

$$+ L(y_k, y_k)v_m(y_k) \int_a^b \frac{1}{|y_k - y|^p} dy \qquad (9.164a)$$

and

$$\int_a^b L(y_k, y)v_m(y)\ell n |y_k - y| dy = \int_a^b [L(y_k, y)v_m(y) - L(y_k, y_k)v_m(y_k)]\ell n |y_k - y| dy$$

$$+ L(y_k, y_k)v_m(y_k) \int_a^b \ell n |y_k - y| dy \qquad (9.164b)$$

With

$$v_m(y_k) = \delta_{km} \qquad (1.19)$$

eq. 9.164 become

$$\int_a^b \frac{L(y_k, y)v_m(y)}{|y_k - y|^p} dy = \int_a^b \frac{[L(y_k, y)v_m(y) - L(y_k, y_k)\delta_{km}]}{|y_k - y|^p} dy$$

$$+ L(y_k, y_k)\delta_{km} \int_a^b \frac{1}{|y_k - y|^p} dy \qquad (9.165a)$$

and

$$\int_a^b L(y_k, y) v_m(y) \ell n|y_k - y| dy = \int_a^b [L(y_k, y) v_m(y) - L(y_k, y_k)\delta_{km}] \ell n|y_k - y| dy$$

$$+ L(y_k, y_k)\delta_{km} \int_a^b \ell n|y_k - y| dy \qquad (9.165b)$$

The first integrals on the right sides of eqs. 9.165 are approximated by quadrature sums. If, for a small parameter ε, a quadrature point y_n satisfies

$$|y_n - y_k| < \varepsilon \qquad (9.166)$$

that term in the quadrature sum is taken to be zero.

The second integrals on the right sides of eqs. 9.165 can be evaluated in closed form as follows:

$$\int_a^b \frac{1}{|x - y|^p} dy = \int_a^x \frac{1}{(x - y)^p} dy + \int_x^b \frac{1}{(y - x)^p} dy = \frac{(x - a)^{1-p} + (b - x)^{1-p}}{1 - p}$$

$$(9.167a)$$

and

$$\int_a^b \ell n|x - y| dy = \int_a^x \ell n(x - y) dy + \int_x^b \ell n(y - x) dy =$$

$$(x - a)\ell n(x - a) + (b - x)\ell n(b - x) - (b - a) \qquad (9.167b)$$

Unless one has a sense of the behavior of $\psi(y)$, and so has a sense of the interpolating function $q(y)$, it is reasonable to use polynomial interpolation by taking $q(y) = y$ and interpolating with

$$v_m(y) \equiv \frac{(y - y_1)...(y - y_{m-1})(y - y_{m+1})...(y - y_N)}{(y_m - y_1)...(y_m - y_{m-1})(y_m - y_{m+1})...(y_m - y_N)} \qquad (1.7)$$

Example 9.15: Solution to a Fredholm equation with a weakly singular kernel using Lagrange interpolation

It is straightforward to show that the solution to

$$\psi(x) = e^x \left[1 - \sqrt{(1 + x)} - \sqrt{(1 - x)}\right] + \frac{1}{2} \int_{-1}^1 \frac{e^{x-y}}{\sqrt{|x - y|}} \psi(y) dy \qquad (9.168)$$

is

$$\psi(x) = e^x \qquad (9.50b)$$

Approximating $\psi(y)$ by

$$\psi(y) = \sum_{m=1}^{M} \psi(y_m) v_m(y) \qquad (9.162)$$

eq. 9.168 becomes

$$\psi(x) = e^x \left[1 - \sqrt{(1+x)} - \sqrt{(1-x)} \right] + \frac{1}{2} e^x \sum_{m=1}^{M} \psi(y_m) \int_{-1}^{1} \frac{e^{-y} v_m(y)}{\sqrt{|x-y|}} dy \qquad (9.169a)$$

To illustrate, we take the somewhat crude interpolation over five points $y = \{-1.0, -0.5, 0.0, 0.5, 1.0\}$. Then, eq. 9.169a becomes

$$\psi(x) = e^x \left[1 - \sqrt{(1+x)} - \sqrt{(1-x)} \right] + \frac{1}{2} e^x \sum_{m=1}^{5} \psi(y_m) \int_{-1}^{1} \frac{e^{-y} v_m(y)}{\sqrt{|x-y|}} dy \qquad (9.169b)$$

With $v_m(y)$ the polynomial given in eq. 1.7, we then set x to each point in the set $\{y_m\} = \{-1.0, -0.5, 0.0, 0.5, 1.0\}$ and evaluate

$$\int_{-1}^{1} \frac{e^{-y} v_m(y)}{\sqrt{|x_k - y|}} dy = \int_{-1}^{1} \frac{[e^{-y} v_m(y) - e^{-x_k} \delta_{km}]}{\sqrt{|x_k - y|}} dy + e^{-x_k} \delta_{km} \int_{-1}^{1} \frac{1}{\sqrt{|x_k - y|}} dy$$

$$\simeq \sum_{n=1}^{N} w_n \frac{[e^{-y} v_m(y_n) - e^{-x_k} \delta_{km}]}{\sqrt{|x_k - y_n|}} + 2 e^{-x_k} \delta_{km} \left[\sqrt{1 + x_k} + \sqrt{1 - x_k} \right] \qquad (9.170)$$

We obtain

$$\begin{pmatrix} \psi(-1.0) \\ \psi(-0.5) \\ \psi(0.0) \\ \psi(0.5) \\ \psi(1.0) \end{pmatrix} = \begin{pmatrix} 0.36751 \\ 0.60603 \\ 1.00058 \\ 1.65099 \\ 2.72063 \end{pmatrix} \qquad (9.171a)$$

which compares well with the exact result e^x given by

$$\begin{pmatrix} \psi_{exact}(-1.0) \\ \psi_{exact}(-0.5) \\ \psi_{exact}(0.0) \\ \psi_{exact}(0.5) \\ \psi_{exact}(1.0) \end{pmatrix} = \begin{pmatrix} 0.36788 \\ 0.60653 \\ 1.00000 \\ 1.64872 \\ 2.71828 \end{pmatrix} \qquad (9.171b)$$

The largest error in these results is 0.13%.

The results given in eq. 9.171a can then be substituted into eq. 9.169b to yield values for $\psi(x)$ at any x. $\square$

As the reader will show in Problem 14, a second approach using Lagrange (or Lagrange-like) interpolation is to write

$$\int_a^b \frac{L(x,y)\psi(y)}{|x-y|^p} dy \simeq \sum_{m=1}^M L(x,y_m)\psi(y_m) \int_a^b \frac{v_m(y)}{|x-y|^p} dy \qquad (9.172a)$$

and

$$\int_a^b L(x,y)\psi(y)\ell n|x-y| dy \simeq \sum_{m=1}^M L(x,y_m)\psi(y_m) \int_a^b v_m(y)\ell n|x-y| dy \qquad (9.172b)$$

If the interpolation functions $v_m(y)$ are polynomials of order M, they can be written as sums of powers of y with $\alpha_M = 1$. That is,

$$\int_a^b \frac{v_m(y)}{|x_k-y|^p} dy = \sum_{r=0}^M \alpha_r \int_a^b \frac{y^r}{|x_k-y|^p} dy \qquad (9.173a)$$

and

$$\int_a^b v_m(y)\ell n|x_k-y| dy = \sum_{r=0}^M \alpha_r \int_a^b y^r \ell n|x_k-y| dy \qquad (9.173b)$$

can be evaluated in closed form.

Writing

$$\int_a^b \frac{y^r}{|x-y|^p} dy = \int_a^x \frac{y^r}{(x-y)^p} dy + \int_x^b \frac{y^r}{(y-x)^p} dy \qquad (9.174)$$

we make the substitutions

$$(x-y)^p = z^q \quad y \in [a,x] \qquad (9.175a)$$

and

$$(y - x)^p = z^q \quad y \in [x, b] \tag{9.175b}$$

to obtain

$$\int_a^b \frac{y^r}{|x - y|^p} dy =$$

$$\frac{q}{p} \left[\int_0^{(x-a)^{p/q}} \left(x - z^{1/p} \right)^r z^{q/p-q-1} dz + \int_0^{(b-x)^{p/q}} \left(x + z^{1/p} \right)^r z^{q/p-q-1} dz \right] \tag{9.176}$$

Another approach using polynomial interpolation is to write

$$\int_a^b \frac{L(x, y)}{|x - y|^p} \psi(y) dy =$$

$$\int_a^b \frac{[L(x, y) - L(x, x)]}{|x - y|^p} \psi(y) dy + L(x, x) \int_a^b \frac{1}{|x - y|^p} \psi(y) dy \tag{9.177}$$

Since the integrand of the first integral is finite (zero) at $y = x$, the first integral is well approximated by the N-point quadrature sum

$$\int_a^b \frac{[L(x, y) - L(x, x)]}{|x - y|^p} \psi(y) dy \simeq \sum_{k=1}^N w_k \frac{[L(x, y_k) - L(x, x)]}{|x - y_k|^p} \psi(y_k) \tag{9.178}$$

To evaluate the second integral, we interpolate $\psi(y)$ over a subset (which can be the entire set) of the quadrature abscissae $\{y_k\}$. That is, we interpolate over the points $\{Y_m\} \, \varepsilon \, \{y_k\}$. Then, the second integral is approximated by

$$\int_a^b \frac{1}{|x - y|^p} \psi(y) dy \simeq \sum_{m=1}^M \psi(Y_m) \int_a^b \frac{v_m(y)}{|x - y|^p} dy \tag{9.179}$$

with $M \leq N$. Writing

$$v_m(y) = \frac{y^M + c_1 y^{M-1} + \dots}{(Y_m - Y_1) \dots (Y_m - Y_{m-1})(Y_m - Y_{m+1}) \dots (Y_m - Y_M)} \tag{9.180}$$

the integrals in eq. 9.179 can be evaluated in closed form as described above.

The approximation to the integral equation then becomes

$$\psi(x) \simeq \psi_0(x) +$$

$$\lambda \sum_{k=1}^N w_k \frac{[L(x, y_k) - L(x, x)]}{|x - y|^p} \psi(y_k) + \lambda L(x, x) \sum_{m=1}^M \psi(Y_m) \int_a^b \frac{v_m(y)}{|x - y|^p} dy \tag{9.181}$$

With $\{\psi(Y_m)\} \, \varepsilon \, \{\psi(y_m)\}$, this can be solved by standard methods.

In practice, we find that it is best to choose a quadrature rule that contains abscissae $\{Y_m\}$ that are approximately equally spaced over $[a, b]$.

Example 9.16: Solution to a Fredholm equation with a weakly singular kernel using Lagrange interpolation

We again consider

$$\psi(x) = e^x\left[1 - \sqrt{(1 + x)} - \sqrt{(1 - x)}\right] + \frac{1}{2}\int_{-1}^{1} \frac{e^{x-y}}{\sqrt{|x - y|}}\psi(y)dy \qquad (9.168)$$

which has solution

$$\psi(x) = e^x \qquad (9.50b)$$

The approximation of eq. 9.181 for this equation is

$$\psi(x) \simeq e^x\left[1 - \sqrt{(1 + x)} - \sqrt{(1 - x)}\right] +$$

$$\frac{1}{2}e^x\sum_{k=1}^{N} w_k\frac{[e^{-y_k} - e^{-x}]}{\sqrt{|x - y|}}\psi(y_k) + \frac{1}{2}e^x\sum_{m=1}^{M} \psi(Y_m)\int_{-1}^{1} \frac{v_m(y)}{\sqrt{|x - y|}}dy \qquad (9.182)$$

The quadrature set used to evaluate the first integral is selected based on the points over which $\psi(y)$ is interpolated in the second integral. If the interval $[-1,1]$ is divided into four segments of approximately equal widths, the five points should be approximately $\{-1.0, -0.5, 0.0, 0.5, 1.0\}$. Since 0.0 is one of these points, we look for an odd order quadrature rule for which 0.0 is one of the abscissae and that has abscissae close ± 1.0 and ± 0.5.

Referring to Stroud, A.H., and Secrest, D., 1966, p. 101, the 17 point Gauss–Legendre quadrature abscissae and weights are shown in Table 9.2.

N = 17	
x	w
0.00000	0.17945
±0.17848	0.17656
±0.35123	0.16800
±0.51269	0.15405
±0.65767	0.13514
±0.78151	0.11188
±0.88024	0.06504
±0.95068	0.05546
±0.99057	0.02415

Table 9.2 Seventeen point Gauss–Legendre quadrature data

We use this quadrature rule to evaluate the first integral, and the points

$$\{Y_m\}_{17} = \{-0.99058, -0.51260, 0.00000, 0.51260, 0.99058\} \qquad (9.183)$$

from this quadrature set to interpolate $\psi(y)$ in the second integral. Then eq. 9.182 becomes

$$\psi(x) \simeq e^x \left[1 - \sqrt{(1+x)} - \sqrt{(1-x)} \right] +$$

$$\frac{1}{2} e^x \sum_{k=1}^{17} w_k \frac{[e^{-y_k} - e^{-x}]}{\sqrt{|x - y|}} \psi(y_k) + \frac{1}{2} e^x \sum_{m=1}^{5} \psi(Y_m) \int_{-1}^{1} \frac{v_m(y)}{\sqrt{|x - y|}} dy \qquad (9.184)$$

where

$$v_m(y) = \frac{(y - Y_1)...(y - Y_{m-1})(y - Y_{m+1})...(y - Y_M)}{(Y_m - Y_1)...(Y_m - Y_{m-1})(Y_m - Y_{m+1})...(Y_m - Y_M)} \qquad (9.185)$$

Results at a sample of the 17 abscissae are

$$\begin{pmatrix} \psi(0.99058) \\ \psi(0.78151) \\ \psi(0.51269) \\ \psi(0.17848) \\ \psi(0.00000) \\ \psi(-0.51269) \\ \psi(-0.88024) \\ \psi(-0.99058) \end{pmatrix} = \begin{pmatrix} 2.69145 \\ 2.18359 \\ 1.66886 \\ 1.19523 \\ 1.00031 \\ 0.59916 \\ 0.41431 \\ 0.37118 \end{pmatrix} \qquad (9.186a)$$

which is a reasonably accurate approximation to

$$\begin{pmatrix} \psi_{exact}(0.99058) \\ \psi_{exact}(0.78151) \\ \psi_{exact}(0.51269) \\ \psi_{exact}(0.17848) \\ \psi_{exact}(0.00000) \\ \psi_{exact}(-0.51269) \\ \psi_{exact}(-0.88024) \\ \psi_{exact}(-0.99058) \end{pmatrix} = \begin{pmatrix} 2.69278 \\ 2.18478 \\ 1.66978 \\ 1.19540 \\ 1.00000 \\ 0.59888 \\ 0.41468 \\ 0.37136 \end{pmatrix} \qquad (9.186b)$$

The largest error in these results is 0.09% with an average error of 0.05%. This indicates that this approach is a bit more accurate than the method described in example 9.15. □

Spline Interpolation Methods

To apply spline interpolation methods, we subdivide $[a, b]$ into several segments, writing eqs. 9.160 as

$$\psi(x) = \psi_0(x) + \lambda \sum_{m=1}^{M} \int_{z_m}^{z_{m+1}} \frac{L(x,y)}{|x-y|^p} \psi(y) dy \qquad (9.187a)$$

and

$$\psi(x) = \psi_0(x) + \lambda \sum_{m=1}^{M} \int_{z_m}^{z_{m+1}} L(x,y) \ell n |x-y| \psi(y) dy \qquad (9.187b)$$

where

$$z_1 \equiv a \qquad (9.188a)$$

and

$$z_{M+1} \equiv b \qquad (9.188b)$$

As before, the segments can be equally spaced, or defined by some other method such as taking a subset of the abscissae of a Gaussian quadrature rule. Then, over each segment, $\psi(y)$ is approximated by a constant. Unlike the approximation of eq. 9.45, here we approximate $\psi(y)$ by

$$\psi(y) \simeq \psi(\alpha z_m + \beta z_{m+1}) \equiv \psi(y_m) \quad z_m \leq y_m \leq z_{m+1} \qquad (9.189)$$

where, unless there is a reason to do otherwise, we take y_m to be the midpoint of the mth segment. Then, setting x to each y_k, eqs. 9.187 become

$$\psi(y_k) \simeq \psi_0(y_k) + \lambda \sum_{m=1}^{M} \psi(y_m) \int_{z_m}^{z_{m+1}} \frac{L(y_k, y)}{|y_k - y|^p} dy \qquad (9.190a)$$

and

$$\psi(y_k) \simeq \psi_0(y_k) + \lambda \sum_{m=1}^{M} \psi(y_m) \int_{z_m}^{z_{m+1}} L(y_k, y) \ell n |y_k - y| dy \qquad (9.190b)$$

which are solved by standard techniques.

Example 9.17: Solution of a Fredholm equation with a weakly singular kernel using spline interpolation

We again consider

$$\psi(x) = e^x \left[1 - \sqrt{(1+x)} - \sqrt{(1-x)} \right] + \frac{1}{2} \int_{-1}^{1} \frac{e^{x-y}}{\sqrt{|x-y|}} \psi(y) dy \qquad (9.168)$$

which has solution

$$\psi(x) = e^x \qquad (9.50b)$$

We define N segments by the points $\{z_1, \ldots, z_{N+1}\}$ from which we determine the points at which we will determine $\psi(y)$ at

$$y_k \equiv \frac{1}{2}(z_k + z_{k+1}) \qquad (9.191)$$

Then eq. 9.168 is approximated by

$$\psi(y_k) = e^{y_k} \left[1 - \sqrt{(1+y_k)} - \sqrt{(1-y_k)} \right] + \frac{1}{2} \sum_{m=1}^{N} \psi(y_m) \int_{z_m}^{z_{m+1}} \frac{e^{y_k-y}}{\sqrt{|y_k-y|}} dy \qquad (9.192)$$

As noted above, we could approximate each of the integrals by "smoothing out" the singularity at $y = y_k$ in the integrands by writing

$$\int_{z_m}^{z_{m+1}} \frac{e^{y_k-y}}{\sqrt{|y_k-y|}} dy = \int_{z_m}^{z_{m+1}} \frac{e^{y_k-y}-1}{\sqrt{|y_k-y|}} dy + \int_{z_m}^{z_{m+1}} \frac{1}{\sqrt{|y_k-y|}} dy \simeq$$

$$\sum_{n=1}^{N} w_n \frac{e^{y_k-y_n}-1}{\sqrt{|y_k-y_n|}} + \int_{z_m}^{z_{m+1}} \frac{1}{\sqrt{|y_k-y|}} dy \qquad (9.193)$$

and easily evaluating the second integral for each of the cases $y_k \geq z_{m+1}$, $z_{m+1} \geq y_k \geq z_{m+1}$, and $y_k \leq z_m$.

To illustrate another approach, we have evaluated the integrals for these three cases by expanding the exponential in a Taylor series to approximate

$$\int_{z_m}^{z_{m+1}} \frac{e^{y_k-y}}{\sqrt{|y_k-y|}} dy \simeq \sum_{n=0}^{4} \frac{1}{n!} \int_{z_m}^{z_{m+1}} \frac{(y_k-y)^n}{\sqrt{|y_k-y|}} dy \qquad (9.194)$$

A sample of the results we obtain are given below.

For $N = 10$ segments, we find

$$
\begin{pmatrix} \psi(-.9) \\ \psi(-.5) \\ \psi(.1) \\ \psi(.9) \end{pmatrix} = \begin{pmatrix} 0.51386 \\ 0.67232 \\ 1.00968 \\ 2.08540 \end{pmatrix} \tag{9.195a}
$$

which is a poor approximation to

$$
\begin{pmatrix} \psi_{exact}(-.9) \\ \psi_{exact}(-.5) \\ \psi_{exact}(.1) \\ \psi_{exact}(.9) \end{pmatrix} = \begin{pmatrix} 0.40657 \\ 0.60653 \\ 1.10517 \\ 2.45960 \end{pmatrix} \tag{9.195b}
$$

These results differ from the exact values by an average of 13.2% with a maximum error of 26.4%.

For $N = 20$ segments, we obtain

$$
\begin{pmatrix} \psi(-.95) \\ \psi(-.55) \\ \psi(.05) \\ \psi(.45) \\ \psi(.85) \end{pmatrix} = \begin{pmatrix} 0.49036 \\ 0.64835 \\ 0.97569 \\ 1.33916 \\ 2.24827 \end{pmatrix} \tag{9.196a}
$$

the exact values of which are

$$
\begin{pmatrix} \psi_{exact}(-.95) \\ \psi_{exact}(-.55) \\ \psi_{exact}(.05) \\ \psi_{exact}(.45) \\ \psi_{exact}(.85) \end{pmatrix} = \begin{pmatrix} 0.38674 \\ 0.57695 \\ 1.05127 \\ 1.56831 \\ 2.58571 \end{pmatrix} \tag{9.196b}
$$

The average and maximum errors of these results are 12.6% and 26.8%, respectively.

These results indicate that this cardinal spline interpolation method converges very slowly. Thus, the results of the Lagrange interpolation method are much more accurate for interpolation over a small set of points. $\square$

Schlitt's method

A method, proposed by Schlitt, D.W., (1968), involves "smoothing out" the singularity by subtracting and adding the unknown function $\psi(x)$ to obtain

$$\psi(x) = \psi_0(x) + \lambda \int_a^b \frac{L(x,y)}{|x-y|^p} [\psi(y) - \psi(x)] dy + \lambda \psi(x) \int_a^b \frac{L(x,y)}{|x-y|^p} dy \qquad (9.197)$$

With

$$J(x) \equiv \int_a^b \frac{L(x,y)}{|x-y|^p} dy \qquad (9.198)$$

eq. 9.197 becomes

$$\psi(x)[1 - \lambda J(x)] = \psi_0(x) + \lambda \int_a^b \frac{L(x,y)}{|x-y|^p} [\psi(y) - \psi(x)] dy \qquad (9.199)$$

If $\psi(x)$ is analytic at all $y \ \varepsilon \ [a, b]$, $\psi(y) - \psi(x)$ can be expanded in a Taylor series, all terms of which contain $(x - y)^n$ with $n \geq 1$. Therefore, the integrand of the integral on the right hand side of eq. 9.199 can be approximated by a quadrature sum. With x set to each point in the abscissae set, we obtain

$$\psi(y_k)[1 - \lambda J(y_k)] = \psi_0(y_k) + \lambda \sum_{\substack{m=1 \\ m \neq k}}^N \frac{L(y_k, y_m)}{|y_k - y_m|^p} [\psi(y_m) - \psi(y_k)] \qquad (9.200a)$$

where, because $p < 1$, all terms in the series for $[\psi(y) - \psi(x)]/|x - y|^p$ have positive powers of $(x - y)$. Therefore, the $m = k$ term in the sum is zero.

This results in the set of equations

$$\psi(y_k) \left[1 - \lambda J(y_k) + \lambda \sum_{\substack{m=1 \\ m \neq k}}^N \frac{L(y_k, y_m)}{|y_k - y_m|^p} \right] - \lambda \sum_{\substack{m=1 \\ m \neq k}}^N \frac{L(y_k, y_m)}{|y_k - y_m|^p} \psi(y_m) = \psi_0(y_k)$$

$$(9.200b)$$

the solution to which is obtained by standard methods. An example of this is left as an exercise for the reader (see Problem 16).

9.6 Fredholm Equations with Kernels Containing a Pole Singularity

Integral equations in which the kernel has a pole (or *Cauchy*) singularity are of the form

$$\psi(x) = \psi_0(x) + \lambda \int_a^b \frac{L(x,y)}{(y-z)_P} \psi(y)dy \quad a \le z \le b \tag{9.201}$$

where the subscript P indicates that the integral is a principal value integral. z can either be a constant or a function of x.

When the singularity is at a constant value of y, it is called a *fixed pole*. An example of such an equation is the *Lippmann–Schwinger* equation for the scattering wave function of a non-relativistic particle (Lippmann, B., and Schwinger, J., 1950). With E a constant related to the energy of the incident particle, the form of this equation is

$$\psi(x) = \psi_0(x) - \frac{2}{\pi} \int_0^\infty \frac{L(x,y)}{(y-E)_P} \psi(y)dy \tag{9.202a}$$

An integral equation with a Cauchy kernel that is developed from quantum field theory is the *Omnes equation* (Omnes, R., 1958)

$$\psi(x) = \psi_0(x) + \frac{1}{\pi} \int_1^\infty \frac{L(y)}{(y-x)_P} \psi(y)dy \tag{9.202b}$$

With x a variable, a pole singularity at $y = x$ is a *variable* or *movable pole*.

Solution to an equation with a fixed pole

Referring to the Lippmann–Schwinger equation, we note that when we approximate integrals by quadrature sums, a semi-infinite interval $[a, \infty]$ suggests that we might use Laguerre quadratures. However, as noted in ch. 4, unless the integrand of an integral explicitly contains e^{-y}, using a Gauss–Laguerre quadrature is less reliable than transforming the range of integration to $[-1, 1]$ and using a Gauss–Legendre quadrature. As such, with $z =$ constant, we consider eq. 9.201 in the form

$$\psi(x) = \psi_0(x) + \lambda \int_{-1}^1 \frac{L(x,y)}{(y-z)_P} \psi(y)dy - 1 \le z \le 1 \tag{9.203}$$

To accurately approximate the solution to this integral equation, we write eq. 9.203 as

$$\psi(x) = \psi_0(x) + \lambda \int_{-1}^{1} \frac{[L(x,y) - L(x,z)]}{(y-z)} \psi(y) dy + \lambda L(x,z) \int_{-1}^{1} \frac{1}{(y-z)_P} \psi(y) dy$$

$$(9.204)$$

Since the integrand of the first integral is no longer singular, the integral is well approximated by a Gauss–Legendre quadrature sum

$$\int_{-1}^{1} \frac{[L(x,y) - L(x,z)]}{(y-z)} \psi(y) dy \simeq \sum_{m=1}^{N} w_m \frac{[L(x,y_m) - L(x,z)]}{(y_m - z)} \psi(y_m) \qquad (9.205)$$

To approximate the second integral, we express $\psi(y)$ by an interpolating polynomial

$$\psi(y) \simeq \sum_{m=1}^{M} v_m(y) \psi(Y_m) \qquad \qquad \boldsymbol{(9.162)}$$

where $M \leq N$ and the set $\{Y_m\}$ over which the interpolation is constructed is a subset (possibly the entire set) of $\{y_m\}$, the quadrature abscissae.

We note that $\{Y_m\}$ being a subset of the abscissae of the N-point quadrature rule are some of the zeros of the Legendre polynomial $P_N(y)$. Thus, referring to eq. 1.15, we can express the interpolating function as

$$v_m(y) = \frac{P_N(y)}{(y - Y_m)P_N'(Y_m)} \qquad (9.206)$$

and the second integral of eq. 9.204 becomes

$$\int_{-1}^{1} \frac{1}{(y-z)_P} \psi(y) dy \simeq \sum_{m=1}^{M} \frac{\psi(Y_m)}{P_N'(Y_m)} \int_{-1}^{1} \frac{P_N(y)}{(y-z)_P(y-Y_m)} dy$$

$$= -\sum_{m=1}^{M} \frac{\psi(Y_m)}{P_N'(Y_m)} \frac{1}{(Y_m - z)} \int_{-1}^{1} \left[\frac{P_N(y)}{(Y_m - y)} - \frac{P_N(y)}{(z-y)_P} \right] dy \qquad (9.207)$$

The advantage of this form of the second integral is that each term can be expressed in terms of *the Legendre function of the second kind*, Q_N, many properties of which are well established (see, for example, Cohen, H., 1992, pp. 299–306 and pp. 370–371). The *Neumann representation* of Q_N is given by

$$Q_N(z) = \frac{1}{2} \int_{-1}^{1} \frac{P_N(y)}{(z-y)} \, dy \tag{9.208}$$

Then, eq. 9.207 can be written as

$$\int_{-1}^{1} \frac{1}{(y-z)_P} \psi(y) dy \simeq -2 \sum_{m=1}^{M} \frac{\psi(Y_m)}{P'_N(Y_m)} \frac{[Q_N(Y_m) - Q_N(z)]}{(Y_m - z)} \tag{9.209}$$

Combining the results given in eqs. 9.204 and 9.208, the approximated integral equation is

$$\psi(x) = \psi_0(x) + \lambda \sum_{m=1}^{N} w_m \frac{[L(x, y_m) - L(x, z)]}{(y_m - z)} \psi(y_m)$$

$$- 2\lambda L(x, z) \sum_{m=1}^{M} \frac{\psi(Y_m)}{P'_N(Y_m)} \frac{[Q_N(Y_m) - Q_N(z)]}{(Y_m - z)} \tag{9.210}$$

which is then solved by standard methods.

If z has a value close to one of the quadrature abscissae y_m, such that for some small ε

$$|y_m - z| < \varepsilon \tag{9.211}$$

that term in the first sum is replaced by

$$\frac{[L(x, y_m) - L(x, z)]}{(y_m - z)} \rightarrow \frac{\partial L(x, y)}{\partial y}\bigg|_{y=z} \tag{9.212}$$

If y_m, is also one of the interpolation points Y_m, that term in the second sum is replaced by

$$\frac{[Q_N(Y_m) - Q_N(z)]}{(Y_m - z)} \rightarrow Q'_N(z) \tag{9.213}$$

From the author's experience, it is found that the most accurate method of approximating the $P_N(y)$, $Q(y)$, $P'_N(y)$, and $Q'_N(y)$ is from the recurrence relations satisfied by these functions. The Legendre polynomials satisfy

$$(\ell + 2)P_{\ell+2}(y) - y(2\ell + 3)P_{\ell+1}(y) + (\ell + 1)P_\ell(y) = 0 \tag{9.214a}$$

for $\ell \geq 0$. Taking one derivative of this expression, we obtain

$$(\ell + 2)P'_{\ell+2}(y) - y(2\ell + 3)P'_{\ell+1}(y) + (\ell + 1)P'_{\ell}(y) = (2\ell + 3)P_{\ell+1}(y) \quad (9.214b)$$

By substituting the expression of eq. 9.214a into the Neumann expression for $Q_N(z)$ we obtain

$$(\ell + 2) \int_{-1}^{1} \frac{P_{\ell+2}(y)}{(z-y)_P} dy - (2\ell + 3) \int_{-1}^{1} \frac{yP_{\ell+1}(y)}{(z-y)_P} dy + (\ell + 1) \int_{-1}^{1} \frac{P_{\ell}(y)}{(z-y)_P} dy = 0$$

$$(9.215a)$$

Referring to eq. 9.210, this can be written as

$$2(\ell + 2)Q_{\ell+2}(z) - (2\ell + 3) \int_{-1}^{1} \frac{(y-z)P_{\ell+1}(y)}{(z-y)_P} dy$$

$$- (2\ell + 3)z \int_{-1}^{1} P_{\ell+1}(y)dy + 2(\ell + 1)Q_{\ell}(z) = 0 \quad (9.215b)$$

The Legendre polynomials satisfy the *orthonormalization condition*

$$\int_{-1}^{1} P_{\ell}(y)P_m(y)dy = \frac{2}{(2m+1)}\delta_{\ell m} \quad (9.216)$$

Therefore, with

$$P_0(z) = 1 \quad (9.217)$$

and $\ell \geq 0$, we obtain

$$\int_{-1}^{1} \frac{(y-z)P_{\ell+1}(y)}{(z-y)_P} dy = -\int_{-1}^{1} P_0(y)P_{\ell+1}(y)dy = 0 \quad (9.218)$$

Thus, from eq. 9.215b, the Legendre function of the second kind satisfies

$$(\ell + 2)Q_{\ell+2}(y) - y(2\ell + 3)Q_{\ell+1}(y) + (\ell + 1)Q_{\ell}(y) = 0 \quad (9.219a)$$

and from the first derivative of this expression,

$$(\ell + 2)Q'_{\ell+2}(y) - y(2\ell + 3)Q'_{\ell+1}(y) + (\ell + 1)Q'_{\ell}(y) = (2\ell + 3)Q_{\ell+1}(y) \quad (9.219b)$$

Example 9.18: Solution to a Fredholm equation with a fixed pole singularity in the kernel using interpolation

(a) The solution to the Lippmann–Schwinger type of equation

$$\psi(x) = 2e^x + \frac{1}{\ell n(3)} \int_{-1}^{1} \frac{e^{(x-y)}}{\left(y - \frac{1}{2}\right)_P} \psi(y) dy \qquad (9.220)$$

is

$$\psi(x) = e^x \qquad (\mathbf{9.50b})$$

We write eq. 9.220 as

$$\psi(x) = 2e^x + \frac{1}{\ell n(3)} \int_{-1}^{1} \frac{\left[e^{(x-y)} - e^{(x-\frac{1}{2})}\right]}{\left(y - \frac{1}{2}\right)} \psi(y) dy + \frac{e^{(x-\frac{1}{2})}}{\ell n(3)} \int_{-1}^{1} \frac{1}{\left(y - \frac{1}{2}\right)_P} \psi(y) dy \quad (9.221)$$

Using the 17-point Gauss–Legendre quadrature rule with abscissae $\{y_m\}$ to approximate the first integral, we obtain

$$\int_{-1}^{1} \frac{\left[e^{(x-y)} - e^{(x-\frac{1}{2})}\right]}{\left(y - \frac{1}{2}\right)} \psi(y) dy \simeq \sum_{m=1}^{17} w_m \frac{\left[e^{(x-y_m)} - e^{(x-\frac{1}{2})}\right]}{\left(y_m - \frac{1}{2}\right)} \psi(y_m) \qquad (9.222)$$

Approximating $\psi(y)$ in the second integral by a polynomial over $\{Y_m\}$, an M-point subset of $\{y_m\}$, we have

$$\int_{-1}^{1} \frac{1}{\left(y - \frac{1}{2}\right)_P} \psi(y) dy \simeq -2 \sum_{m=1}^{M} \frac{\psi(Y_m)}{P'_{17}(Y_m)} \frac{\left[Q_{17}(Y_m) - Q_{17}\left(\frac{1}{2}\right)\right]}{\left(Y_m - \frac{1}{2}\right)}$$

$$= -2 \sum_{m=1}^{17} C_m \frac{\psi(Y_m)}{P'_{17}(Y_m)} \frac{\left[Q_{17}(Y_m) - Q_{17}\left(\frac{1}{2}\right)\right]}{\left(Y_m - \frac{1}{2}\right)} \qquad (9.223)$$

where C_m is 0 if y_m is not in the subset $\{Y_m\}$ and is 1 if y_m is in $\{Y_m\}$. Setting x to each quadrature point in the set $\{y_k\}$, eq. 9.221 is approximated by

$$\psi(y_k) = 2e^{y_k} + \frac{1}{\ln(3)} \left[\sum_{m=1}^{17} w_m \frac{\left[e^{(y_k - y_m)} - e^{(y_k - \frac{1}{2})} \right]}{\left(y_m - \frac{1}{2} \right)} \right.$$

$$\left. - \sum_{m=1}^{17} C_m \frac{\psi(y_m)}{P'_{17}(y_m)} \frac{\left[Q_{17}(y_m) - Q\left(\frac{1}{2}\right) \right]}{\left(y_m - \frac{1}{2} \right)} \right]$$

(9.224)

With $\{Y_m\} = \{y_m\}$, so that all $C_m = 1$, we solve eq. 9.224 by matrix inversion. We obtain results that are essentially the exact values of $\psi(x)$. Each of the computed values of $\psi(x)$ differ from that value of e^x by a difference of $1.3 \times 10^{-13}\%$. We do find that the accuracy of the results are sensitive to the points over which $\psi(y)$ is interpolated. For example, if $\{Y_m\} = \{y_1, y_3, \ldots, y_{15}, y_{17}\}$ (the odd index points of $\{y_m\}$), a sample of typical results are

$$\psi(y_k) \simeq 2e^{y_k} + \frac{1}{\ln(3)} \left[\sum_{m=1}^{17} w_m \frac{\left[e^{(y_k - y_m)} - e^{(y_k - \frac{1}{2})} \right]}{(y_m - \frac{1}{2})} \psi(y_m) \right.$$

$$\left. - 2e^{(y_k - \frac{1}{2})} \sum_{m=1}^{5} \frac{1}{P'_{17}(Y_m)} \frac{\left[Q_{17}(Y_m) - Q_{17}(\frac{1}{2}) \right]}{(Y_m - \frac{1}{2})} \psi(Y_m) \right]$$

$$= 2e^{y_k} + \frac{1}{\ln(3)} \left[\sum_{m=1}^{17} \left[w_m \frac{\left[e^{(y_k - y_m)} - e^{(y_k - \frac{1}{2})} \right]}{(y_m - \frac{1}{2})} \psi(y_m) \right. \right.$$

$$\left. \left. - 2e^{(y_k - \frac{1}{2})} C_m \frac{1}{P'_{17}(y_m)} \frac{\left[Q_{17}(y_m) - Q_{17}(\frac{1}{2}) \right]}{(y_m - \frac{1}{2})} \psi(y_m) \right] \right]$$

(9.225)

Solving this by standard methods, we obtain

$$\begin{pmatrix} \psi(0.99058) \\ \psi(0.51269) \\ \psi(0.00000) \\ \psi(-0.65767) \\ \psi(-0.95068) \end{pmatrix} = \begin{pmatrix} 2.08635 \\ 1.29373 \\ 0.77479 \\ 0.40139 \\ 0.29944 \end{pmatrix} \qquad (9.226a)$$

which differs from

$$\begin{pmatrix} \psi_{exact}(0.99058) \\ \psi_{exact}(0.51269) \\ \psi_{exact}(0.00000) \\ \psi_{exact}(-0.65767) \\ \psi_{exact}(-0.95068) \end{pmatrix} = \begin{pmatrix} 2.69278 \\ 1.66978 \\ 1.00000 \\ 0.51806 \\ 0.38648 \end{pmatrix} \qquad (9.226b)$$

by approximately 22.5%. This example suggests that accurate results are only obtained by interpolating over the entire quadrature set. □

Solution to an equation with a variable pole

We consider an integral equation with a kernel that has a variable pole by setting $z = x$ in eq. 9.201 to obtain

$$\psi(x) = \psi_0(x) + \lambda \int_{-1}^{1} \frac{L(x,y)}{(y-x)_P} \psi(y)dy \quad -1 \leq x \leq 1 \qquad (9.227a)$$

As we did with the integral equation with a kernel with a fixed pole, we write eq. 9.227a as

$$\psi(x) = \psi_0(x) + \lambda \int_{-1}^{1} \frac{[L(x,y) - L(x,x)]}{(y-x)} \psi(y)dy + \lambda L(x,x) \int_{-1}^{1} \frac{1}{(y-x)_P} \psi(y)dy \qquad (9.227b)$$

then approximate the first integral by the quadrature sum

$$\int_{-1}^{1} \frac{[L(x,y) - L(x,x)]}{(y-x)} \psi(y)dy \simeq \sum_{m=1}^{N} w_m \frac{[L(x,y_m) - L(x,x)]}{(y_m - x)} \psi(y_m) \qquad (9.228)$$

For the second integral, we refer to the results of example 9.18 and interpolate $\psi(y)$ over the entire set of quadrature abscissae $\{y_m\}$ as

$$\psi(y) \simeq \sum_{m=1}^{N} v_m(y)\psi(y_m) \tag{9.162}$$

We again construct the interpolating functions in terms of Legendre polynomials as

$$v_m(y) = \frac{P_N(y)}{(y - y_m)P'_N(y_m)} \tag{9.206}$$

Then,

$$\int_{-1}^{1} \frac{1}{(y-x)_P}\psi(y)dy \simeq \sum_{m=1}^{N} \frac{\psi(y_m)}{P'_N(y_m)} \int_{-1}^{1} \frac{P_N(y)}{(y-x)_P(y-y_m)}dy$$

$$= -\sum_{m=1}^{N} \frac{\psi(y_m)}{P'_N(y_m)} \frac{1}{(y_m - x)} \int_{-1}^{1} \left[\frac{P_N(y)}{(y_m - y)} - \frac{P_N(y)}{(x-y)_P} \right] dy$$

$$= -2\sum_{m=1}^{M} \frac{\psi(y_m)}{P'_N(y_m)} \frac{[Q_N(y_m) - Q_N(x)]}{(y_m - x)} \tag{9.229}$$

Combining the results given in eqs. 9.228 and 9.229, the approximated integral equation is

$$\psi(x) \simeq \psi_0(x) + \lambda \sum_{m=1}^{N} w_m \frac{[L(x,y_m) - L(x,x)]}{(y_m - x)}\psi(y_m)$$

$$- 2\lambda L(x,x) \sum_{m=1}^{N} \frac{\psi(y_m)}{P'_N(y_m)} \frac{[Q_N(y_m) - Q_N(x)]}{(y_m - x)} \tag{9.230}$$

By setting x to each y_k of the quadrature abscissae, we obtain a set of N linear equations for $\psi(y_k)$ which is solved by standard methods. Again, the values of the Legendre functions are found from the recurrence relations of eqs. 9.214 and 9.219. The $m = k$ terms in the two sums must be replaced by

$$\frac{[L(y_k, y_m) - L(y_k, y_k)]}{(y_m - y_k)} \rightarrow \left. \frac{\partial L(y_k, y)}{\partial y} \right|_{y=y_k} \tag{9.231a}$$

and

$$\frac{[Q_N(y_m) - Q_N(y_k)]}{(y_m - y_k)} \rightarrow Q'_N(y_k) \tag{9.231b}$$

Example 9.19: Solution to a Fredholm equation with a variable pole singularity in the kernel using interpolation

The solution to the Omnes type of equation

$$\psi(x) = e^x \left[1 + \ell n \left(\frac{1+x}{1-x} \right) \right] + \int_{-1}^{1} \frac{e^{(x-y)}}{(y-x)_P} \psi(y) dy \qquad (9.232)$$

is

$$\psi(x) = e^x \qquad \qquad (9.50b)$$

We write eq. 9.232 as

$$\psi(x) = e^x \left[1 + \ell n \left(\frac{1+x}{1-x} \right) \right] + \int_{-1}^{1} \frac{\left[e^{(x-y)} - 1 \right]}{(y-x)} \psi(y) dy + \int_{-1}^{1} \frac{1}{(y-x)_P} \psi(y) dy$$

$$(9.233a)$$

Using the 17-point Gauss–Legendre quadrature rule again, this integral equation is approximated by

$$\psi(y_k) \simeq e^{y_k} \left[1 + \ell n \left(\frac{1+y_k}{1-y_k} \right) \right]$$

$$+ \sum_{m=1}^{17} \left[w_m \frac{\left[e^{(y_k - y_m)} - 1 \right]}{(y_m - y_k)} - 2 \frac{1}{P'_{17}(y_m)} \frac{\left[Q_{17}(y_m) - Q_{17}(y_k) \right]}{(y_m - y_k)} \right] \psi(y_m)$$

$$(9.233b)$$

We again obtain highly accurate results. The average difference of the 17 values from the corresponding exact values is $5.5 \times 10^{-13}\%$ and the individual difference of each value of $\psi(y_k)$ is of this order of magnitude. $\square$

Problems

1. The homogeneous Fredholm integral equation of the first kind $\frac{1 - e^{(1+x)}}{(1+x)} = \int_0^1 e^{-xy} \psi(y) dy$ has solution $\psi(x) = e^{-x}$
   Transform the range of integration to $[-1, 1]$ and expand $\psi_0(x)$, $\psi(x)$, and $K(x,y)$ in series over Legendre polynomials. From this, find the approximate solution to the above equation by approximating $\psi(x) \simeq \sum_{m=0}^{2} \beta_m P_m(x)$ at $x = \{0.20,$ $0.60, 1.00\}$. Compare this approximate solution to the exact values at these points

2. The Fredholm integral equation of the first kind $\sqrt{\dfrac{\pi}{1+x^2}} = \displaystyle\int_{-\infty}^{\infty} e^{-x^2 y^2} \psi(y)\,dy$
   has solution $\psi(x) = e^{-x^2}$

   (a) Expand each of $\psi_0(x)$, $\psi(x)$, and $K(x,y)$ in three term series over approximations Hermite polynomials. From this, find the approximate solution for $\psi(x)$ at $x = \{2, 10, 50\}$.
   (b) Transform the range of integration to $[-1, 1]$ and expand $\psi_0(x)$, $\psi(x)$, and $K(x,y)$ in series over Legendre polynomials. From this, find the approximate solution to the above equation for a three term series approximation to $\psi(x)$
   at $x = \{2, 10, 50\}$. At these points, determine the values of $1 - x^2 + \dfrac{x^4}{2!}$ the first three terms in the MacLaurin series for the solution. Compare these values to the results of parts (a) and (b).

3. The solution to the integral equation $\psi(x) = 1 - \dfrac{\sin(\pi x)}{\pi} + \displaystyle\int_0^1 x\cos(\pi xy)\psi(y)\,dy$
   is $\psi(x) = 1$
   Find an approximate solution to this equation using

   (a) a 3-point Gauss–Legendre quadrature rule
   (b) a three term Neumann series in λ, then setting $\lambda = 1$
   (c) a 3-point interpolation of $x\cos(\pi xy)$ of the form $x\cos(\pi xy) \simeq$
   $\displaystyle\sum_{k=1}^{3} x\cos(\pi xy_k)v_k(y)$ over the points $y = \{0.0,\ 0.5,\ 1.0\}$. With $q(y) =$
   $\cos(\pi y)$ interpolate using $v_k(y) = \dfrac{(q(y) - q(y_r))(q(y) - q(y_s))}{(q(y_k) - q(y_r))(q(y_k) - q(y_s))}$ with k,
   r, and s taking the values 1, 2, and 3, and with $k \neq r \neq s$.

4. The solution to the integral equation $\psi(x) = 1 - \dfrac{2\sin(\pi x)}{\pi} + \displaystyle\int_{-1}^1 x\cos(\pi xy)\psi(y)\,dy$
   is $\psi(x) = 1$
   Find an approximate solution to this equation by expanding the functions ψ and ψ_0 and the kernel $K(x, y)$ in series over the three lowest order Legendre polynomials as in example 9.5.

5. Find the exact solution to $\psi(x) = x + \int_0^1 (3x + y)\psi(y)\,dy$ using the fact that the kernel is degenerate.

6. Find the two exact eigensolutions to $\psi(x) = \lambda \int_0^1 \sin[\pi(x + y)]\psi(y)\,dy$

7. Find approximations to the eigensolutions of the homogeneous Fredholm equation of the second kind given in Problem 6 by

   (a) approximating the integral by a 3-point Gauss–Legendre quadrature rule.
   (b) Interpolating the kernel over the points $y = \{0.0, 0.5, 1.0\}$ as described in example 9.9, using $q(y) = \sin(\pi y)$

8. Find the Volterra integral equation of the second kind $\psi(x) =$
   $\psi_0(x) + \displaystyle\int_0^x e^{(x-y)}\psi(y)\,dy$ that has the same solution as the Volterra equation of the first kind $\displaystyle\int_0^x e^{(x-y)}\psi(y)\,dy = e^x(e^x - 1)$

9. Find the first three non-zero terms in the Taylor series for $\psi(x)$ which satisfies
$\psi(x) = x^2 + \int_0^x x^2 y^2 \psi(y)dy$ which has solution $\psi_{exact}(x) = xe^{\frac{1}{3}x^5}$

10. Use the method of approximating $\psi(y)$ by the constant $\alpha\psi(x_k) + \beta\psi(x_{k+1})$ for
the Volterra integral equation $\psi(x) = 1 + x + \int_0^x e^{(x-y)}\psi(y)dy$ the solution to
which is $\psi(x) = \dfrac{1}{4} + \dfrac{x}{2} + \dfrac{3}{4}e^{2x}$
Find the approximate solution to this equation at $x = 0.15$ for

(a) $\alpha = 0, \beta = 1$
(b) $\alpha = \beta = 1/2$
(c) $\alpha = 1/3, \beta = 2/3$
(d) $\alpha = 2/3, \beta = 1/3$

11. (a) Show that the solution to the Volterra equation $\psi(x) = x^2 - \int_1^x \dfrac{x}{y}\psi(y)dy$ is
$\psi(x) = x$
    (b) By approximating $\psi(y)$ by the constant $[\psi(x_k) + \psi(x_{k+1})]/2$, estimate this
solution at $x = 1.15$ with $\Delta x = 0.05$.

12. Approximate the solution to $\psi(x) = \left[x - \sqrt{1+x}(8x^2 - 4x + 3) - \sqrt{1-x}\right.$
$\left.(8x^2 + 4x + 3)\right] + \dfrac{15}{2}\int_{-1}^{1}\dfrac{xy}{\sqrt{|x-y|}}\psi(y)dy$ by approximating $\psi(x)$ by a polynomial Lagrange interpolation. Compare your result to the analytic solution
$\psi(x) = x$.

13. One can show that the solution to $\psi(x) = e^x[3 - (1+x)\ell n(1+x) - (1+x)$
$\ell n(1+x)] + \int_{-1}^{1} e^{x-y}\ell n|x - y|\psi(y)dy$ is $\psi(x) = e^x$
Using the Lagrange polynomial interpolation of $\psi(y)$ over the three points $\{-1,$
$0, 1\}$, obtain an approximate solution to the above equation. Use these approximate values to estimate $\psi(-0.5)$ and $\psi(0.5)$.

14. By approximating $e^{-y}\psi(y)$ by a Lagrange polynomial interpolation over the
three points $\{-1, 0, 1\}$, obtain an approximate solution to $\psi(x) =$
$e^x[3 - (1+x)\ell n(1+x) - (1+x)\ell n(1+x)] + \int_{-1}^{1} e^{x-y}\ell n|x - y|\psi(y)dy$
Compare your results to the exact values $\psi(x) = e^x$ at these three points.

15. Approximate the solution to $\psi(x) = \left[x - \sqrt{1+x}(8x^2 - 4x + 3) - \sqrt{1-x}\right.$
$\left.(8x^2 + 4x + 3)\right] + \dfrac{15}{2}\int_{-1}^{1}\dfrac{xy}{\sqrt{|x-y|}}\psi(y)dy$ by writing this equation as $\psi(x) =$
$\left[x - \sqrt{1+x}(8x^2 - 4x + 3) - \sqrt{1-x}(8x^2 + 4x + 3)\right] + \frac{15}{2}\int_{-1}^{1}\dfrac{(xy-x^2)}{\sqrt{|x-y|}}\psi(y)dy+$
$\dfrac{15}{2}x^2\int_{-1}^{1}\dfrac{1}{\sqrt{|x-y|}}\psi(y)dy$ then estimating the value of the first integral by a
Gauss–Legendre quadrature rule, and interpolating $\psi(y)$ in the second integral
by a polynomial Lagrange interpolation over four points that are a subset of

the quadrature abscissae and are close to $\{-1, -1/3, 1/3, 1\}$. Compare your approximate results to the analytic solution $\psi(x) = x$.

16. Use the Schlitt method described above to approximate the solution to $\psi(x) = e^x\left[1 - \sqrt{(1+x)} - \sqrt{(1-x)}\right] + \frac{1}{2}\int_{-1}^{1}\frac{e^{x-y}}{\sqrt{|x-y|}}\psi(y)dy$

Compare your results to the analytic solution $\psi(x) = e^x$.

17. The integral equation $\psi(x) = \dfrac{\ell n(3)}{4}x + \displaystyle\int_{-1}^{1}\frac{xy}{(y - \frac{1}{2})_P}\psi(y)dy$ has solution $\psi(x) = x$

Approximate the solution to this integral equation by interpolating $\psi(y)$ as

$\psi(y) = \displaystyle\sum_{m=1}^{N}\psi(y_m)v_m(y)$ with $v_m(y) = \dfrac{P_N(y)}{P'_N(y_m)(y-y_m)}$ where $P_N(y)$ is the Legendre polynomial of order N. Take $N = 5$.

18. The integral equation $\psi(x) = x\left[1 - 2x + x^2\ell n\left(\dfrac{1+x}{1-x}\right)\right] + \displaystyle\int_{-1}^{1}\frac{xy}{(y - x)_P}\psi(y)dy$ has solution $\psi(x) = x$

Approximate the solution to this integral equation by interpolating $\psi(y)$ as

$\psi(y) = \displaystyle\sum_{m=1}^{5}\psi(y_m)v_m(y)$ with $v_m(y) = \dfrac{P_5(y)}{P'_5(y_m)(y-y_m)}$ where $P_N(y)$ is the Legendre polynomial of order N.

Appendix 1
PADE APPROXIMANTS

When $f(x)$ is defined at all x in a specified domain, it is said to be *analytic* in that domain. Then, $f(x)$ can be represented by an infinite *Taylor* series

$$f(x) = \sum_{k=0}^{\infty} \frac{1}{k!} \left(\frac{d^k f}{dx^k}\right)_{x=x_0} (x - x_0)^k \tag{A1.1}$$

where x_0 is some point within the domain of analyticity.

In the domain of analyticity, $f(x)$ can also be represented by an *infinitely continued fraction*:

$$f(x) = A_0 + \cfrac{A_1(x - x_0)}{B_0 + \cfrac{B_1(x - x_0)}{C_0 + \cfrac{C_1(x - x_0)}{D_0 + \cfrac{D_1(x - x_0)}{\ddots}}}} \tag{A1.2}$$

This infinitely continued fraction can be approximated by truncating the process at some level. Such a truncated continued fraction is identical to a *Pade* (pronounced "Pah-day") *Approximant* (Pade, H., 1892).

For example, if the process in eq. A1.2 is truncated at the 2nd (or B) level, then $f(x)$ can be approximated by

$$\begin{aligned} f(x) &\simeq A_0 + \frac{A_1(x - x_0)}{B_0 + B_1(x - x_0)} = \frac{A_0 B_0 + (A_0 B_1 + A_1)(x - x_0)}{B_0 + B_1(x - x_0)} \\ &\equiv \frac{p_0 + p_1(x - x_0)}{q_0 + q_1(x - x_0)} \end{aligned} \tag{A1.3}$$

H. Cohen, *Numerical Approximation Methods*, DOI 10.1007/978-1-4419-9837-8,
© Springer Science+Business Media, LLC 2011

We note that truncating the infinitely continued fraction at the B-level results in the ratio of two first-order polynomials. This is called the [1,1] *Pade Approximant* of f (x). It is denoted by $f^{[1,1]}(x)$. If the process in eq. A1.2 is truncated at the 3rd (C-) level, then

$$f(x) \simeq A_0 + \cfrac{A_1(x - x_0)}{B_0 + \cfrac{B_1(x - x_0)}{C_0 + C_1(x - x_0)}}$$

$$= \frac{p_0 + p_1(x - x_0) + p_2(x - x_0)^2}{q_0 + q_1(x - x_0)} \equiv f^{[2,1]}(x) \qquad (A1.4)$$

and truncating the infinitely continued fraction at the 4th (D-) level yields the [2,2] Pade Approximant

$$f(x) \simeq A_0 + \cfrac{A_1(x - x_0)}{B_0 + \cfrac{B_1(x - x_0)}{C_0 + \cfrac{C_1(x - x_0)}{D_0 + D_1(x - x_0)}}}$$

$$= \frac{p_0 + p_1(x - x_0) + p_2(x - x_0)^2}{q_0 + q_1(x - x_0) + q_2(x - x_0)^2} \equiv f^{[2,2]}(x) \qquad (A1.5)$$

and so on. (One can easily verify that truncating the continued fraction at the 5th (E-) level results in the [3,2] Pade Approximant.)

Thus, a truncated continued fraction results in a ratio of polynomials such that if the order of the numerator polynomial is N, the order of the polynomial in the denominator is either N or $N - 1$. If the continued fraction is truncated at the nth level, we obtain a $[n/2, n/2]$ Pade Approximant if n is even (called the *diagonal Pade Approximant*) or a $[(n + 1)/2, (n-1)/2]$ Pade Approximant (called the *off-diagonal Pade Approximant*) if n is odd.

The definition of the Pade Approximant can be generalized as the ratio of any two polynomials. Let M and N be any two positive integers. Then

$$f^{[N,M]}(x) \equiv \frac{p_0 + p_1(x - x_0) + p_2(x - x_0)^2 + ... + p_N(x - x_0)^N}{q_0 + q_1(x - x_0) + q_2(x - x_0)^2 + ... + q_M(x - x_0)^M} \qquad (A1.6)$$

is the $[N, M]$ Pade Approximant. If $M = N$ or $M = N-1$, $f^{[N, M]}$ is equivalent to a truncated continued fraction. If $M \neq N$ and $M \neq N-1$, this *non-diagonal Pade Approximant* does not represent a truncated continued fraction and is less accurate an approximation to $f(x)$ than the diagonal or off-diagonal Approximant.

(For a discussion of the convergence of a Pade Approximant, see Chisholm, J. S. R., 1973, pp. 11–21.) Therefore, within the domain of analyticity, the most accurate Pade Approximants are

$$f^{[N,N]}(x) = \frac{p_0 + p_1(x - x_0) + p_2(x - x_0)^2 + \dots + p_N(x - x_0)^N}{q_0 + q_1(x - x_0) + q_2(x - x_0)^2 + \dots + q_N(x - x_0)^N} \qquad (A1.7a)$$

and

$$f^{[N,N-1]}(x) = \frac{p_0 + p_1(x - x_0) + p_2(x - x_0)^2 + \dots + p_N(x - x_0)^N}{q_0 + q_1(x - x_0) + q_2(x - x_0)^2 + \dots + q_{N-1}(x - x_0)^{N-1}} \qquad (A1.7b)$$

We will, therefore, focus our attention on these.

There are M zeros in the Mth order denominator polynomial. These zeros are singularities of $f^{[N,M]}(x)$, but since $f^{[N,M]}(x)$ is an approximation to a function that is analytic in some domain of analyticity, these zeros must be outside that domain. (For discussions about Pade Approximants in the domain in which $f(x)$ is singular, see Baker, G., 1975, pp. 123–126 or Nuttall, J., 1977, pp. 101–109.) To specify the Pade Approximant, we must determine the coefficients of the various powers of $x - x_0$. We note that since $f^{[N,M]}(x)$ is required not to have a singularity at $x = x_0$, q_0 cannot be zero. Therefore, we can divide numerator and denominator by q_0 to obtain

$$f^{[N,M]}(x) = \frac{\dfrac{p_0}{q_0} + \dfrac{p_1}{q_0}(x - x_0) + \dfrac{p_2}{q_0}(x - x_0)^2 + \dots + \dfrac{p_N}{q_0}(x - x_0)^N}{1 + \dfrac{q_1}{q_0}(x - x_0) + \dfrac{q_2}{q_0}(x - x_0)^2 + \dots + \dfrac{q_M}{q_0}(x - x_0)^M}$$

$$\equiv \frac{p_0' + p_1'(x - x_0) + p_2'(x - x_0)^2 + \dots + p_N'(x - x_0)^N}{1 + q_1'(x - x_0) + q_2'(x - x_0)^2 + \dots + q_M'(x - x_0)^M} \qquad (A1.8)$$

Because these undetermined constants can be denoted by p' and q' or by p and q, we can ignore the "prime" marks and denote the constants by p and q. Thus, eq. A1.8 is equivalent to eqs. A1.7 with $q_0 = 1$. Therefore, the most general form of the $[N, M]$ Pade Approximant is

$$f^{[N,M]}(x) = \frac{p_0 + p_1(x - x_0) + p_2(x - x_0)^2 + \dots + p_N(x - x_0)^N}{1 + q_1(x - x_0) + q_2(x - x_0)^2 + \dots + q_M(x - x_0)^M} \equiv \frac{P_N(x)}{Q_M(x)} \qquad (A1.9)$$

Methods for determining the p and q coefficients will be described at the places in the text where application of the Pade Approximant is presented.

Appendix 2
INFINITE SERIES CONVERGENCE TESTS

We divide a series $S(z)$ into a finite sum, $S_1(z)$ and a remainder series $S_2(z)$ such that, with n_0 and N finite integers and $n_0 < N-1$,

$$S(z) = \sum_{n=n_0}^{N-1} \sigma_n(z) + \sum_{n=N}^{\infty} \sigma_n(z) \equiv S_1(z) + S_2(z) \tag{A2.1}$$

We choose N such that

$$|\sigma_{n+1}(z)| < |\sigma_n(z)| \tag{A2.2}$$

for all $n \geq N$.

Cauchy integral test for an absolute series

Let $S(z)$ be an absolute series. For the sake of discussion, we take all $\sigma_n(z)$ to be positive. We note that for any two successive terms,

$$\Delta n \equiv (n+1) - n = 1 \tag{A2.3}$$

Therefore, we can write $S_2(z)$ as

$$S_2(z) = \sum_{n=N}^{\infty} \sigma_n(z)\Delta n \tag{A2.4}$$

Then, for some fixed z, Fig. A2.1a illustrates rectangles of height $\sigma_{N+1}(z)$, $\sigma_{N+2}(z)$, ... each of width $\Delta n = 1$. The solid line represents the envelope curve for $\sigma(n,z)$ treating n as a variable.

451

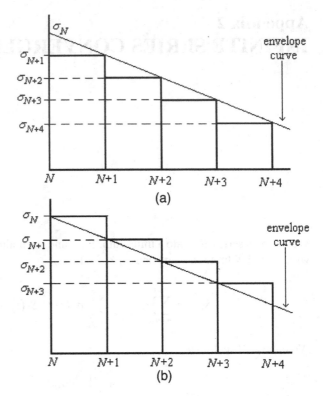

The area under the envelope curve is given by

$$A_{envelope}(z) = \int_N^\infty \sigma(n,z)dn \qquad (A2.5)$$

The area under the rectangles of Fig. A2.1a is

$$A_{rectangles_a}(z) = \sigma_{N+1}(z)\Delta n + \sigma_{N+2}(z)\Delta n + \ldots$$

$$= \sum_{n=N+1}^\infty \sigma_n(z) = S_2(z) - \sigma_N(z) \qquad (A2.6)$$

We now consider the rectangles shown in Fig. A2.1b. We note that the rectangles are of heights $\sigma_N(z)$, $\sigma_{N+1}(z)$, ... and widths $\Delta n = 1$. The envelope curve is the same as that of Fig. A2.1a.

The area of the rectangles of Fig. A2.1b are given by

$$A_{rectangles_b}(z) = \sigma_N(z)\Delta n + \sigma_{N+1}(z)\Delta n + \ldots$$

$$= \sum_{n=N}^\infty \sigma_n(z) = S_2(z) \qquad (A2.7)$$

We see that in Fig. A2.1a, the area of the rectangles is less than the area under the envelope curve. Thus,

$$S_2(z) - \sigma_N(z) < \int_N^\infty \sigma(n, z)dn \qquad (A2.8a)$$

It is clear from Fig. A2.1b that the area of the rectangles is greater than the area under the envelope curve. Therefore,

$$S_2(z) > \int_N^\infty \sigma(n, z)dn \qquad (A2.8b)$$

These inequalities place bounds on $S_2(z)$

$$\int_N^\infty \sigma(n, z)dn < S_2(z) < \sigma_N(z) + \int_N^\infty \sigma(n, z)dn \qquad (A2.9)$$

Therefore, if the envelope integral is finite, $S_2(z)$ is finite and $S(z)$ converges. If the envelope integral is infinite, $S_2(z)$ is infinite and the series diverges.

Limit test for an alternating series

For an alternating series, we must start the sum for $S_2(z)$ with N an even integer. This guarantees that the first term in the remainder series is positive. Then, $S_2(z)$ can be written as

$$S_2(z) = |\sigma_N(z)| - |\sigma_{N+1}(z)| + |\sigma_{N+2}(z)| - |\sigma_{N+3}(z)| + \cdots \qquad (A2.10a)$$

which we group as

$$S_2(z) = [|\sigma_N(z)| - |\sigma_{N+1}(z)|] + [|\sigma_{N+2}(z)| - |\sigma_{N+3}(z)|] + \cdots \qquad (A2.10b)$$

Since N has been chosen so that

$$|\sigma_{n+1}(z)| < |\sigma_n(z)| \qquad (A2.2)$$

for all $|\sigma_n(z)|$ in $S_2(z)$, the difference in each bracket is positive. Therefore,

$$S_2(z) > |\sigma_N(z)| - |\sigma_{N+1}(z)| \qquad (A2.11)$$

We next group the terms in $S_2(z)$ as

$$S_2(z) = |\sigma_N(z)| - [|\sigma_{N+1}(z)| - |\sigma_{N+2}(z)|] - [|\sigma_{N+3}(z)| - |\sigma_{N+4}(z)|] - \cdots \qquad (A2.12)$$

Again, the differences in the brackets are positive, so that

$$S_2(z) < |\sigma_N(z)| \tag{A2.13}$$

Therefore, the bounds on $S_2(z)$ are given by

$$|\sigma_N(z)| - |\sigma_{N+1}(z)| < S_2(z) < |\sigma_N(z)| \tag{A2.14}$$

Since a necessary condition for $S(z)$ to converge is

$$\lim_{n \to \infty} \sigma_n(z) = 0 \tag{A2.2}$$

and this condition will restrict $S_2(z)$ to be between two small (ultimately zero) bounds, this is the only condition necessary for convergence of an alternating series.

Cauchy ratio test for any series

The Cauchy ratio is defined by

$$\rho_n \equiv \left| \frac{\sigma_{n+1}(z)}{\sigma_n(z)} \right| \tag{A2.15a}$$

and its limiting value is defined by

$$\rho \equiv \lim_{n \to \infty} \rho_n \tag{A2.15b}$$

Taking N large enough that the Cauchy ratio is essentially its limiting value, we write

$$\begin{aligned}
S_2(z) &= \sigma_N(z) + \sigma_{N+1}(z) + \sigma_{N+2}(z) + \sigma_{N+3}(z) + \ldots \\
&= \sigma_N(z) + \rho\sigma_N(z) + \rho^2\sigma_N(z) + \rho^3\sigma_N(z) + \ldots \\
&= \sigma_N(z)\left[1 + \rho + \rho^2 + \rho^3 + \ldots\right]
\end{aligned} \tag{A2.16}$$

If $\rho < 1$, the series in the bracket converges to $1/(1 - \rho)$, and $S_2(z)$ converges. If $\rho > 1$, the series in the bracket diverges, and if $\rho = 1$, the convergence or divergence of the series in the bracket is unknown. This is the Cauchy ratio test.

Appendix 3
GAMMA AND BETA FUNCTIONS

A3.1 Definitions of the Gamma and Beta Functions

Gamma function

A common definition of the Γ function is given by

$$\Gamma(p+1) \equiv \int_0^\infty z^p e^{-z} dz \quad \mathrm{Re}(p) > -1 \qquad (A3.1)$$

Integrating this by parts, we have

$$\Gamma(p+1) = -z^p e^{-z}\big|_0^\infty + p \int_0^\infty z^{p-1} e^{-z} dz = p\Gamma(p) \qquad (A3.2)$$

where the integrated term is zero for $\mathrm{Re}(p) > 0$. Eq. A3.2 is referred to as the *iterative property* of the Γ function.

For p a positive integer N, we use this iterative property of $\Gamma(p)$ to obtain

$$\Gamma(N+1) = N(N-1)(N-2)...3 \times 2 \times 1 \times \Gamma(1) \qquad (A3.3a)$$

Setting $p = 0$ in eq. A3.1 we obtain

$$\Gamma(1) = \int_0^\infty e^{-z} dz = 1 \qquad (A3.3b)$$

Therefore, eq. A3.3a becomes

$$\Gamma(N+1) = N! \qquad (A3.4)$$

From this result, $\Gamma(p)$ has been defined as the *generalized factorial*. That is, when p is not an integer, the meaning of $p!$ is $\Gamma(p+1)$.

Gamma function of half integer order

Referring to eq. A3.1, we define

$$\left(\frac{1}{2}\right)! = \Gamma\left(\frac{3}{2}\right) = \int_0^\infty z^{\frac{1}{2}} e^{-z} dz \qquad (A3.5a)$$

Substituting $z = x^2$, this becomes

$$\left(\frac{1}{2}\right)! = 2 \int_0^\infty x^2 e^{-x^2} dz \qquad (A3.5b)$$

We integrate by parts, setting $u = x$ and $dv = xe^{-x} dx$. Then

$$\left(\frac{1}{2}\right)! = \int_0^\infty e^{-x^2} dx \qquad (A3.6)$$

To evaluate this integral, we note that the integral is unchanged if x is replaced by y. Thus,

$$\left(\frac{1}{2}\right)! = \int_0^\infty e^{-y^2} dy \qquad (A3.7)$$

The product of the integrals in eqs. A3.6 and A3.7 is

$$\left[\left(\frac{1}{2}\right)!\right]^2 = \int_0^\infty \int_0^\infty e^{-(x^2+y^2)} dx dy \qquad (A3.8)$$

The region of integration of eq. A3.8 is the first quadrant of the x–y plane. As seen in Fig. A3.1, $dA_{xy} = dx dy$ is the area element in Cartesian coordinates. Transforming to circular coordinates r and θ, we write

$$x = r\cos\theta \qquad (A3.9a)$$

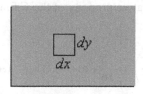

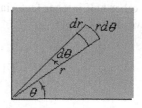

Fig. A3.1 The first quadrant of the x–y plane
and a Cartesian area element

Fig. A3.2 Area element in circular
coordinates

and

$$y = r \sin \theta \qquad (A3.9b)$$

Referring to Fig. A3.2, the infinitesimal area element in circular coordinates
is given by

$$dA_{r\theta} = (dr)(rd\theta) = rdrd\theta \qquad (A3.10)$$

and the first quadrant is described by

$$0 \leq r \leq \infty, 0 \leq \theta \leq \frac{\pi}{2} \qquad (A3.11)$$

Therefore,

$$\left[\left(\frac{1}{2} \right)! \right]^2 = \int_0^{\pi/2} d\theta \int_0^\infty e^{-r^2} rdr \qquad (A3.12)$$

which can be integrated easily to obtain

$$\left[\left(\frac{1}{2} \right)! \right]^2 = \frac{\pi}{4} \qquad (A3.13a)$$

Therefore,

$$\left(\frac{1}{2} \right)! = \frac{\sqrt{\pi}}{2} \qquad (A3.13b)$$

For other half integer factorials, we use the iterative property of the Γ function. We have, for example,

$$\left(\frac{5}{2}\right)! = \Gamma\left(\frac{7}{2}\right) = \frac{5}{2}\frac{3}{2}\frac{1}{2}\Gamma\left(\frac{1}{2}\right) = \frac{15}{16}\sqrt{\pi} \qquad (A3.14)$$

and

$$\left(\frac{1}{2}\right)! = \frac{\sqrt{\pi}}{2} = \frac{1}{2}\left(-\frac{1}{2}\right)\left(-\frac{3}{2}\right)\left(-\frac{5}{2}\right)! \qquad (A3.15a)$$

from which

$$\left(-\frac{5}{2}\right)! = \frac{4\sqrt{\pi}}{3} \qquad (A3.15b)$$

Beta function

From the integral representation of the Γ function given in eq. A3.1, we consider

$$\Gamma(p)\Gamma(q) = \int_0^\infty \int_0^\infty u^{p-1}v^{q-1}e^{-(u+v)}dudv \quad \text{Re}(p)>0,\ \text{Re}(q)>0 \qquad (A3.16)$$

With

$$x = \sqrt{u} \qquad (A3.17a)$$

and

$$y = \sqrt{v} \qquad (A3.17b)$$

eq. A3.16 becomes

$$\Gamma(p)\Gamma(q) = 4\int_0^\infty \int_0^\infty x^{2p-1}y^{2q-1}e^{-(x^2+y^2)}dxdy \qquad (A3.18)$$

which we view as an integral over the first quadrant of the x–y plane in Cartesian coordinates. Referring to eq. A3.9a and A3.10 and Fig. A3.2, we transform to circular coordinates to obtain

$$\Gamma(p)\Gamma(q) = 4 \int_0^{\frac{\pi}{2}} \cos^{2p-1}\theta \sin^{2q-1}\theta \, d\theta \int_0^\infty r^{2p+2q-1} e^{-r^2} dr \qquad \text{(A3.19a)}$$

Substituting $z = r^2$, this becomes

$$\Gamma(p)\Gamma(q) = 2 \int_0^{\frac{\pi}{2}} \cos^{2p-1}\theta \sin^{2q-1}\theta \, d\theta \int_0^\infty z^{p+q-1} e^{-z} dz$$
$$= 2\Gamma(p+q) \int_0^{\frac{\pi}{2}} \cos^{2p-1}\theta \sin^{2q-1}\theta \, d\theta \qquad \text{(A3.19b)}$$

The beta function is defined by

$$\beta(p,q) = 2 \int_0^{\frac{\pi}{2}} \cos^{2p-1}\theta \sin^{2q-1}\theta \, d\theta \qquad \text{(A3.20)}$$

and we see from (A3.19b) that

$$\beta(p,q) = \frac{\Gamma(p)\Gamma(q)}{\Gamma(p+q)} \qquad \text{(A3.21)}$$

Clearly,

$$\beta(p,q) = \beta(q,p) \qquad \text{(A3.22)}$$

Substituting

$$u = \cos^2\theta \qquad \text{(A3.23)}$$

eq. A3.20 becomes

$$\beta(p,q) = \int_0^1 u^{p-1}(1-u)^{q-1} du \qquad \text{(A3.24)}$$

A third representation of the β function is obtained by setting

$$u = \frac{x}{1+x} \qquad \text{(A3.25)}$$

The result is

$$\beta(p,q) = \int_0^\infty \frac{x^{p-1}}{(1+x)^{p+q}} dx \tag{A3.26}$$

where, in eqs. A3.20, A3.21, A3.24 and A3.25, it is understood that these definitions of $\beta(p, q)$ are valid for $\text{Re}(p) > 0$ and $\text{Re}(q) > 0$.

Let $N \geq 1$ be an integer and $p < N$ be noninteger. If $q = N - p$, then

$$\beta(p, N - p) = \int_0^\infty \frac{x^{p-1}}{(1+x)^N} dx \tag{A3.27}$$

Using contour methods, as shown in Cohen, H., 2007, pp. 208–211

$$\beta(p, N - p) = \frac{\Gamma(p)\Gamma(N - p)}{\Gamma(N)}$$

$$= (-1)^{N+1} \frac{\Gamma(p)}{(N - 1)!\Gamma(p - N + 1)} \frac{\pi}{\sin(\pi p)} \tag{A3.28}$$

from which we have

$$\Gamma(N - p)\Gamma(p - N + 1) = (-1)^{N+1} \frac{\pi}{\sin(\pi p)} \tag{A3.29}$$

Thus,

$$\lim_{p \to 0} \Gamma(N - p)\Gamma(p - N + 1) = \Gamma(N)\Gamma(1 - N) = (-1)^{N+1} \infty \tag{A3.30}$$

Setting $N = 1, 2, \ldots$, we see that gamma functions of zero and negative integers are $\pm\infty$ as shown in Fig. A3.3.

Legendre duplication formula

From eqs. A3.21 and A3.24, we have

$$\beta(p, p) = \int_0^1 x^{p-1}(1 - x)^{p-1} dx = \frac{[\Gamma(p)]^2}{\Gamma(2p)} \tag{A3.31}$$

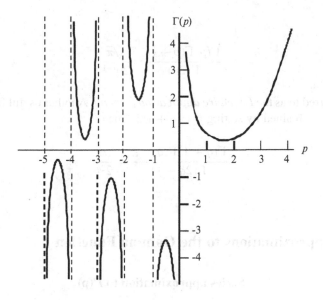

Fig. A3.3 $\Gamma(p)$ for $-\infty \le p \le \infty$

Substituting

$$x = \frac{(1+y)}{2} \tag{A3.32}$$

this becomes

$$\begin{aligned}
\frac{[\Gamma(p)]^2}{\Gamma(2p)} &= \frac{1}{2^{2p-1}} \int_{-1}^{1} (1-y^2)^{p-1} dy \\
&= \frac{1}{2^{2p-1}} \left[\int_{-1}^{0} (1-y^2)^{p-1} dy + \int_{0}^{1} (1-y^2)^{p-1} dy \right] \\
&= \frac{1}{2^{2p-2}} \int_{0}^{1} (1-y^2)^{p-1} dy
\end{aligned} \tag{A3.33}$$

We then set

$$y^2 = u \tag{A3.34}$$

to obtain

$$\begin{aligned}
\frac{[\Gamma(p)]^2}{\Gamma(2p)} &= \frac{1}{2^{2p-1}} \int_{0}^{1} u^{-1/2}(1-u)^{p-1} du = \frac{1}{2^{2p-1}} \beta\left(p, \frac{1}{2}\right) \\
&= \frac{\Gamma(p)\Gamma(\frac{1}{2})}{\Gamma(p+\frac{1}{2})} = \frac{\Gamma(p)}{\Gamma(p+\frac{1}{2})} \sqrt{\pi}
\end{aligned} \tag{A3.35}$$

or

$$\frac{\Gamma(p)\Gamma(p+\frac{1}{2})}{\Gamma(2p)} = \frac{\sqrt{\pi}}{2^{2p-1}} \qquad (A3.36a)$$

This is referred to as the *Legendre duplication formula*. Another useful form for this expression is obtained by setting $p = q + 1/2$. Then

$$\frac{\Gamma(q+1)\Gamma(q+\frac{1}{2})}{\Gamma(2q+1)} = \frac{\sqrt{\pi}}{2^{2q}} \qquad (A3.36b)$$

A3.2   Approximations to the Gamma Function

Series approximation to Γ(p)

When p is not an integer or half integer, the value of the Γ function cannot be determined exactly. If p is small, one method is to develop a MacLaurin series in powers of p and approximate the Γ function by truncating this series.

To develop this series, we begin with the *Euler limit representation* of the Γ function,

$$\Gamma(p) = \lim_{m\to\infty} \left[m^p \frac{m!}{p(p+1)...(p+m)} \right] \qquad (A3.37)$$

(For a derivation of this, the reader is referred to Cohen, H., 1992, p. 268.)
With $\Gamma(1) = 1$, we begin with

$$\ell n[\Gamma(p+1)] = \ell n[\Gamma(1)] + \sum_{n=1}^{\infty} \left[\frac{d^n \ell n[\Gamma(p+1)]}{dp^n} \right]_{p=0} \frac{p^n}{n!} \qquad (A3.38a)$$

Since $\Gamma(1) = 1$, this becomes

$$\ell n[\Gamma(p+1)] = \sum_{n=1}^{\infty} \left[\frac{d^n \ell n[\Gamma(p+1)]}{dp^n} \right]_{p=0} \frac{p^n}{n!} \equiv \sum_{k=1}^{\infty} \psi^{(n)}(1) \frac{p^n}{n!} \qquad (A3.38b)$$

The functions

$$\psi^{(n)}(p+1) = \frac{d^n \ell n(p+1)}{dp^n} \qquad (A3.39)$$

are called the *digamma function* ($n = 1$) and *polygamma functions* ($n > 1$).

From the Euler limit representation of eq. A3.37

$$
\ell n[\Gamma(p+1)] = \lim_{m\to\infty} \left[\ell n\left(m^{p+1} \frac{m!}{(p+1)(p+2)...(p+m+1)} \right) \right]
$$

$$
= \lim_{m\to\infty} \left[(p+1)\ell n(m) + \ell n(m!) - \sum_{k=0}^{m} \ell n(p+k+1) \right]
$$

$$
= \lim_{m\to\infty} \left[(p+1)\ell n(m) + \ell n(m!) - \sum_{k=1}^{m} \ell n(p+k) \right] \tag{A3.40}
$$

Thus, the digamma function is given by

$$
\psi^{(1)}(p+1) = \frac{d\ell n[\Gamma(p+1)]}{dp} = \lim_{m\to\infty} \left[\ell n(m) - \sum_{k=1}^{m} \frac{1}{(p+k)} \right] \tag{A3.41a}
$$

from which

$$
\psi^{(1)}(1) = \lim_{m\to\infty} \left[\ell n(m) - \sum_{k=1}^{m} \frac{1}{k} \right] \equiv -\gamma = -0.5772156 \tag{A3.41b}
$$

The constant γ is called the *Euler–Mascheroni constant*.

To see that γ is finite, we note that the series is an absolute series. Therefore, by taking

$$
\lim_{m\to\infty} \sum_{k=1}^{m} \frac{1}{k} = \sum_{k=1}^{N-1} \frac{1}{k} + \lim_{m\to\infty} \sum_{k=N}^{m} \frac{1}{k} \equiv S_1 + S_2 \tag{A3.42}
$$

S_2 is bounded by

$$
\lim_{m\to\infty} \int_N^m \frac{1}{k} dk < S_2 < \frac{1}{N} + \lim_{m\to\infty} \int_N^m \frac{1}{k} dk \tag{A3.43a}
$$

As noted in ch. 3, eq. 3.71, by taking the average of its upper and lower bounds, S_2 can be approximated by

$$
S_2 \simeq \frac{1}{2N} + \lim_{m\to\infty} \ell n(m) - \ell n(N) \tag{A3.43b}
$$

Table A3.1 Values of polygamma functions

$-\gamma$	$\psi^{(2)}(1)/2!$	$\psi^{(3)}(1)/3!$	$\psi^{(4)}(1)/4!$	$\psi^{(5)}(1)/5!$
-0.57722	0.82247	-0.40069	0.27058	-0.27039
$\psi^{(6)}(1)/6!$	$\psi^{(7)}(1)/7!$	$\psi^{(8)}(1)/8!$	$\psi^{(9)}(1)/9!$	$\psi^{(10)}(1)/10!$
0.16956	-0.14405	0.12551	-0.11133	0.10001

Therefore,

$$\psi^{(1)}(1) \simeq \lim_{m\to\infty}\left[\ell n(m) - \sum_{k=1}^{N-1}\frac{1}{k} - \frac{1}{2N} + \ell n(N) - \ell n(m)\right]$$

$$= -\left[\sum_{k=1}^{N-1}\frac{1}{k} + \frac{1}{2N} - \ell n(N)\right] \equiv -\gamma \qquad (A3.44)$$

Its value, which can be approximated by taking a "large" finite value for N in eq. A3.44. For $N = 500$ we obtain

$$\gamma = 0.5772153.... \qquad (A3.45)$$

which is an accurate approximation to γ. (An aid to remembering the value of the Euler–Mascheroni constant is that $\tan(30^\circ) = 0.5773503....$)

The polygamma functions are found by taking various derivatives of eq. A3.41a. Referring to eq. A3.39 for $n \geq 2$, we obtain

$$\psi^{(n)}(p+1) = (-1)^{n+1}n!\sum_{k=1}^{\infty}\frac{1}{(p+k)^n} \qquad (A3.46)$$

which is finite for $p \geq 0$, $n \geq 2$.

Using the approximation of ch. 3, eq. 3.71, the digamma functions are approximated as

$$\psi^{(1)}(1) = -\gamma \qquad (A3.47a)$$

and

$$\frac{\psi^{(n)}(1)}{n!} \simeq (-1)^{n+1}\left[\sum_{k=1}^{N-1}\frac{1}{k^n} + \frac{1}{2N} + \frac{1}{(n-1)N^{n-1}}\right] \quad n>1 \qquad (A3.47b)$$

A small sample of values of $\psi^{(n)}(1)/n!$ is given in Table A3.1.

Stirling's approximation to $\Gamma(p)$

If p is large, an approximation to $\Gamma(p)$ can be obtained using a technique called *the method of steepest descent*. Starting with

$$\Gamma(p+1) = p! = \int_0^\infty z^p e^{-z} dz \qquad\qquad (A3.1)$$

we make the substitution

$$x = \frac{z}{p} \qquad\qquad (A3.48)$$

to obtain

$$p! = p^{p+1} \int_0^\infty x^p e^{-px} dx = p^{p+1} \int_0^\infty e^{-p[x-\ell n(x)]} dx \qquad\qquad (A3.49)$$

Since $x - \ell n(x) > 0$ for all $x > 0$,

$$e^{-p[x-\ell n(x)]}$$

will be small for large values of p. Therefore, the largest contribution to the integral comes from the region around the minimum of $x - \ell n(x)$ and the contribution of the integrand to the integral will be small at points x away from this minimum.
  Defining

$$x = 1 + \delta \qquad\qquad (A3.50)$$

so that $\delta \varepsilon\, [-1, \infty]$. It is straightforward to show that the minimum of the exponent is at $x = 1$ or at $\delta = 0$. In terms of δ this exponent is

$$x - \ell n(x) = (1+\delta) - \ell n(1+\delta) = (1+\delta) - \left[\delta - \frac{\delta^2}{2} + \frac{\delta^3}{3} - \ldots\right]$$

$$= 1 + \frac{\delta^2}{2} + O(\delta^3) \qquad\qquad (A3.51)$$

Since p is large, we assume that the terms of order δ^3 will contribute very little to the integral compared with the contribution from the δ^2 term. Therefore, we ignore the contribution from $O(\delta^3)$ and approximate eq. A3.49 by

$$p! \simeq p^{p+1} \int_{-1}^\infty e^{-p[1+\delta^2/2]} d\delta = p^{p+1} e^{-p} \int_{-1}^\infty e^{-p\delta^2/2} d\delta \qquad\qquad (A3.52a)$$

In addition, since p is large, the contribution to the integral from values of δ in the interval $-\infty \leq \delta < -1$ is negligible. Thus, without significant error, we extend the range of integration, writing

$$p! \simeq p^{p+1}e^{-p} \int_{-\infty}^{\infty} e^{-p\delta^2/2} d\delta \qquad (A3.52b)$$

Substituting

$$\delta\sqrt{\frac{p}{2}} = y \qquad (A3.53)$$

eq. A3.52b becomes

$$p! \simeq p^{p+1}e^{-p}\sqrt{\frac{2}{p}} \int_{-\infty}^{\infty} e^{-y^2} dy = p^{p+1}e^{-p}\sqrt{\frac{2}{p}}2\int_{0}^{\infty} e^{-y^2} dy = p^p e^{-p}\sqrt{2\pi p} \quad (A3.54)$$

This is *Stirling's approximation* to $\Gamma(p)$ for large p.

Namias' extension of the stirling approximation

A more accurate approximation, also attributed to Stirling, is derived from the Legendre duplication formula given in eq. A3.36a. As shown by Namias, V., 1986

$$p! \simeq p^p e^{-p}\sqrt{2\pi p}\, e^{\sum\limits_{k=1}^{\infty} \frac{a_k}{p^k}} \qquad (A3.55a)$$

where the coefficients a_k satisfy the recurrence relation

$$\left(2^{k+1} - 1\right)a_k + (k-1)! \sum_{m=1}^{k-1} \frac{2^m}{(k-m)!\,(m-1)!}a_m = \frac{1}{2(k+1)} \qquad (A3.55b)$$

For $k = 1$,

$$\sum_{m=1}^{0} \frac{2^m}{(k-m)!\,(m-1)!}a_m = 0 \qquad (A3.56)$$

Therefore,

$$a_1 = \frac{1}{12} \qquad (A3.57a)$$

Then, with $k = 2$, we obtain

$$a_2 = \frac{1}{7}\left(\frac{1}{6} - 2a_1\right) = 0 \qquad \text{(A3.57b)}$$

In this way, we find

$$a_3 = -\frac{1}{360} \qquad \text{(A3.57c)}$$

$$a_4 = 0 \qquad \text{(A3.57d)}$$

$$a_5 = \frac{1}{1260} \qquad \text{(A3.57e)}$$

and so on. Thus, the modified Stirling approximation, developed by Namias, is

$$p! \simeq p^p e^{-p} \sqrt{2\pi p}\, e^{\left(\frac{1}{12p} - \frac{1}{360p^3} + \frac{1}{1260p^5} - \ldots\right)} \qquad \text{(A3.58)}$$

The Stirling approximation of eqs. A3.54 and A3.58 can also be used to approximate $\Gamma(1 + p)$ for small values of p. To illustrate, let N be an integer large enough that $\Gamma(N + 1 + p)$ is well approximated by eq. A3.54. Then,

$$\Gamma(N + 1 + p) \simeq (N + p)^{N+p} e^{-(N+p)} \sqrt{2\pi(N + p)} \qquad \text{(A3.59a)}$$

Using the iterative property of the Γ function, we have

$$\begin{aligned} &(N + p)^{N+p} e^{-(N+p)} \sqrt{2\pi(N + p)} \\ &\simeq (N + p)(N + p - 1)\ldots(1 + p)\Gamma(1 + p) \end{aligned} \qquad \text{(A3.59b)}$$

from which

$$\Gamma(1 + p) \simeq \frac{(N + p)^{N+p} e^{-(N+p)} \sqrt{2\pi(N + p)}}{(N + p)(N + p - 1)\ldots(1 + p)} \qquad \text{(A3.60)}$$

Appendix 4
PROPERTIES OF DETERMINANTS

Some of the properties we will use to develop method of evaluating a determinant are given below. Derivations of these properties are presented elsewhere in the literature; see, for example, Cohen, H., 1992, pp. 423, 424.

- If all elements in any two rows are interchanged, or if all elements in any two columns are interchanged, the resulting determinant is $-A_N$. Thus, if

$$
\begin{vmatrix}
a_{11} & a_{12} & \bullet\bullet & a_{1N} \\
a_{21} & a_{22} & \bullet\bullet & a_{2N} \\
\bullet & & & \bullet \\
\bullet & & & \bullet \\
a_{(N-1)1} & a_{(N-1)2} & \bullet\bullet & a_{(N-1)N} \\
a_{N1} & a_{N2} & \bullet\bullet & a_{NN}
\end{vmatrix} = A_N
\qquad (A4.1)
$$

then, with the interchange $row_1 \leftrightarrow row_{N-1}$

$$
\begin{vmatrix}
a_{(N-1)1} & a_{(N-1)2} & \bullet\bullet & a_{(N-1)N} \\
a_{21} & a_{22} & \bullet\bullet & a_{2N} \\
\bullet & & & \bullet \\
\bullet & & & \bullet \\
a_{11} & a_{12} & \bullet\bullet & a_{1N} \\
a_{N1} & a_{N2} & \bullet\bullet & a_{NN}
\end{vmatrix} = -A_N
\qquad (A4.2a)
$$

and with the interchange $col_2 \leftrightarrow col_N$

$$
\begin{vmatrix}
a_{11} & a_{1N} & \bullet\bullet & a_{12} \\
a_{21} & a_{2N} & \bullet\bullet & a_{22} \\
\bullet & & & \bullet \\
\bullet & & & \bullet \\
a_{(N-1)1} & a_{(N-1)N} & \bullet\bullet & a_{(N-1)2} \\
a_{N1} & a_{NN} & \bullet\bullet & a_{N2}
\end{vmatrix} = -A_N
\qquad (A4.2b)
$$

- If any two rows are identical or if any two columns are identical, the determinant is zero. Thus

$$
\begin{vmatrix}
a_{11} & a_{12} & \bullet\bullet & a_{1N} \\
a_{21} & a_{22} & \bullet\bullet & a_{2N} \\
\bullet & & & \bullet \\
\bullet & & & \bullet \\
a_{11} & a_{12} & \bullet\bullet & a_{1N} \\
a_{N1} & a_{N2} & \bullet\bullet & a_{NN}
\end{vmatrix} = 0 \tag{A4.3a}
$$

and

$$
\begin{vmatrix}
a_{11} & a_{11} & \bullet\bullet & a_{1N} \\
a_{21} & a_{21} & \bullet\bullet & a_{2N} \\
\bullet & & & \bullet \\
\bullet & & & \bullet \\
a_{(N-1)1} & a_{(N-1)1} & \bullet\bullet & a_{(N-1)N} \\
a_{N1} & a_{N1} & \bullet\bullet & a_{NN}
\end{vmatrix} = 0 \tag{A4.3b}
$$

- If a determinant is multiplied by a constant, it is the same as multiplying every element is any row by that constant, or every element in any column by that constant. Thus,

$$
CA_N = \begin{vmatrix}
a_{11} & a_{12} & \bullet\bullet & a_{1N} \\
Ca_{21} & Ca_{22} & \bullet\bullet & Ca_{2N} \\
\bullet & & & \bullet \\
\bullet & & & \bullet \\
a_{(N-1)1} & a_{(N-1)2} & \bullet\bullet & a_{(N-1)N} \\
a_{N1} & a_{N2} & \bullet\bullet & a_{NN}
\end{vmatrix} \tag{A4.4a}
$$

and

$$
CA_N = \begin{vmatrix}
a_{11} & a_{12} & \bullet\bullet & Ca_{1N} \\
a_{21} & a_{22} & \bullet\bullet & Ca_{2N} \\
\bullet & & & \bullet \\
\bullet & & & \bullet \\
a_{(N-1)1} & a_{(N-1)2} & \bullet\bullet & Ca_{(N-1)N} \\
a_{N1} & a_{N2} & \bullet\bullet & Ca_{NN}
\end{vmatrix} \tag{A4.4b}
$$

A corollary to this is that if there is a common multiple of all the elements in any row, or a common multiple of all the elements in any column, that common multiple can be factored out of the determinant. Thus,

$$
\begin{vmatrix}
Ca_{11} & Ca_{12} & \bullet\bullet & Ca_{1N} \\
a_{21} & a_{22} & \bullet\bullet & a_{2N} \\
\bullet & & & \bullet \\
\bullet & & & \bullet \\
a_{(N-1)1} & a_{(N-1)2} & \bullet\bullet & a_{(N-1)N} \\
a_{N1} & a_{N2} & \bullet\bullet & a_{NN}
\end{vmatrix}
= C
\begin{vmatrix}
a_{11} & a_{12} & \bullet\bullet & a_{1N} \\
a_{21} & a_{22} & \bullet\bullet & a_{2N} \\
\bullet & & & \bullet \\
\bullet & & & \bullet \\
a_{(N-1)1} & a_{(N-1)2} & \bullet\bullet & a_{(N-1)N} \\
a_{N1} & a_{N2} & \bullet\bullet & a_{NN}
\end{vmatrix}
\tag{A4.5a}
$$

and

$$
\begin{vmatrix}
a_{11} & Ca_{12} & \bullet\bullet & a_{1N} \\
a_{21} & Ca_{22} & \bullet\bullet & a_{2N} \\
\bullet & & & \bullet \\
\bullet & & & \bullet \\
a_{(N-1)1} & Ca_{(N-1)2} & \bullet\bullet & a_{(N-1)N} \\
a_{N1} & Ca_{N2} & \bullet\bullet & a_{NN}
\end{vmatrix}
= C
\begin{vmatrix}
a_{11} & a_{12} & \bullet\bullet & a_{1N} \\
a_{21} & a_{22} & \bullet\bullet & a_{2N} \\
\bullet & & & \bullet \\
\bullet & & & \bullet \\
a_{(N-1)1} & a_{(N-1)2} & \bullet\bullet & a_{(N-1)N} \\
a_{N1} & a_{N2} & \bullet\bullet & a_{NN}
\end{vmatrix}
$$
$$\tag{A4.5b}$$

- Let two $N \times N$ determinants have $(N - 1)$ rows in common and have one row in which one or more of the elements are different. Or let two $N \times N$ determinants have $(N - 1)$ columns in common and have one column in which one or more of the elements are different. The sum of these determinants is a determinant with the $(N - 1)$ common rows, or the $(N - 1)$ common columns, and the row or column that is not the same replaced by the sums of the elements in that row or column. Thus, if

$$
A_N =
\begin{vmatrix}
a_{11} & a_{12} & \bullet\bullet & a_{1N} \\
a_{21} & a_{22} & \bullet\bullet & a_{2N} \\
\bullet & & & \bullet \\
\bullet & & & \bullet \\
a_{(N-1)1} & a_{(N-1)2} & \bullet\bullet & a_{(N-1)N} \\
a_{N1} & a_{N2} & \bullet\bullet & a_{NN}
\end{vmatrix}
\tag{A4.1}
$$

and

$$
B_N =
\begin{vmatrix}
a_{11} & a_{12} & \bullet\bullet & a_{1N} \\
b_{21} & b_{22} & \bullet\bullet & b_{2N} \\
\bullet & & & \bullet \\
\bullet & & & \bullet \\
a_{(N-1)1} & a_{(N-1)2} & \bullet\bullet & a_{(N-1)N} \\
a_{N1} & a_{N2} & \bullet\bullet & a_{NN}
\end{vmatrix}
\tag{A4.6}
$$

then,

$$A_N + B_N = \begin{vmatrix} a_{11} & a_{12} & \bullet\bullet & a_{1N} \\ (a_{21} + b_{21}) & (a_{22} + b_{22}) & \bullet\bullet & (a_{2N} + b_{2N}) \\ \bullet & & & \bullet \\ \bullet & & & \bullet \\ a_{(N-1)1} & a_{(N-1)2} & \bullet\bullet & a_{(N-1)N} \\ a_{N1} & a_{N2} & \bullet\bullet & a_{NN} \end{vmatrix} \qquad (A4.7)$$

If

$$B_N = \begin{vmatrix} a_{11} & b_{12} & \bullet\bullet & a_{1N} \\ a_{21} & b_{22} & \bullet\bullet & a_{2N} \\ \bullet & & & \bullet \\ \bullet & & & \bullet \\ a_{(N-1)1} & b_{(N-1)2} & \bullet\bullet & a_{(N-1)N} \\ a_{N1} & b_{N2} & \bullet\bullet & a_{NN} \end{vmatrix} \qquad (A4.8)$$

then

$$A_N + B_N = \begin{vmatrix} a_{11} & (a_{12} + b_{12}) & \bullet\bullet & a_{1N} \\ a_{21} & (a_{22} + b_{22}) & \bullet\bullet & a_{2N} \\ \bullet & & & \bullet \\ \bullet & & & \bullet \\ a_{(N-1)1} & (a_{(N-1)2} + b_{(\bar{N}-1)2}) & \bullet\bullet & a_{(N-1)N} \\ a_{N1} & (a_{N2} + b_{N2}) & \bullet\bullet & a_{NN} \end{vmatrix} \qquad (A4.9)$$

- A determinant is unchanged if each element in a row is replaced by the sum of
  that element and a constant multiple of the corresponding element in any other
  row. The determinant is also unchanged if each element in a column is replaced
  by the sum of that element and a constant multiple of the corresponding element
  in any other column. Thus, if

$$A_N = \begin{vmatrix} a_{11} & a_{12} & \bullet\bullet & a_{1N} \\ a_{21} & a_{22} & \bullet\bullet & a_{2N} \\ \bullet & & & \bullet \\ \bullet & & & \bullet \\ a_{(N-1)1} & a_{(N-1)2} & \bullet\bullet & a_{(N-1)N} \\ a_{N1} & a_{N2} & \bullet\bullet & a_{NN} \end{vmatrix} \qquad (A4.1)$$

then

$$
A_N = \begin{vmatrix}
a_{11} & a_{12} & \bullet\bullet & a_{1N} \\
(a_{21} + Ca_{11}) & (a_{22} + Ca_{12}) & \bullet\bullet & (a_{2N} + Ca_{1N}) \\
\bullet & & & \bullet \\
\bullet & & & \bullet \\
a_{(N-1)1} & a_{(N-1)2} & \bullet\bullet & a_{(N-1)N} \\
a_{N1} & a_{N2} & \bullet\bullet & a_{NN}
\end{vmatrix} \tag{A4.10a}
$$

and

$$
A_N = \begin{vmatrix}
a_{11} & (a_{12} + Ca_{11}) & \bullet\bullet & a_{1N} \\
a_{21} & (a_{22} + Ca_{21}) & \bullet\bullet & a_{2N} \\
\bullet & & & \bullet \\
\bullet & & & \bullet \\
a_{(N-1)1} & (a_{(N-1)2} + Ca_{(N-1)1}) & \bullet\bullet & a_{(N-1)N} \\
a_{N1} & (a_{N2} + Ca_{N1}) & \bullet\bullet & a_{NN}
\end{vmatrix} \tag{A4.10b}
$$

Appendix 5
PROOF OF THE SINGULARITY OF A MATRIX

First Proof

Let A be a matrix of L rows and columns:

$$A \equiv \begin{pmatrix} a_{11} & a_{12} & \bullet\bullet & a_{1L} \\ a_{21} & a_{22} & \bullet\bullet & a_{2L} \\ \bullet & \bullet & \bullet & \bullet \\ \bullet & \bullet & \bullet & \bullet \\ a_{L1} & a_{L2} & \bullet\bullet & a_{LL} \end{pmatrix} \tag{A5.1}$$

such that the elements in each row (denoted by k) have the property that

$$\sum_{\ell=1}^{L} a_{k\ell} = 0 \tag{A5.2}$$

To show that such a matrix is singular, we use the well-known property that the value of a determinant is unchanged when the elements in any one column are replaced by the original elements in that column added to a constant multiple of any other column. (See, for example, Cohen, H., 1992, pp. 424–425.) Thus, the value of a determinant of order $L \times L$ is unchanged when we perform the following operations on the first column of the determinant:

$$col_1^{(1)} = col_1 + col_2 \tag{A5.3a}$$

then

$$col_1^{(2)} = col_1^{(1)} + col_3 = col_1 + col_2 + col_3$$
$$\bullet \tag{A5.3b}$$
$$\bullet$$

475

and finally

$$col_1^{(L-1)} = col_1^{(L-2)} + col_L = col_1 + col_2 + ... + col_{L-1} + col_L \qquad (A5.3c)$$

Thus, for each row k of the determinant, the element in the first column, a_{k1}, is replaced by

$$a'_{k1} = a_{k1} + a_{k2} + ... + a_{kL} = 0 \qquad (A5.4)$$

Therefore, the original determinant has the same value as

$$\begin{vmatrix} 0 & a_{12} & \bullet\bullet & a_{1(L-1)} & a_{1L} \\ 0 & a_{22} & \bullet\bullet & a_{2(L-1)} & a_{2L} \\ \bullet & \bullet & & \bullet & \bullet \\ \bullet & \bullet & & \bullet & \bullet \\ 0 & a_{(L-1)2} & \bullet\bullet & a_{(L-1)(L-1)} & a_{(L-1)L} \\ 0 & a_{L2} & \bullet\bullet & a_{L(L-1)} & a_{LL} \end{vmatrix} = 0 \qquad (A5.5)$$

As such, the original matrix is singular.

Second Proof

We note the following properties of a matrix:

- In diagonal form, all off-diagonal elements of a matrix are zero, and the elements along the diagonal are its eigenvalues.
- If a matrix is diagonalized by a similarity transformation, the determinant of the diagonalized form of the matrix (the product of its eigenvalues) is the same as the determinant of the original matrix (see, for example, Cohen, H., 1992, p. 455).

Let V be a column vector with 1 for each of its L components:

$$V \equiv \begin{pmatrix} 1 \\ 1 \\ \vdots \\ 1 \end{pmatrix} \qquad (A5.6)$$

Then

$$
AV = \begin{pmatrix} \sum_{\ell=1}^{L} a_{1\ell} \\ \sum_{\ell=1}^{L} a_{2\ell} \\ \bullet \\ \bullet \\ \sum_{\ell=1}^{L} a_{L\ell} \end{pmatrix} = \begin{pmatrix} 0 \\ 0 \\ \bullet \\ \bullet \\ 0 \end{pmatrix} \qquad (A5.7)
$$

This can be written as

$$
A \begin{pmatrix} 1 \\ 1 \\ \vdots \\ 1 \end{pmatrix} = \lambda \begin{pmatrix} 1 \\ 1 \\ \vdots \\ 1 \end{pmatrix} \qquad (A5.8)
$$

with $\lambda = 0$. That is, V is an eigenvector of A with eigenvalue of zero. Since at least one of its eigenvalues is zero, the determinant of A is zero.

References

Abramowitz, M. and Stegun, I., 1964, *Handbook of Mathematical Functions with Formulas, Graphs and Mathematical Tables*, National Bureau of Standards, Washington, DC.

Baker, G., 1975, *Essentials of Pade Approximants*, Academic Press, New York.

Baker, C. T. H., 1977, *The Numerical Treatment of Integral Equations*, Clarendon Press, Oxford.

Baker, C., Fox, L., Mayers, D, and Wright, K., 1964, *Numerical Solution of Fredholm Integral Equations of the First Kind*, The Computer Journal, v. 7, no. 2, p. 141.

Chio, F., 1853, *Memoire sur les Fonctions Connue sous le Nom des Resultant ou des Determinants*, Turin.

Chisholm, J. S. R., 1973, *Convergence of Pade Approximants* in *Pade Approximants and their Applications*, edited by P. R. Graves-Morris, Academic Press, New York.

Cohen, H., 1992, *Mathematics for Scientists and Engineers*, Prentice-Hall Publishing Co., Englewood Cliffs.

Cohen, H., 2007, *Complex Analysis with Applications in Science and Engineering*, Springer Publishing Co., New York.

Conkwright, N., 1941, *Introduction to the Theory of Equations*, Ginn and Co., New York.

Crank, J. and Nicolson, P., 1947, *A Practical Method for Numerical Integration of Solutions of Partial Differential Equations of the Heat-Conduction Type*, Proceedings of the Cambridge Philosophical Society, v. 43, pp, 50–67.

Forsythe, G. and Wasow, W., 1960, *Finite Difference Methods for Partial Differential Equations*, John Wiley and Sons, Inc., New York.

Fox, L. 1965, *An Introduction to Numerical Linear Algebra with Exercises*, Oxford University Press, Oxford.

Francis, J., 1961 and 1962, *The QR Transformation*, The Computer Journal, v. 4, Part I, pp. 265–271, Part II, pp. 332–345.

Fuller, L. and Logan, J., 1975, *On the Evaluation of Determinants by Chio's Method*, The Two-Year College Mathematics Journal, v. 6, No. 1, pp. 8–10.

Graeffe, H. K., 1837, *Die Aufloesung der hoeheren numerischen Gleichungen (The Resolution of Higher [Order] Numerical Equations)*, Zurich.

Hansen, P., 1992, *Numerical Tools for Analysis and Solution of Fredholm Integral Equations of the First Kind*, Inverse Problems, v. 8, p. 849.

Hanson, R., 1971, *A Numerical Method for Solving Fredholm Integral Equations of the First Kind Using Singular Values*, SIAM Journal on Numerical Analysis, v. 8, no. 3, p. 616.

Hochstadt, H., 1973, *Integral Equations*, John Wiley & Sons, Inc., New York.

Kells, L. 1954, *Elementary Differential Equations*, 4th ed, McGraw-Hill Book Co., New York.

Kirkwood, J. G. and Riseman, J., 1948, Journal of Chemical Physics, v. 16, p. 565.

Kunz, K., 1957, *Numerical Analysis*, McGraw-Hill Publishing Co., New York.

Lippmann, B. and Schwinger, J., 1950, *Physical Review*, v. 79, p. 469.

Mathews, J., 1992, *Numerical Methods for Mathematics, Science and Engineering*, 2nd ed. Prentice-Hall Publishing Co., Englewood Cliffs.

Mathews, J. and Walker, R. L., 1970, *Mathematical Methods of Physics*, 2nd ed., W. A. Benjamin, Inc., Menlo Park.

Mikhlin, S. G. and Smolitskiy, K. L., 1967, *Approximate Methods for Solution of Differential and Integral Equations*, edited by Bellman, R. and Kalaba, R. E. in *Modern Analytic and Computational Methods in Science and Mathematics*, American Elsevier Publishing Co., New York.

Mitchell, A. R. and Griffiths, D. F., 1980, *The Finite Difference Method in Partial Differential Equations*, John Wiley and Sons, New York.

Namias, V., 1985, *Simple Derivation of the Roots of a Cubic Equation*, American Journal of Physics, v. 53, p 775.

Namias, V., 1986, *A Simple Derivation of Stirling's Asymptotic Series*, American Mathematical Monthly, v. 93, No. 1, p. 25.

Nuttall, J., 1977, *The Convergence of Pade Approximants to Functions with Branch Points* in *Pade and Rational Approximation, Theory and Application*, edited by E. B. Saff and R. S. Varga, Academic Press, New York.

Omnes, R., 1958, *Il Nuovo Cimento*, v. 8, p. 316.

Pade, H., 1892, *Sur la Representation Approchee d'une Fonction par les Fractions Rationalles, (On the Approximate Representation of a Function by Rational Fractions)* Doctoral Thesis.

Patel, V., 1994, *Numerical Analysis*, Saunders College Publishing, Fort Worth.

Pesic, P., 2003, *Abel's Proof: An Essay on the Sources and Meaning of Mathematical Insolvability*, MIT Press, Cambridge, pp. 155–180.

Schlitt, D. W., 1968, *Journal of Mathematical Physics*, v. 9, p. 436.

Smith, G.D., 1965, *Numerical Solution of Partial Differential Equations*, Oxford University Press, London.

Solomon, R., 2008, *The Little Book of Mathematical Principles, Theories and Things*, Metro Books, New York.

Spiegel, M., 1973, *Fourier Series with Applications to Boundary Value Problems*, Schaum's Outline Series, McGraw-Hill Publishing Co., New York.

Strikwerda, J., 1989, *Finite Difference Schemes and Partial Differential Equations*, Wadsworth & Brooks/Cole, Belmont.

Stroud A. H. and Secrest, D., 1966, *Gaussian Quadrature Formulas*, Prentice-Hall, Inc., Englewood Cliffs.

Twomey, S., 1963, *On the Numerical Solution of Fredholm Integral Equations of the First Kind by the Inversion of the Linear System Produced by Quadrature*, Journal of the ACM, v. 10, no. 1, p. 97.

Ullman, R., 1964, Journal of Chemical Physics, v. 40, p. 2193.

Wilkinson, J. H., 1965, *The Algebraic Eigenvalue Problem*, Clarendon Press, Oxford.

Index